T0241533

LIFE AND SCIENTIFIC WORK
OF
PETER GUTHRIE TAIT

Yours truly
P. G. Tait.

LIFE AND SCIENTIFIC WORK

OF

PETER GUTHRIE TAIT

SUPPLEMENTING THE TWO VOLUMES OF SCIENTIFIC PAPERS PUBLISHED IN 1898 AND 1900

by

CARGILL GILSTON KNOTT

D.Sc.; one of the Secretaries R.S.E.; Order of the Rising Sun, Empire of Japan (Class IV); Assistant to Professor Tait from 1879 to 1883; Professor of Physics, Imperial University of Japan from 1883 to 1891; Lecturer on Applied Mathematics in Edinburgh University from 1892

Cambridge: at the University Press 1911

CAMBRIDGE
UNIVERSITY PRESS

University Printing House, Cambridge CB2 8BS, United Kingdom

Cambridge University Press is part of the University of Cambridge.

It furthers the University's mission by disseminating knowledge in the pursuit of education, learning and research at the highest international levels of excellence.

www.cambridge.org
Information on this title: www.cambridge.org/9781107494923

© Cambridge University Press 1911

This publication is in copyright. Subject to statutory exception and to the provisions of relevant collective licensing agreements, no reproduction of any part may take place without the written permission of Cambridge University Press.

First published 1911
First paperback edition 2015

A catalogue record for this publication is available from the British Library

ISBN 978-1-107-49492-3 Paperback

Cambridge University Press has no responsibility for the persistence or accuracy of URLs for external or third-party internet websites referred to in this publication, and does not guarantee that any content on such websites is, or will remain, accurate or appropriate.

PREFACE

AT the time of his death in 1901 Professor P. G. Tait had just finished editing the Second Volume of his Collected Scientific Papers. The series is now completed by this Memorial Volume whose preparation I undertook at the request of Mrs Tait, who kindly placed a great deal of material at my disposal, and who, together with the other members of the family, has been closely in touch with the work as it proceeded.

Professor Crum Brown, the late Professor's brother-in-law and colleague for over 30 years, closely associated himself with the work. His knowledge and judgement were always at my service.

Lord Kelvin at the outset afforded me much useful information generally about events of an early date, especially certain facts connected with the preparation of "The Treatise on Natural Philosophy," a work unfortunately never completed.

The proofs have been read by Dr A. W. Ward, Master of Peterhouse, and Mr J. D. Hamilton Dickson, Fellow and Tutor of Peterhouse, to both of whom I am deeply indebted for many valuable criticisms and suggestions; and for similar helpful services my sincere thanks are also due to Professor J. G. MacGregor and Professor W. Peddie.

The interest expressed by others among Professor Tait's friends and students has greatly encouraged me in my work. Their reminiscences of the Natural Philosophy Class Room or Laboratory, and their memory of the stimulating character of the teaching, will be found reflected in the pages which follow.

In arranging the material I have been influenced largely by one consideration—the convenience of the reader. The opening chapter, including the description of Professor Tait on holiday in St Andrews, for which I am indebted to Mr J. L. Low, gives simply the main facts of the Life. The various aspects of the Scientific Work are taken up, in more or less detail, in the succeeding chapters.

The care with which Professor Tait preserved the letters he received from his scientific correspondents has enabled me greatly to enrich the pages of the Memoir by the inclusion of letters from Sir William Rowan Hamilton, Professor Cayley, Lord Kelvin, and Professor Clerk Maxwell. Introduced as far as possible in its immediate setting, the correspondence brings out interesting points of history, and shows how heartily all these great men helped one another in their scientific investigations. It is much to be regretted that Professor Tait's own letters to Clerk Maxwell are not now available.

Professor Tait's foreign correspondence was carefully arranged and annotated by Dr J. S. Mackay, to whom I am greatly indebted for thus enabling me rapidly to choose what was serviceable for the purposes of the Memoir.

Several of the old students having suggested that the controversy between Professor Tait and Mr Herbert Spencer would prove interesting, I have given the details at some length. It seemed advisable to bring the real points at issue clearly before the reader's mind, more especially as Mr Spencer had given his own views at great length in a published pamphlet and in the appendix to subsequent editions of his *First Principles*. On looking into the matter I found myself forced to begin with what preceded Professor Tait's share in the controversy; and in this connection I wish to thank Lord Justice Fletcher Moulton for his help in presenting an accurate account of the stages of a lively debate which had its origin in his review of Mr Spencer's Work.

The original photograph of Professor Tait writing a note in his retiring-room is the property of the Rev. L. O. Critchley, M.A., who most willingly granted the inclusion of the portrait in the present volume. To him also special thanks are due.

I wish also to record my thanks to the Editors of *Nature*, of the *Philosophical Magazine*, and of the *Badminton Magazine*, for permission to reprint articles contributed by Professor Tait; and to the Council of the Royal Society of Edinburgh for certain diagrams and figures which have been reproduced.

CARGILL GILSTON KNOTT

EDINBURGH UNIVERSITY
February 1911

TABLE OF CONTENTS

CHAPTER I

MEMOIR—PETER GUTHRIE TAIT

CHAPTER II

EXPERIMENTAL WORK

CHAPTER III

MATHEMATICAL WORK

CONTENTS

CHAPTER VII

ADDRESSES, REVIEWS, AND CORRESPONDENCE

CHAPTER VIII

POPULAR SCIENTIFIC ARTICLES

PORTRAITS

ADDENDA AND CORRIGENDA

p. 22, footnote.—In connection with early laboratories, those begun by Professor Clifton in Oxford, and by Professor Grylls Adams in King's College, London, should also have been mentioned.

p. 48, l. 2 from foot—*for* "rauschend" *read* "rauchend."

p. 96, l. 6 from top—*for* "Reader in" *read* "Professor of."

CHAPTER I

MEMOIR

PETER GUTHRIE TAIT

Of all human activities and developments none are more characteristic of the Victorian Era than those clustering round the word Science. Scientific theory and its application to the growing needs of mankind advance hand in hand. On the one side are the developments of steam power, and the practical creations of Electric Telegraphy, Telephony and Dynamo-electric machinery; on the other the framing of new theories of Heat and Electricity. Practical engineers and scientific men of all types and degrees of ability and talent have had their share in this great development, which within two generations has transformed the whole aspect of human life.

But of far greater import to the philosophical student than the dovetailed features of this development is the apprehension of the broad principle of Energy which has unified the various branches of science. The biography of any of the outstanding natural philosophers of the latter half of the Nineteenth Century must, indeed, be to a large extent a history of Energetics, to use Rankine's convenient nomenclature. These minds, trained under masters of an older school who knew of no such guiding principle, grew with the scientific environment which they were themselves creating. It is not easy for us, who are the heirs of the rich legacy of thought which our immediate predecessors bequeathed to us, fully to realise the greatness of the transformation which they effected.

We may be able to note here and there the subtle manner in which, not always consciously to themselves, they acted and reacted one upon the other; but we are perhaps too near the age of transition to see clearly the interplay of all that made for progress. Each of us has had his own peculiar training, his own personal contact with the mighty ones of the immediate past; and this forms as it were a telescopic tube determining limits to our field of vision. No doubt we may range the whole horizon; but after all we look from our own point of vantage. What may appear as a towering peak to one may seem but an ordinary eminence to another. Nevertheless, incomplete and historically partial though it must be, a sketch of the career of a leader of scientific thought who lived his strenuous mental life through this formative time cannot be without its value as a contribution to the history of the growth of ideas.

Such a one, pre-eminently, was Professor Tait of Edinburgh University. He was the personal friend of Hamilton, Andrews, Stokes, Joule, Kelvin, Maxwell, Stewart, Helmholtz, Cayley, Sylvester—to name a few of the more outstanding of those who have passed away. These contemporaries were to him personalities and not mere writers of papers or of books. He got much from them and he gave much to them. As a historian of contemporary developments he takes high rank; and to him we owe in a manner which can only now be clearly recognised the very existence of Thomson and Tait's *Natural Philosophy* and of Hamilton's *Elements of Quaternions*.

In tracing his career I have received every help possible from Mrs Tait and the other members of the family. My own recollections of his tales of earlier days have been corroborated and supplemented by evidence from letters written contemporaneously with the events they describe. His Scrap Book, a fascinating collection of all kinds of letters and cuttings bearing upon his own work and the work of others that touched him closely, has been of unique value.

I feel it a great honour to have had confided to me the privilege of preparing this memorial volume. My sole endeavour has been to give a faithful picture of Professor Tait as teacher, investigator, author, and friend. To this end I have reproduced a few of his more popular scientific articles as well as numerous quotations from letters, addresses, and reviews.

The picturesque account of the St Andrews holiday life of Professor Tait is from the pen of Mr John L. Low, the author of *F. G. Tait, a Record*, being the biography of Professor Tait's soldier son, Lieutenant in the Black Watch, who lost his life in the South African War.

EDINBURGH. 1837–48

Peter Guthrie Tait was born at Dalkeith on 28 April, 1831. He was educated in his very early years at the Dalkeith Grammar School. On his father's death his mother came to Edinburgh with her young family of two girls and one boy; and after a year or two at Circus Place School, Tait entered the Academy at the age of ten. He and his sisters finally lived with their uncle, John Ronaldson, in an old-fashioned roomy house called Somerset Cottage, which is still occupied by the Misses Tait. Mr Ronaldson was a banker by profession, but was keenly interested in many scientific pursuits. He would take his nephew geological rambles in the long summer days, and study the planets and stars through his telescopes during the dark winter nights ; or he would dabble in the mysteries of photography which had just been invented by Daguerre and Talbot. There is little doubt that the receptive mind of the young lad must have been greatly influenced by his uncle's predilection for scientific study. A small room on the left of the hall as one enters Somerset Cottage contains to this day the stand and tube of a Newtonian reflector, and a good serviceable refractor of two-inch aperture. The room has been long used by Miss Tait for storing her canvasses and artistic materials; but the scientific contents of the apartment have never been disturbed since 1854, when P. G. Tait definitely made his home in Belfast. On his return to Edinburgh in 1860 his interests were in other directions than observational astronomy, and the old telescopes and theodolite were left in undisturbed possession. Nevertheless, his early appreciation of astronomical instruments declared itself from time to time when he purchased a beautiful speculum or a complete reflector for the Natural Philosophy Museum. In his Scrap Book Tait preserved a neatly constructed chart of date 1844, showing graphically the positions of Jupiter's satellites on successive nights from Sept. 18 to Sept. 31. These "Observations on Jupiter" were made by himself when he was a little over thirteen years of age. Probably they were interrupted by bad weather.

The environment amid which Tait spent his schooldays is well described in the Chronicles of the Cumming Club, a remarkable book printed for private circulation in 1887. Written by the late Lt.-Col. Alexander

Fergusson, it places on record the life history of a class of boys which began its corporate existence in the winter of 1841.

Peter Guthrie Tait was one of this class, which at the start numbered some sixty lads all about ten years of age. The reason for this great gathering of the first year or "Geits[1]" class was the popularity of the master, James Cumming. According to the custom then holding in Edinburgh Academy, each master began in rotation with the first year's scholars and carried them on for four years under his exclusive instruction in classical studies. For the remaining three years of the regular curriculum the boys, although coming directly under the care of the Rector, still continued to spend some hours of tuition with the master who had trained them from the first. When in accordance with the routine of the school the time came for Mr Cumming to start the new first year, his fame as a teacher drew an unusually large number of boys.

Of the members of this particular Cumming class as many as twenty-seven entered "the Services at an important juncture in the history of our country," and won thirty-nine military honours including six British and Foreign Knightly Orders. This was the class in which Tait was throughout his schooldays the "permanent dux." In 1850 the surviving members of the class formed themselves into a club called the Cumming Club, which met for good fellowship year by year.

In Colonel Fergusson's brightly written chronicle we find a perfect picture of the school life in Edinburgh during the early part of last century. Especially are we introduced to the masters who helped to mould the mind of P. G. Tait. Tait himself had many reminiscences of his schoolmasters; and for James Cumming, the classical master, and James Gloag, who gave him his first acquaintance with mathematics, he retained always the greatest admiration and respect. So thoroughly was Tait taught the classics that (as he once told me) he never required to turn up a Greek Lexicon all the time he was at school. This no doubt was largely due to the pupil's own extraordinary verbal memory; but the master who could teach with such results must have been to the manner born.

Gloag was a teacher of strenuous character and quaint originality—a type familiar enough in Scotland before School Boards and Leaving Certificates cooperated to mould teachers after the same type. With him mathematics

[1] In Jamieson's Dictionary of the Scottish Language, *geit, gett, gyte,* variously spelt, is defined as "a contemptuous name for a child." Compare modern "kid."

was a mental and moral discipline. How keenly Gloag enjoyed exposing the superficial knowledge of a boy who thought he knew! A very characteristic story is told in the Chronicles of the way in which, in the presence of the Rector, Gloag demanded a proof from one of the Rector's classical pets. After the Rector in a foolish assumption of knowledge had for some time encouraged the boy with such remarks as "Why, my boy, don't you see it? Think a moment! It's quite easy, don't you know—perfectly simple!" Gloag in a moment of supreme triumph exclaimed

"Naw, Mr Ractor, Sir, it's nott easy—the thing's impōssible, Sir—it's gross nonsense, Sir!"

Such was the teacher who first led Tait's mind in the paths in which ere long he was to gain the highest distinction.

Lewis Campbell and James Clerk Maxwell were also Edinburgh Academy boys; and in Campbell's *Life of Maxwell* an interesting account is given of the school. They were a year ahead of Tait and were not therefore members of the Cumming Club. Fleeming Jenkin, the first Professor of Engineering in Edinburgh University, was a classmate of Tait, as were also Sir Patrick Heron Watson the eminent surgeon, Sir Edward Harland of Harland and Wolff, Belfast, A. D. Stewart, C.E., who selected the plans for the Forth Bridge, Andrew Wilson, traveller and author of *The Abode of Snow*, General Cockburn, General Sherriff, Frederick Pitman, W.S., one of the early Secretaries of the Cumming Club, Dr Thomas Wright Hall, a well-known physician for many years resident in Brazil, and many others whose careers are sketched in the *Roll Call* of the Chronicles of the Cumming Club.

Tait himself preserved in printed form the result of the examination held in 1846 to determine the winner of the Edinburgh Academical Club Prize. The competition was open to all the Rector's classes, namely, the Fifth to the Seventh. Lewis Campbell came out first over all and gained the prize. Tait was third, being the only Fifth Class boy who was named in the list, and Maxwell was sixth. In the department of mathematics, however, the order of merit was Tait, Campbell, Maxwell, the others named being far behind. On the classical and linguistic side Tait naturally fell behind the more widely read scholars of the higher classes.

In the competition for the Academical Club Prize in 1847, Tait was again third, but Maxwell, now in the Seventh Class, was second on the whole. In mathematics, Maxwell was first and Tait was second.

Tait's skill in Latin verses is specially recorded in the School Reports, and a good specimen of his efforts in versification will be found by the curious in the Edinburgh Academy Report for 1845. To the end of his life he remembered hundreds of lines of Greek and Latin poetry. His children remember how he used to declaim Odes of Horace and long passages of Homer when the fancy struck him. German ballads also were among his stock in trade for apt quotation. A favourite time for such outpourings was on St Andrews Links before breakfast, when he was still young enough to cover the ground without trouble at a good five miles an hour. It may be doubted if anyone whose classical studies ended when he was little more than fifteen years old ever carried away such a store of poetry, or found in it such a never-failing source of pleasure. He frequently spoke of Archdeacon Williams, the Rector of the Academy with whom he read Homer, as a born teacher. "A gentleman, every inch of him," was his emphatic verdict a few weeks before his death.

In the Rector's report for the year 1851–2, when Tait's position as Senior Wrangler added glory to his old school, it is stated that Tait gained eight medals, six as dux of his class for the successive years 1841–47, and two for mathematical excellence in the Fifth and Sixth classes.

Tait left the Academy in 1847, and then spent a session at Edinburgh University under the tutelage of Kelland and Forbes.

He enrolled himself in the two highest of Kelland's three mathematical classes and attended all the examinations. He secured high positions in both, but was distanced in the competition by several of his fellow students. In the highest class he was third in the honours list.

There was only one class in Natural Philosophy; but this was divided by Forbes into three divisions. All members of the class attended the same lectures, on the subject matter of which they were periodically examined. The home reading, on which there were special examinations, varied with the division. A student usually entered the third or lowest division, passing into the higher divisions if he enrolled himself in the class more than once. Tait boldly entered himself for the first division. There is a tradition that Forbes in his most dignified manner tried to induce Tait to be content with the second division. This was the course Clerk Maxwell took, in spite of the fact that he was certainly as advanced in his mathematical studies as Tait, and had moreover already published a mathematical paper of distinct originality. Neither Maxwell nor Tait *markedly* excelled in comparison with the best of their fellow students. Tait was third in the honours list of the five men who formed the first division. The Gold Medal, which was awarded to the student who made most marks in the special examinations in the highest division, was gained by James Sime, one of the most brilliant students of his day, and well known in Edinburgh educational circles throughout a long and active life. In the examinations on Newton's *Principia* (first three sections) and Airy's *Tracts* (probably that on the undulatory theory of light), Sime gained twice as many marks as Tait. In the ordinary examinations on the Class Lectures Tait had a slight advantage, although a wrong addition in the class book makes him a mark or two behind Sime. The prize was, however, gained by Maxwell. It is not a little curious that the Gold Medal was not won by Balfour Stewart in 1846, nor by Tait in 1848, nor by Maxwell in 1849; and yet Edinburgh University can claim no greater names in physical science than these three.

An interesting fact which I learned from Tait himself is worth recording. On one occasion when, in preparation for a lecture on statics, I was arranging and admiring the models of catenaries of various forms which belong to the Natural Philosophy Museum of Edinburgh University, Tait remarked, " I helped Forbes to make these when I was a young student here." The models were constructed of beautifully turned disks of wood of suitable form, suitably strung together, and represented the common catenary, the circular arc catenary and the catenaries of parabolic form and of uniform strength. I pointed to the last word "strength" which was misspelled, the penultimate letter being dropped probably from want of room, and said in joke, " Is this an example of your accuracy ? " " Ah," he rejoined, " I was responsible only for the calculations of the sizes of the disks, not for anything else."

Clerk Maxwell spent three sessions in Edinburgh University before he decided to go to Cambridge; but Tait was content with one session, and began his mathematical training in Cambridge before he was eighteen.

CAMBRIDGE. 1848–54

It was a curious fate which brought to Peterhouse in 1848 the two young mathematicians, P. G. Tait and W. J. Steele, the one from Scotland, the other from Ireland by way of Glasgow[1]. They "coached" with the famous private mathematical tutor of those days, also a Peterhouse man, William Hopkins, another of whose pupils a few years earlier was William Thomson, afterwards Tait's lifelong friend. Tait and Steele at once became marked out as future high wranglers; but one would hardly have dared to prophesy that they would come out respectively first and second in the Tripos.

Tait's method of preparing for the great contest is preserved in his own hand-writing on three quarto sheets afterwards pasted into the Scrap Book. From Dec. 16, 1851, to Jan. 5, 1852, each day (Sundays excepted) is marked off for revision of definite subjects of study, morning and evening. When the work is accomplished, the subject is scored out and the time taken marked in the margin. Four hours are the most he gives at one sitting, and on no day does his time of study exceed $6\frac{1}{2}$ hours, usually much less. Opposite Jan. 6, Tuesday, is printed by hand the words "Senate House." Then comes an irrelevant note of a lunar eclipse which occurred on Jan. 7, and below this appears in large letters right across the sheet the word "Porgatorio." The three days of Purgatory past, the time schedule begins again on Jan. 8 (evening) with "Brief Respite from Torment"; and during the succeeding eight working days the morning and evening tasks are again portioned out. But the work is more serious now. Tait never gives less than $5\frac{1}{2}$ hours a day, and on one occasion reaches $7\frac{1}{2}$ hours. Beneath the last date "January 19, Monday and subsequent" he prints across the page in huge capital letters "L'ENFER!" The guiding principle seems to have been not greatly to exceed in sustained work during any one day the time allotted for the examination.

[1] In Kelvin's early paper on the Absolute Thermometric Scale (*Cambridge Phil. Trans.*, June 1848, *Phil. Mag.*, Oct. 1848) William Steele is mentioned as having assisted in comparing the proposed scale with that of the air thermometer (see *Math. and Phys. Papers*, Vol. I, p. 105).

Steele seems to have been generally ahead of Tait in the College examinations, so that Tait's winning of the Senior Wranglership came somewhat as a surprise to those who deemed they knew. The story of this day, famous in the annals of Peterhouse, is well told by J. D. Hamilton Dickson in the Magazine of the Peterhouse Sexcentenary Club for the Michaelmas Term, 1902.

"How the old gyp's face used to light up as he told the story of that January morning when the Tripos list was read. One gyp was in the Senate House to hear the list, and as soon as Steele's name came out as Senior Wrangler he was to rush out and make a signal by stretching out his arms like a big T; another gyp near the 'Bull' was to repeat the signal; and a third at the College gate was to rush in with the news. When that list was read and Tait's name came first the gyp nearly collapsed, but hearing Steele's name next he recovered, and noting only that Peterhouse was first, rushed out, made the signal, and fled with all speed to College to correct the pardonable error he had telegraphed."

Tait telegraphed home "Tait Senior, Steele second, tell Gloag." How Gloag received the news is told in a footnote in the Chronicles of the Cumming Club.

"When intelligence reached the Academy of the great event, Gloag was 'raised' and out of himself with excitement. 'Have ye hard the news aboot Tait?' he asked of everybody he met, M— among others. 'No,' answered M—, 'he's got a Bishopric, I suppose, or something of that sort.' 'No, Sir, it's not Archibald Cam'ell Tait it's Peter Guthrie Tait, a vara different parson[1]—Senior Wrangler, Sir,' and off he went to spread the news."

Through the kindness of Sir Doyle Money Shaw, at that time president of the Cumming Club, Mr Beatson Bell, for many years Secretary of the Club, was able to show me the brief note in which Tait told of his success.

COLL: DIV: PET: CANT.
Jany. 31st 1852.

My dear Doyle,
      I'm all in a flutter
      I scarcely can utter, &c., as
the song has it:—
      I AM SENIOR WRANGLER!

Tell it to the Cumming Club—&c.
&c. and believe me
      yours very sincerely
        PETER GUTHRIE TAIT, B.A.

[1] So Gloag pronounced "person."

Tait's achievement was made the occasion of a special meeting of the Cumming Club.   It was (to quote from the Chronicles)

"felt to be an honour conferred on the Academy, the Masters—Gloag in particular —the Class, and the Club. Consequently they could do no less than offer to their old friend and Dux a banquet specially designed to do him worship.   And right well they did it.......

"For once the exclusive rule of the Club was broken through, and invitations scattered with a lavish hand amongst those—and they were many—who beyond the limits of the Class, held kindly memories of Tait and of the Academy.......

"It was a high occasion for them all.   Gloag could hardly divest himself of the idea that he was the hero of the occasion, such credit did he take to himself.......

"Festive conversation was at fullest swing—that is to say, many talkers, few listeners—when suddenly the scene of revelry was broken in upon by an ominous 'boom.' Tongues were still for a moment, but only for a moment.

"Then once again, clearer, deadlier than before, the 'boom' is heard above the clatter of tongues.

"In a moment the mystery is solved.   The President, Doyle Shaw, ever active for good, or evil, from his end of the table as it approached the gallery, had observed peeping over the edge of this gallery, at an inviting angle, the rim of a big drum.   Straightway the idea arose that by well directed vertical fire this tempting object might be reached.   The first orange discharged hit the mark unobserved by the company, but the second 'boom' discovered all.

"The idea was hailed as a brilliant one that only needed development.   The entire dessert, oranges and apples, was soon expended.   Then the thought occurred to Doyle Money Shaw to improve on his original idea.   While the practice was still going on he managed cleverly to 'swarm' up one of the pillars with the intention of capturing the big drum.   But on arriving at the spot and with a shout of ecstasy he announced to those below that the entire band instruments were there.   Without a moment's loss of time these were handed down, and from hand to hand; and nothing would serve these festive spirits but the 'Conquering Hero' in Tait's honour."

Steele was evidently a man after Tait's own heart.   They were close friends throughout their College life, and when Fellows of the same college they collaborated in the production of a treatise on the *Dynamics of a Particle*. The book was planned and to some extent written during a holiday they spent together after they took their degree.   Unfortunately Steele's health gave way, and his early death left his portion of the work unfinished. With the true chivalry of his nature Tait issued the book in 1856 under the joint names of Tait and Steele; and "Tait and Steele" is still its familiar title.   The character of the book will be discussed later.   The MS was presented to Peterhouse by Mrs Tait, and is now preserved in the College

Library. The accompanying picture of the group containing Tait and Steele, who are respectively first and third reckoning from the left, has been reproduced from a somewhat faded photograph. Its probable date is 1852.

Having taken his degree as Senior Wrangler and First Smith's Prizeman, Tait was elected a Fellow of his College and began to establish himself as a "coach." To quote from an address he gave to the Edinburgh Graduates fourteen years later, he became one of those who,

"eagerly scanning examination papers of former years, and mysteriously finding out the peculiarities of the Moderators and Examiners under whose hands their pupils are doomed to pass, spend their lives in discovering which pages of a text-book a man ought to read and which will not be likely to 'pay.' The value of any portion as an intellectual exercise is never thought of; the all-important question is—Is it likely to be set? I speak with no horror of or aversion to such men; I was one of them myself, and thought it perfectly natural, as they all do. But I hope that such a system may never be introduced here."

His hopes, it is to be feared, are being only partially realised.

Tait's experience as a coach was fortunately very limited. During the two and a half years he continued to reside at Peterhouse he had hardly time to establish a reputation. There is indeed a story[1] of "Tait's one Pupil," who had begun to read with Hopkins. So unsatisfactory was his progress that Hopkins advised him to seek another tutor. Naturally the pupil protested and said he would do his utmost not to keep the others back. But Hopkins was obdurate. Accordingly the aspirant to Wrangler honours became Tait's one pupil, and was taught to such good purpose that when the Tripos list came out he was one place above Hopkins' best man. When congratulated upon the success of his pupil Tait is said to have remarked, "Oh, that's nothing—I could coach a coal scuttle to be Senior Wrangler."

Tait, however, was not a man to let time hang on his hands. He read widely and thoroughly in all branches of mathematical physics. During these years also he learned to read Italian with ease and made himself master of the French and German languages.

---

[1] The story is given with full details in a letter from W. A. Porter, whose authority was C. B. Clarke, 3rd Wrangler in 1856, and Mathematical Lecturer in Queens', 1857–65.

BELFAST. 1854–60

On September 14, 1854, P. G. Tait was appointed Professor of Mathematics in Queen's College, Belfast. Among his colleagues were Thomas Andrews, the famous experimenter on the liquefaction of gases, Wyville Thomson, afterwards of Edinburgh and the scientific leader of the Challenger Expedition, James Thomson (Lord Kelvin's brother), subsequently professor of Engineering in Glasgow and the discoverer of the lowering of the melting point of ice by pressure, and James McCosh, afterwards President of Princeton.

The Right Hon. Thomas Sinclair, of Belfast, who as senior scholar in mathematics in 1857 assisted Tait in tutoring the junior men, mentions that in addition to conducting his official classes in mathematics Tait supplemented Professor Stevelly's lectures in Natural Philosophy by starting a voluntary class for Honours men in the more advanced treatment of dynamics. This was a great boon to those studying for honours. The voluntary class is mentioned in a footnote in the Calendar, but there is no indication that the class was carried on by the professor of mathematics. We can well imagine the delight with which Tait would escape from the comparative dreariness of Pure mathematics into the satisfying realities of Applied. Tait proved an admirable teacher, clear and systematic in his treatment of the various branches taught. In addition to the regular lectures, he gave tutorial instruction to his pupils, setting them exercises and problems and helping each individually in turn.

In these years he continued to practise on the flute on which he was a skilled performer. In Cambridge he had been a member of the amateur orchestra, and we hear of him appearing at a concert in Belfast to play a flute obligato to a distinguished local soprano singer.

The two great scientific facts of his life in Belfast were his association with Dr Andrews in experimental work and his study of Hamilton's calculus of Quaternions. Often in conversation Tait expressed his indebtedness to Andrews for initiating him into certain lines of experimentation. Their joint papers on Ozone are published in Andrews' memorial volume. The original conception of the investigation was due to the older man who had already published important work on the same subject. Tait gave efficient aid, more particularly in the calculations involved, and in the construction of much of the apparatus used. He proved such an apt pupil in the art of glass blowing that ere long Andrews gave that part of the manipulation over to his eager and energetic companion. Tait used to speak with intense admiration of the extreme care and patience with which Andrews carried out all his researches. Each difficulty or discrepancy as it arose had to be disposed of before progress could be reported and the investigation advanced a stage. At times indeed the patient care of the skilled experimenter must have chafed somewhat the brilliant young mathematician ever eager to get to the heart of things; but no amount of argument or theorising on Tait's part could move the master from the steady tenor of his way. Years after when Andrews in his failing health visited Edinburgh Physical Laboratory to inspect a set of his own apparatus for the liquefaction of gases it was at once a privilege and an inspiration to witness the deep affection and admiration with which Tait regarded his whilom colleague.

In his letter to Mrs Andrews immediately after the death of her husband, Tait expresses his feelings and regard in these words:

"It does not become me to speak of the irreparable loss which you and your family have suffered. But it may bring some consolation to you to be assured that there are many, in many lands, whose sympathies are sincerely with you;—and who lament, with you, the loss of a great man and a good man.

"For my own part, I feel that I cannot adequately express my obligation to him whether as instructor or example. I have always regarded it as one of the most important determining factors in my own life (private as well as scientific) and one for which I cannot be sufficiently thankful, that my appointment to the Queen's College at the age of 23 brought me for six years into almost daily association with such a friend."

Hamilton's first book, *Lectures on Quaternions*, was published in 1853. We learn from the inscription on the title page of Tait's copy that he bought it the same year while still a resident at Peterhouse. As he explained in the preface to his own Treatise (1st edition, 1867) Tait was attracted to the study of quaternions by the promise of usefulness in physical applications. Yet in Hamilton's Lectures very few pages indeed touch upon dynamical problems. Tait used to tell how his faith in the new calculus was put to a severe test as he read through these remarkable so-called lectures of Hamilton. Lecture after Lecture he carefully perused, wearied though he was with Hamilton's extraordinary prolixity in laying strong and deep the foundations of his calculus. He seemed to be making no progress. Did the fault lie with the author, or with Tait's own inability to understand the system? Such were his feelings through the first six "Lectures." But perseverance had its reward when he came to Lecture VII. Here, after a few sections of recapitulation, Hamilton revels in the wealth of geometrical applications fitted to display the power of the calculus. This so-called Seventh Lecture occupies 356 pages in a book of which the other six Lectures occupy 380!

Tait was one of very few who really appreciated the immense value of Hamilton's work. Many who with gay confidence began to read the Lectures lost heart and fell back from Quaternion heights into Cartesian valleys, where the paths seemed easier in their artificial symmetry. Now, however, the early hopes of Hamilton and Tait are being realised in the growing use of vector methods and symbolism, especially in their physical applications. Hamilton's and Tait's theorems have been rediscovered by later workers, some of whom, under the domination of new notations for the quantities and functions which Hamilton made familiar, think the novelty extends to the functions and quantities themselves!

During his undergraduate days Tait made the acquaintance of William Archer Porter and James Porter, brothers from Belfast. William, Third Wrangler in 1849, was for a time Tutor of Peterhouse, and after being called to the English bar became Principal of Combaconum College in India, and subsequently Tutor and Secretary to the Maharajah of Mysore. James Porter was Seventh Wrangler in 1851 and was elected a Fellow immediately after graduating. He was for some years mathematical professor of the Collegiate Institute in Liverpool, but returned ere long to Peterhouse, first as Tutor then as Master (1876–1901). He was endowed with a great activity both mental and physical, which found expression on the one hand in a keen participation in athletic sports, and on the other in whole-hearted efforts to promote the highest interests of the University. In Dr T. A. Walker's *History of Peterhouse* (1906) the Rev. James Porter is described as a "man of notable business qualifications and of a rare generosity of spirit."

When Tait went to Belfast he became closely intimate with the Porter family, and on October 13, 1857, he married one of the sisters of his Peterhouse friends. As Kelvin expressed it: "During these bright years in Belfast he found his wife and laid the foundation of a happiness which lasted as long as his life."

The youngest brother, John Sinclair Porter, was one of Tait's most distinguished students at Queen's College. He entered the Indian Civil Service in 1861 and retired in 1889.

There is a good story told of how Tait saved valuable personal property of his colleague Wyville Thomson from the process of arrestment executed upon the landlord's house and goods. When the bailiffs took possession Tait came on the scene and after some conversation got permission for Wyville Thomson and his wife, who were simply lodgers, to fill two boxes with their purely personal goods. The men of law retired to the kitchen to be refreshed for their labours. They looked out occasionally and always saw the two boxes in the hall being filled. But they did not realise that as soon as one box was filled another took its place, a process of substitution which continued for some little time. Meanwhile the landlord's family thought they might be doing similar deeds of saving, and began to pitch things out of the window. A feather bed happened to fall on an onlooker. The consequent excitement roused the bailiffs from their ease, but not until all the valuables of the Thomsons had been removed.

Although Tait was professor of pure mathematics in Queen's College, his real interest lay towards the physical side. Writing to his uncle, John Ronaldson, in 1858 he says:

"I have got the contoured map of Knocklayd from the Ordnance Office and have done a rough calculation which shows 10".28 as the effect on the plumb line, a very hopeful indication. If Thomson reports as well of the geology we shall commence in earnest next summer."

Knocklayd is a conspicuous hill of conical form in County Antrim, and evidently Tait contemplated using it after the manner of the Schiehallion Experiment to measure the mass of the earth. In one of his quarto note books there are tabulations of stars convenient for zenith observations which he purposed making with suitable instruments both at Belfast and at Knocklayd. Beyond these preparations, nothing more definite seems to have been done. Other problems had to be dealt with and the proposed book on Quaternions pushed on; and before two more summers had passed Tait had bidden farewell to Ireland and had begun his great career in Edinburgh.

EDINBURGH. 1860–1901

In 1860 the Chair of Natural Philosophy in Edinburgh University became vacant owing to the retirement of James David Forbes, and Tait offered himself as a candidate. The other candidates were Professor Fuller, King's College, Aberdeen; the Rev. Cosmo Reid Gordon, Manchester; Professor Clerk Maxwell, Marischal College, Aberdeen; E. J. Routh, Peterhouse, Cambridge; Edward Sang, Edinburgh; and Professor Swan, St Andrews. There is no difficulty now about placing these men in their appropriate niches in the Temple of Fame; but in 1860, when the best work of most of them was still to do, it could not have been an easy matter to discriminate among them. In the Edinburgh *Courant* of the day we find a remarkably sane and prescient discussion of the choice which the Curators had made. Some of the sentences are well worth quoting as showing that even in these days the characteristics of some of the men had been clearly diagnosed. After noting the distinction already gained by Fuller and Routh as eminently successful teachers, the writer disposes of their claims in comparison with those of Maxwell and Tait by the remark that neither "had as yet acquired a reputation for powers of original scientific investigation." With regard to Maxwell and Tait the writer continues

"it will be no disrespect to the warmest friends of the successful candidate, and we do not mean to dispute the decision of the curators, by saying, that in Professor Maxwell the curators would have had the opportunity of associating with the University one who is already acknowledged to be one of the remarkable men known to the scientific world. His original investigations on the nature of colours, on the mechanical condition of stability of Saturn's Rings, and many similar subjects, have well established his name among scientific men; while the almost intuitive accuracy of his ideas would give his connection with a chair of natural philosophy one advantage, namely, that of a sure and valuable guide to those who came with partial knowledge requiring direction and precision. But there is another power which is desirable in a professor of a University with a system like ours, and that is, the power of oral exposition proceeding upon the supposition of a previous imperfect knowledge, or even total ignorance, of the study on the part of pupils. We little doubt that it was the deficiency of this power in Professor Maxwell principally that made the curators prefer Mr Tait......With a clear understanding, and talents only second in order to genius, cultivated by persevering industry, he has attained to great and solid scientific acquirements, and to very much of that habitual accuracy which his rival, Mr Maxwell, possesses by a sort of intuition. We have never heard Mr Tait lecture, but we should augur from all we can learn that he will have great powers of impressing and instructing an audience such as his class will consist of, combined with that conscientious industry which is so necessary in a successful professor."

Whoever wrote these words or supplied the underlying thoughts had formed a just estimate of the respective strengths of the candidates. Fuller was certainly one of the greatest mathematical teachers any Scottish University ever possessed; Routh was unsurpassed in Cambridge as a trainer of Senior Wranglers and has, moreover, left his mark on dynamical science; Maxwell towers as one of the creative geniuses of all time, curiously lacking though he was in the power of oral exposition; Tait, who possessed, also by intuition, the clearest physical conceptions, has left behind him a great record of research both in mathematics and physics, while, as a teacher and clear exponent of physical laws and principles, he took a foremost place among his contemporaries.

He had all the gifts of a born lecturer. His tall form and magnificent head at once impressed the student audiences which gathered year after year on the opening day of the session. The impression was deepened as with easy utterance, clear enunciation, and incisive phrase, he proceeded to indicate the nature of the subject of study.

J. M. Barrie in *An Edinburgh Eleven* gives a graphic picture of Tait lecturing:

"Never, I think, can there have been a more superb demonstrator. I have his burly figure before me. The small twinkling eyes had a fascinating gleam in them; he could concentrate them until they held the object looked at; when they flashed round the room he seemed to have drawn a rapier. I have seen a man fall back in alarm under Tait's eyes, though there were a dozen benches between them. These eyes could be merry as a boy's, though, as when he turned a tube of water on students who would insist on crowding too near an experiment......"

This is good; but in some other respects Barrie's pen portrait is unsatisfactory if not misleading. For example in the succeeding paragraph he states that "Tait's science weighed him to the earth"—a remark almost too grotesque to need refutation. With regard to the real character of the man whose eyes could flash rapier-like glances or scintillate with heartiest merriment Barrie had, indeed, little chance of intimate knowledge. Tait used to speak of himself as a "lecturing machine" appointed by the University to instruct the youth of our country in the "common sense view of the universe we live in." Students were invited to send in their difficulties in writing before the lecture; but conditions were not favourable for personal intercourse between teacher and pupil.

Tait let nothing interfere with his official duties towards his class, declining on principle to make mention of anything but what had a direct connection with University regulations or College work. Once an enthusiastic secretary approached him with the request that he would announce a meeting of the highly important society represented by the petitioner. Tait opened his lecture with the remark that in this class room they met to discuss Natural Philosophy and that he made it a rule to speak only of what concerned the work of the class. A few mornings later there appeared in the public prints the announcement of the birth of his youngest son. As Tait appeared on the platform behind the lecture table he was greeted with a burst of applause, which lasted several minutes. In grim silence he waited till the noise subsided; then, with a quizzical glance round the full benches, he remarked—"Gentlemen, I said the other day that I make it a rule to take notice here only of what affects directly the work of the class." This pertinent sally was received with laughter and a ringing cheer, and then the students settled down to listen attentively to the lecture of the day.

To the student who passed through the general class of Natural Philosophy on the way to the ordinary degree Tait was the superb lecturer and nothing more. Those who entered the optional laboratory course or who took the Advanced Class with a view to honours were better able to appreciate his varied gifts; but a full revelation of the great personality came only to the privileged few who acted as his assistants, or who worked with him or for him in the laboratory. The sterling honesty of the man shone through all he did. As Sir Patrick Heron Watson once said, the charm of Tait was his naturalness—and he had known Tait from their boyhood's days. Sincerity was to him the touchstone of a man's character. Strong in his likes he was also strong in his dislikes. With true chivalry he fought for the claims of his friends if these were challenged by others. It was this indeed which led him into controversy. Thus arose the controversies with Tyndall concerning the history of the modern theory of heat and Forbes' glacier work, and the discussion with Clausius in reference to the thermo-dynamic discoveries of Kelvin. His passage at arms with Herbert Spencer Tait himself never regarded as anything else than a big joke.

As a lecturer Tait was probably unsurpassed by any of his contemporaries. His lecture notes were merely jottings of headings with the experiments indicated and important numerical values interspersed. In the original note book, which was in use till 1881, these headings were entered with intervening spaces so as to allow for additions as time went on. In 1881 he rewrote the greater part of the notes in a smaller octavo book, and this he continued to use to the end.

These lecture notes had to do with the properties of matter, which largely occupied the attention of the class for the first half of the winter session. Tait regarded this part of the course as a general introduction to the study of Natural Philosophy. He devoted the first few days to a discussion of the nature of the subject and of the means by which we gain knowledge of the physical universe. His treatment of the subjective and objective from the point of view of the natural philosopher was always clear and reasonable. I remember going back with a former classmate to hear Tait's opening lecture. Since we had first sat together in the benches of the Natural Philosophy Class room my friend had pondered deeply on metaphysical themes; and, as we listened again to Tait's exposition of objective and subjective, he whispered to me "Beautiful, Berkeley couldn't have done it better."

The conceptions of time and space, and the realities known as matter and energy, were introduced and placed in their right setting from the physical standpoint. These preliminaries disposed of, Tait began his systematic lectures on the properties of matter. His aim was to build a truly philosophical body of connected truths upon the familiar experiences of the race. In ordered sequence the various obvious properties of matter were considered, first, in themselves, then in their theoretical setting and their practical applications. Thus, to take but one example, the discussion of the divisibility of matter led to the consideration of mechanical sub-division and of the elementary principles of the diffraction and interference of light, illustrated by colours of soap films, halos and supernumerary rainbows. The fuller explanation of these was, however, reserved for a later date when the laws of physical optics were taken up in more detail. In this way the intelligent student was able during the first two months to gain a general outlook upon physical science. The nature of the course may be inferred from the contents of his book *The Properties of Matter*; but no written page could teach like the living voice of the master.

After the first few weeks the systematic lectures on the properties of matter were given during not more than three hours each week, Tuesdays and Thursdays being devoted to elementary dynamics. These were supplemented by some tutorial lectures by the assistant. The properties of matter having been disposed of, the subjects of heat, sound, light and electricity were taken up in turn, the amount of time given to each varying with different years. With the exception of heat Tait's lecture notes on these branches were not prepared with the same affectionate care as had been bestowed upon those dealing with the properties of matter. He had a few systematic notes on geometrical optics but none on physical optics or electricity. Indeed, as time went on, the properties of matter, like the Arab's Camel, encroached more and more on the limited time of the session. This was inevitable. Tait was always adding to his notes either new facts or new illustrations, and he never dropped any part out. His experiments hardly ever failed. They were chosen because they were instructive and elucidated the physical principle under discussion—not merely because they were beautiful or sensationally striking.

To the intelligent student who had worked through the earlier part of the course—namely, dynamics and properties of matter—the comparatively meagre treatment of physical optics and electricity was not perhaps of great consequence. He had been guided along a highway from which all parts of the great domain could be sighted and some information gained of each secluded region. He had been taught how to look and how to appreciate the view. He had been warned that the senses alone were untrustworthy guides; that he must illuminate the dark places with the light of reason, with the search light of a scientific imagination. To those of us who came with some knowledge of physical science, Tait's whole method was a revelation. But the great majority of those students who knew nothing of natural philosophy till they came under the fascination of his lectures were hardly in a position to appreciate the majestic beauty of the whole presentation.

In addition to the task of digesting the lectures the students were expected to do some extra reading on which they were specially examined. The junior division, that is, nearly the whole class, read Herschel's *Astronomy*; and the senior division, consisting of a few enthusiasts who were strong enough in mathematics, studied the first three sections of Newton's *Principia*. This home work was however purely voluntary even when, under the later regulations, the attendance of students at the examinations on the Class Lectures became compulsory.

To the advanced student able to follow him Tait was not merely a superb lecturer but was also a great natural philosopher and mathematician. The more abstruse the subject the more clearly did Tait seem to expound it. The listener felt that here was a master who could open the secrets of the universe to him. Unfortunately, when deprived of the aid of Tait's lucid exposition, in the easiest of English speech, of the knottiest mathematical or physical problems, the student, now left to himself, felt that his original ignorance was doubled.

In the Advanced Class Tait treated dynamical science in the manner of " Thomson and Tait." He does not seem to have kept notes of his course, but simply to have prepared his ideas the night before the lecture. In the earlier days down to about 1876 he used as a guide the elementary treatise known as "Little T and T'." Following the sequence of ideas there set down he developed the subject by use of the calculus. After 1876 he used for lecture notes a set neatly written out by his assistant, now Professor Scott Lang of St Andrews; but later he found his *Britannica* article on Mechanics with interleaved blank sheets more suitable for his purpose. In the end he lectured along the lines of his own book on Dynamics, which was largely a reprint of the Mechanics article with important additions on Elasticity and Hydrodynamics.

One outstanding feature of Tait's style of lecturing was its calm, steady, emphatic strength. He never seemed to hurry; and yet the ground covered was enormous. Was he for example establishing the general equations of hydrodynamics? Bit by bit the expressions were formed, each added item being introduced and fitted on with the clearest of explanations, until by a process almost crystalline in its beauty the whole formula stood displayed. All was accomplished with the minimum of chalk, but with sufficient slowness to allow of the student adding the running commentary to his copy of the formulae. The equations only and their necessary transformations were put on the black board, the student being credited with sufficient alertness of mind and agility of hand to supply enough of the explanation to make his notes remain intelligible to himself.

Though broadly the same, his advanced course varied in detail from year to year. For certain parts he had a particular affection, such as, applications of Fourier analysis, Green's theorem, and especially the theory of strains. The last named was, indeed, a subject peculiarly his own, and many of his demonstrations, although given in ordinary Cartesian coordinates, were suggested by the quaternion mode of attack.

An important feature of the Natural Philosophy Department since 1868 was the Physical Laboratory, for which Tait had secured a money grant as early as April 1867 but was unable at the time to find accommodation[1]. Lying quite outside any recognised course of study this purely voluntary course of practical physics offered no inducement to the ordinary student intent on getting his Degree. Tait's idea was to attract men who wished to familiarise themselves with methods of research. This he did by giving every encouragement to the man who had thought of some physical question worthy of investigation, or (as was more frequent) by suggesting some line of research to the eager student. Whoever showed real aptitude had all the resources of the Department placed at his disposal; and beyond the initial fee of two guineas for the first winter session no other charge was made, no matter how long the student continued to work in the laboratory. Those students whose interest in the subject brought them back after the first session of their enrolment, were nicknamed "veterans"; and on their enthusiastic help Tait largely depended for the successful carrying out of his many ideas. This will be brought out in Chapter II on Tait's Experimental Work.

Having given a broad outline of Tait's method of instruction I propose now to sketch briefly the main scientific events of his life, the more important of which will, however, be discussed in detail in later chapters.

On taking up the duties of the Edinburgh Chair Tait gave his first care to the preparation of his class lectures; and we get glimpses of the early development of his ideas from his letters to Andrews, for access to which I am indebted to the kindness of the Misses Andrews.

The following is his own description of his first lecture given on November 5, 1860.

"The Lecture (that is, the formal inaugural lecture) has not yet appeared in public. I began to-day, but, fancying that a dry technical lecture to commence with might perhaps keep off rather than attract amateur students, I gave a set of

---

[1] Thomson's laboratory in Glasgow began about 1850; and Carey Foster's in University College, London, was established in 1866.

experiments—the most striking I could muster—professedly without any explanation —in fact gave them as examples of the objects of Nat. Phil.......I gave a 20 m. lecture on the nature of the study, and the arrangements for the present session, and then plunged into the paradoxes. I reserved as the last the beautiful one of balls and egg shells suspended on a vertical jet of water, as they cannot be shown without some risk of a wetting to the performer and the nearest of the audience. To-morrow I bring into play the large American induction coil, and show the rotation of a stream of violet light in vacuo round a straight electromagnet. I shall also show an inch spark in air......and the discharge by it about 10 times per second of a jar with about 3 square feet of tin foil. There is no self acting break—for safety the interruption is made by a toothed wheel worked by hand—which for short experiments is much preferable. I shall also show the huge Cöln magnet (made under Plücker's direction) which took six of us to heave it up a gently inclined plane into the class room this afternoon......[1]"

Outside his official University work his tireless energies were finding other fields for exercise. He wrote most of the longer and more important physical articles as well as the article Quaternions for the first edition of *Chambers' Encyclopaedia* (1859–68) edited by Dr Findlater. His friendship with Findlater had important consequences; for it was he who first took Tait out to learn the game of golf on the Bruntsfield Links, where they played frequently together.

In 1861 he began the writing of Thomson and Tait's *Natural Philosophy*, while at the same time he was busy strengthening himself in the use of quaternions and preparing his book on the subject.

Together with Kelvin he communicated to *Good Words* in 1862 an article on Energy, which was intended as a corrective to Tyndall's statements regarding the historical development of the modern theory of heat. This led to two important articles in the *North British Review* which finally took shape as his admirable *Sketch of Thermodynamics* (1868).

Some curious speculations by Balfour Stewart as to the thermal equilibrium within an enclosure of a number of radiating bodies moving with different velocities led Balfour Stewart and Tait to plan a series of experiments on the heating of a disk by rapid rotation in vacuo. The results were communicated to the Royal Society of London; but no definite conclusion

[1] In these days a roomy platform a few steps above the floor both of the class room and retiring room lay behind the long curving table on which the experiments were arranged. About 1880 the rapidly increasing number of students compelled the addition of two new benches, and this addition was managed by removing the platform, lowering the table and setting it back nearer the wall. The old Natural Philosophy lecture room is now used by the Logic and Psychology departments.

could be drawn from them. The outstanding difficulty was the uncertainty that all possible sources of heating had been taken account of. After some years of laborious experimenting the research was finally abandoned.

When *Nature* was started in 1869 the Editor, (Sir) Norman Lockyer, secured the services of Tait not only as a reviewer of books but also as a contributor of articles; and, especially during the seventies, Tait supplied many valuable and at times very racy discussions of scientific developments.

In 1871 as President of Section A at the Edinburgh meeting of the British Association, Tait gave a characteristic address (*Scientific Papers*, Vol. 1, p. 164), in which Hamilton's Quaternions and Kelvin's Dissipation of Energy are held up to admiration.

The publication in 1873 of Tyndall's *Forms of Water*, in which the work done by J. D. Forbes in the elucidation of Glacier motion was somewhat belittled, roused Tait's indignation and led to a controversy of some bitterness (see *Nature*, Vol. VIII, pp. 381, 399, 431). Tyndall defended himself in the *Contemporary Review*; and Tait's final reply, in which Tyndall's quotations from the writings of Forbes are shown to be so incomplete as to lead the reader to a false conclusion, appeared in the English translation of Rendu's *Glaciers of Savoy* edited by Professor George Forbes (Macmillan and Co., 1874).

To *Good Words* of 1874 Tait contributed a series of most readable articles on Cosmical Astronomy, which embodied his lectures delivered to the Industrial Classes in the Museum of Science and Art, now known as the Royal Scottish Museum. At the same time Balfour Stewart and he launched their *Unseen Universe* upon an astonished world.

Before 1880 the Editors of the new edition of the *Encyclopaedia Britannica* secured Tait as the contributor of the articles Hamilton, Light, Mechanics, Quaternions, and Radiation. The two longest articles—namely Light and Mechanics—were afterwards published, with additions, as separate books; while the article on Radiation is practically embodied in his book on Heat.

While all this literary work was going on, he was studying the errors of the Challenger Thermometers, writing an elegant paper on Mirage, investigating the intricacies of knots, pushing on his quaternion investigations when leisure permitted and putting together his *Properties of Matter* (1885).

Throughout these years he also took a very practical interest in actuarial mathematics. A great believer in the benefits of Life Assurance he was for a lengthened period a Director of the Scottish Provident Institution. The Directors of this Company were divided into two standing Committees of Agency and of Investment. Tait naturally served on the former; but he was never happier than when engaged with James Meikle, the well-known actuary, in solving actuarial problems. The two men had, each of them, the greatest confidence in the other's capacity. Very often after Board meetings, Meikle would way-lay the Professor and draw him into his sanctum to discuss some knotty question.

The last heavy piece of mathematical investigation which fascinated Tait was the Kinetic Theory of Gases. Prompted by Kelvin, he wrote four important memoirs which by simplifying the mathematical treatment have greatly helped to clear up the difficulties inherent to the theory[1].

Before this work was well off his hands he was mastering the intricacies of the flight of the golf ball and planning experiments in impact and ballistics to elucidate some of the problems requiring solution. Not only did Tait in the end solve the main problem but it was he who first discovered that there was a problem to be solved. For hundreds of years Scotsmen had driven their balls over the historic links of St Andrews, Musselburgh, and Prestwick; but no one had ever put the question to himself, why does a well driven ball "carry" so far and remain so long in the air? The adept knew by experience that it was not a question of mere muscle, but largely of knack. It was reserved for Tait, however, to find in it a dynamical problem capable of exact statement and approximate solution. From his earliest initiation into Scotland's Royal Game, he began to form theories and make experiments with different forms of club and various kinds of ball; but not until late in the eighties did he begin to get at the heart of the mystery. Golf had now become a popular British sport, played wherever the English speech was prevalent; and Tait's second youngest son, Freddie, was rapidly coming to the front as one of the most brilliant of amateur golfers. While the son was surprising and delighting the world by his strong straight driving, his remarkable recoveries from almost unplayable "lies," and his brilliant all-round play with every kind of club, the father was applying his mathematical and physical knowledge to explain the prolonged flight of the golf ball. The practical golfer at first

---

[1] It is interesting to note that the first and second memoirs were translated into Russian by Captain J. Gerebiateffe and published with annotations expanding Tait's mathematical processes in the Russian *Review of Artillery* (1894).

smiled in a superior way at this new science of the game; and Tait was scoffed at when he enunciated the truth that underspin was the great secret in long driving.

It is interesting to see how step by step he advanced to the final elucidation of the whole problem or rather set of problems. Not until he had made definite calculations, did Tait or anyone else for a moment imagine that the flight of a golf ball could not be explained in terms of initial speed of projection, initial elevation or direction of projection, and the known resistance of the air. By means of ingenious experiments on the firing of guns, Bashforth had completely worked out the law of resistance of the air to the passage of projectiles through it. When however Tait tried to make use of the data supplied by Bashforth's tables he found that it was impossible to reach even an approximate agreement between his theoretically calculated path and the path as observed. Two facts were known with fair accuracy—the distance travelled by a well-driven ball, and the time it remained in the air; and a third fact was also with some measure of certainty known, namely, the angle of projection or the elevation. But no reasonable combination of elevation, speed of projection, and resistance of air could give anything like the combined time and "carry" as observed daily on the links. Tait also showed that, on this obvious theory of projection and resistance, very little extra "carry" could be secured by extra effort on the part of the player in giving a stronger stroke with a correspondingly higher speed of projection. The resistance of the air rapidly cut down the initial high velocities. When therefore Freddie Tait on January 11, 1893, exceeded far all his previous efforts by a glorious drive of 250 yards' "carry" on a calm day, he deemed that his father's dynamical theory was at fault.

How often has the tale been told on Golf links and in the Club-house that Freddie Tait disproved his father's supposed dictum, by driving a ball many yards farther than the maximum distance which mathematical calculation had proved to be possible! It is no doubt a good story, but very far indeed from the mark, as a glance at Tait's writings on the subject will at once prove.

On August 31, 1887, Tait communicated to the *Scotsman* newspaper an article called "The unwritten Chapter on Golf," reproduced a few weeks later in *Nature* (Vol. XXXVI, p. 502). In that article he shows clearly that the evils of "slicing," "pulling" and "topping" were all due to the same dynamical cause, namely rotation of the travelling golf ball about a particular axis and in a particular way. The explanation was based on the fact, established experimentally by Magnus in 1852 but already made clear by Newton in 1666, that, when a spherical ball is rotating and at the same time advancing in still air, it will deviate from a straight path in the same direction as that in which the front side is being carried by the rotation. Thus (to quote Tait) "in topping, the upper part of the ball is made to move forward faster than does the centre, consequently the front of the ball descends in virtue of the rotation, and the ball itself skews in that direction. When a ball is undercut it gets the opposite spin to the last, and, in consequence, it tends to deviate upwards instead of downwards. The upward tendency often makes the path of a ball (for a part of its course) concave upwards in spite of the effects of gravity...."

This last sentence contains the germ of the whole explanation; but it was not developed by Tait till four or five years later. Neither here nor in any of his writings on the subject is any rash statement made as to the greatest possible distance attainable by a well-driven golf ball. In his first article "On the Physics of Golf" (*Nature*, Vol. XLII, August 28, 1890) Tait calculates by an approximate formula the range of flight of a golf ball for a particular elevation and various speeds of projection, *the ball being assumed to have no rotation.* In this way by comparison with known lengths of "carry" he finds a probable value for the initial speed of projection. He also points out that, to double the "carry," the ball because of atmospheric resistance must set out with nearly quadruple energy. About a year later (Sept. 24, 1891, *Nature*, Vol. XLIV, p. 497), he treats more particularly of the time of flight. He finds that, although we may approximate to the observed value of the range of a well-driven ball by proper assumptions as to speed and elevation, it is impossible, along those lines, to arrive at anything like the time of flight. The non-rotating golf ball will according to calculation remain in the air a little more than half the time the ball is known from experience to do. "The only way of reconciling the results of calculation with the observed data is to assume that for some reason the effects of gravity are at least partially counteracted. This, in still air, can only be a rotation due to undercutting."

Thus he comes back to the rotation of the ball as the feature which not only explains the faults of slicing, pulling and topping, but is the great secret of long driving. When the rotation is properly applied as an underspin about a truly horizontal axis, the ball goes unswervingly towards its goal; but, when owing to faulty striking the axis of rotation is tilted from the horizontal one way or the other, there is a component spin about a vertical axis and the ball swerves to right or left according as the axis of rotation tilts down to the right or to the left. The clue was found, and the rest of the investigation was merely a question of overcoming the mathematical difficulties of the calculation. Thus undoubtedly before his son's brilliant drive of 250 yards' "carry," Tait knew well the influence of the underspin in prolonging both the range and time of flight; and before the summer of 1893 he had calculated the effect of the underspin sufficiently to establish the truth of his theory as a complete explanation of the flight of a golf ball. The results are given in the third article "On the Physics of Golf" (*Nature*, Vol. XLVIII, June 29, 1893), which is an abridgment of his first paper "On the Path of a Rotating Spherical Projectile" (*Trans. R. S. E.* Vol. XXXVII, *Sci. Pap.* Vol. II, p. 356). The theory is stated in popular language in an article on "Long Driving" communicated to the *Badminton Magazine* (March 1896) and reprinted below with slight additions and alterations made by Tait himself.

Following up the indications of his theory Tait attempted to improve the driving power of a "cleek" or "iron" by furrowing its face with a number of fine parallel grooves, which by affording a better grip on the ball might be expected to produce a greater amount of underspin. He got several clubs constructed on this principle; and four form part of the Tait collection of apparatus in Edinburgh University, having been presented to the Natural Philosophy Department by Mrs Tait. One of these is a "universal iron," in which the iron head in addition to being grooved is adjustable to all possible inclinations. The idea was to supply the golfer with *one* club having a degree of "loft" which could be varied at will. Tait himself found the weapon serviceable enough; but Freddie would have none of it.

The elucidation of the golf ball problem led Tait to another line of research, namely, the investigation of the laws of impact. These experiments and their bearing on the manner of projection of the ball are discussed in a later chapter, and in the article on "Long Driving" already referred to.

Outside his University duties Tait's energies were devoted mainly to the interests of the Royal Society of Edinburgh. Elected a Fellow in 1860 he became one of the Secretaries to the ordinary meetings in 1864, and in 1879 succeeded Professor J. H. Balfour as General Secretary. This important post he continued to hold till his last illness. With the exception of his early mathematical papers, his conjoint papers with Andrews and with Balfour Stewart, and a few mathematical notes communicated in later years to the

Mathematical Society of Edinburgh, all Tait's original contributions to Science are to be found either in his own books or in the publications of the Royal Society of Edinburgh. For many a year hardly a month passed without some communication from him bearing on a physical or mathematical problem. But whether he himself had a communication to make or not, he was always in his place to the right of the Chairman guiding the business of the Society and frequently taking part in the discussions.

The Royal Society of Edinburgh is no longer tenant under Government of the building in Princes Street known as the Royal Institution, the west wing of which had been planned for the Society when the building was erected. The need of more accommodation for the Society's unique library and for the National Art Galleries of Scotland demanded some change; and finally, in 1907, by Act of Parliament the Royal Institution was wholly given up to Art and the Royal Society was assigned a more commodious home in George Street. A description of the old Meeting Room, of which now only the outer wall remains, is not inappropriate in the memoir of one who was for fully thirty years the most conspicuous of the Society's permanent officials, and the most active contributor to its literature.

The arrangement of the room in which the meetings of the Society were held was certainly not convenient for modern requirements, such as experimental demonstrations or lantern exhibitions; but there was a peculiar dignity and old-world flavour about it which will long linger in the memory. It is easily pictured—an oblong room with doors at the ends flanked by crowded book-shelves. Along the east wall were two low book-cases, separated by fire-place and blackboards, and surmounted by portraits of illustrious Fellows such as Sir Walter Scott, Principal Forbes, Sir Robert Christison, and Professor Tait himself; and along the west wall were five windows looking towards the Castle. The President's Chair stood on a slightly raised platform in the very centre of the west wall before the central curtained window, and in front, running fully half across the width of the room towards the reader's desk, was a large oblong table, round which the members of Council were expected to sit. On this table the reader of the paper of the evening would place his microscopes or specimens or objects of interest. With the exception of the President and the leading officials, the Fellows occupied cushioned benches looking towards the large central table. The three secretaries sat invariably on the right of the Chairman, with their eyes towards the north door through which the members entered the room.

Occasionally, when the meetings were very full, part of the audience had to cross between the reader's desk and the Council Table and take their seats behind the secretaries' chairs.   For modern lecture purposes a worse arrangement could hardly have been devised; and yet it was quite in keeping with the fundamental idea of a Society whose Fellows met to communicate and discuss subjects of literary and scientific interest.   At any rate, the reader or lecturer from his position in front of the blackboard looked across his small table towards the President at the far end of the long table, and addressed the Chair in reality, not contenting himself with the formal phrase which has largely lost its significance.

It seems but yesterday when Piazzi Smythe with the peculiar hesitation in his speech uttered his *éloge* of Leverrier in the quaintly wrought involved sentences of a bygone century.   Or it was Kelvin moving eagerly on the soft carpet and putting his gyrostats through their dynamical drill; or Fleeming Jenkin amusing and instructing the audience with the sounds of the first phonograph which was used scientifically to analyse human speech; or Lister quaffing a glass of milk which had lain for weeks simply covered up by a lid under which no air germs could creep; or Turner demonstrating the characteristics of whales or of human skulls; or Tait himself talking in easy English about strains and mirage, golf ball underspin or kinetic theory of gases.

With the exception of the last two years of his life Tait hardly ever failed throughout his long tenure of the Secretaryship to be at the meetings of the Royal Society.   There he sat listening courteously—it might be to the most wearisome of readers who knew not how to give the broad lines without the details—or on the alert for the next bit of inimitable humour with which Lord Neaves when presiding used to delight the Society.   No one could enjoy a joke better than Tait; and who could resist being infected with his whole-hearted laugh or the merry twinkle in the eye which some humorous situation called forth?   To many of the frequenters of the meetings in the seventies and eighties, Tait was in fact the Royal Society; and there is no doubt that he guided its affairs with consummate skill.   At the Council meetings which occurred regularly twice a month during the working session all matters of business were carefully presented by him in due order.   It was his duty to conduct the correspondence of the Society, which during his Secretaryship grew steadily with the progress of the years.

Lord Kelvin, especially during his various terms of office as President, attended the meetings of the Royal Society of Edinburgh with fair regularity; and on the morning following the Monday evening meetings paid a visit to Tait's Laboratory immediately after the conclusion of Tait's lecture. It was then that we laboratory "veterans" had an opportunity of coming into closer touch with the great Natural Philosopher, who would occasionally pass round the laboratory and inspect the experiments which were in progress. Most instructive discussions would at times arise, Kelvin's mind branching off into some line of thought suggested by, but not really intimately connected with, the experiment. At other times the conversation between Kelvin and Tait turned on the papers which had been communicated the evening before. I remember a lively discussion arising on the statistical effect of light impressions on the eye. The argument was reminiscent of the old tale of the two knights and the shield; for while Tait was laying stress on the time average, Thomson was looking at it from the point of view of the space average.

For many years Tait's successive assistants reported the Meetings of the Royal Society to *Nature*; and this duty fell to me during the years 1879–83. At one of these meetings Sir William, as he then was, had in his well-known discursive but infinitely suggestive manner so talked round the subject of the communication that I had some difficulty in quite understanding its real essence. Next morning I tried to get enlightenment from Tait. He laughed and said "I had rather not risk it; but the great man is coming at twelve—better tackle him himself." When in due time Sir William was "tackled," he fixed his gaze at infinity for a few moments and then, a happy thought striking him, he said, with a quick gesture betokening release from burden, "Oh, I'll tell you what you should do. Just wait till the *Nature* Report is published—that fellow always reports me well." Tait's merriment was immense as he unfolded the situation, and he chaffed Thomson as to his obvious inability to explain his own meaning. Not a few of both Kelvin's and Tait's communications to the Royal Society of Edinburgh were never written out by them; they appear as reports only in the columns of *Nature*.

Another scene, in which Thomson and Tait were the main agents, rises in the memory. Once on a Saturday morning in summer when two of us were working with electrometer and galvanometer in the Class room Tait arrived in some excitement and said "Thomson will be here in half an hour on his way to London. He wishes to try some experiments with our Gramme machine and will need your cooperation with electrometer and galvanometer." Sir William soon appeared, and we were immediately commandeered into his service. And then followed the wildest piece of experimenting I ever had the delight of witnessing. The Gramme machine was run at various rates with various resistances introduced, and simultaneous readings of the quadrant electrometer and a shunted mirror galvanometer were taken. The electrometer light-spot danced all over the scale, and I had to bring it to reason by frequent changes in its sensitiveness demanding a continual retesting with a standard cell so as to be able to reduce to the same scale. Full of impatience and excitement Thomson kept moving to and fro between the slabs on which the instruments stood, suggesting new combinations and jotting down in chalk on the blackboard the readings we declared. Tait stood by, assisting and at the same time criticising some of the methods. At length Sir William went to the further side of the lecture table and copied into his note book the columns of figures on the blackboard. After a few hasty calculations he said:—"That will do, it is just what I expected." Then off he hurried for a hasty lunch at Tait's before the start for London where during the next week he was to give expert evidence in a law case. As they withdrew Tait looked back at us with a laugh and said " *There's* experimenting for you!" Early on Monday morning we were startled by a message from Tait who had just received a telegram asking for the numbers on the blackboard. Thomson had mislaid his note book! Also the original record had been obliterated! Fortunately for a man of Thomson's profound physical intuitions the loss would not be irreparable. He had in fact tested his theory as the experiments were in progress.

Tait's official position combined with his high reputation as mathematician and physicist brought him into touch with many of the great scientific men of the day. More especially was his verdict on questions of scientific history regarded with interest and respect, in spite of the fact that in several instances his views and those of his correspondents diverged considerably. I have quoted in a later chapter from both Helmholtz and Verdet in illustration of this point.

Many instances are to be found in his correspondence expressive of the esteem in which he was held by his contemporaries on the Continent of Europe. The letters display a friendliness of tone and a frankness of utterance which show that the writers, one and all, recognised his unfailing honesty of purpose and looked upon him as one whose opinion was worth the asking. The subjects discussed were chiefly scientific, but occasionally matters of purely personal interest were touched upon.

Up to the last year of his busy life, Tait's mind was for ever thinking out some new line of attack on the elusive laws of nature or on the properties of quaternion functions, while with ready utterance and facile pen he was teaching hundreds and thousands the grand principles as well as some of the mysteries of his science. As the years increased, he mingled less and less with general society. In his own home he was the most hospitable of hosts, full of story and jest, and alive to all the passing humours of the moment. Possessed of a verbal memory of unusual accuracy he could often suit the occasion with a quotation from one of his favourite authors,—Horace, Cervantes, Shakespeare, Scott, Byron, Dumas, Thackeray, Dickens, etc. It mattered not on what he was engaged, he had a ready welcome for his friends in the small study which looked out south across the Meadows. His shaded gas-lamp which stood on the table cast a shadow round the walls, somewhat further dimmed by the wreaths of tobacco smoke which stole slowly from his pipe—for though a steady he was not a rapid smoker. There he would sit when alone and work the long night through, rising occasionally to fill his pipe—as he once remarked " it is when you are filling your pipe that you think your brilliant thoughts." But let a visitor enter, then, unless there was a batch of examination papers to finish off before a certain early date, he would lay his work aside and clear decks for a social or scientific chat as the case might be.

In that den walled with book shelves and furnished with a few chairs, the table littered with journals, with proofsheets and manuscript, with books waiting to be reviewed, or with the most recent gifts of original papers from scientific men in every centre of life and civilisation—in that den Tait had entertained the greatest mathematicians and physicists of the age; Kelvin, Maxwell, Stokes, Helmholtz, Newcomb, Cayley, Sylvester, Clifford, Bierens de Haan, Cremona, Hermite, to name only some of those who are no more with us. There only was it possible to find him at leisure to discuss a scientific question. At college matters were different,—the lecture was just about to begin, or it had just ended and some other University work called for attention.

There were three outstanding occasions on which Professor Tait and Mrs Tait made their home a lively centre of science and fun,—namely, the British Association Meetings of 1871 and 1892 and the University Tercentenary Celebration of 1884. At the later meeting of the British

Association, the Natural Philosophy class room was the haunt of Section A; and Stokes, Helmholtz, Kelvin, and Tait sat side by side on the platform through most of the forenoon sederunts. The afternoons, however, were frequently given up to less formal gatherings. On one such occasion Mrs Tait's drawing room was converted into a lecture hall with lantern and screen; and C. V. Boys gave a seance of his flash photographs of the aerial disturbance produced by a bullet shot from a pistol. At another gathering which was purely social Mrs Tait, to make sure of the tea being absolutely perfect, had a kettle "singing" merrily on the open fire. Stokes and Kelvin were seated on a couch conversing diligently with a lady whose knowledge of Japan and Japanese students was interesting them—when suddenly a sharp hissing sound was heard above the talk and laughter which filled the room. "See" said the calm contemplative Stokes pointing with his finger, "the kettle is boiling over"; but Kelvin, who was furthest from the fire, leaped forward in his alert eager way, drawing out his handkerchief as he went, and lifted the kettle off just as Mrs Tait herself reached the hearth rug.

On another occasion, when the meeting of Section A was in full swing, Tait, wishing to show Helmholtz and Kelvin some of the experimental work which was in progress at the time, led them out quietly through the door into the apparatus room behind the platform and then down to the basement. Here in the large cellar containing the Admiralty Hydraulic Press he had some compression experiments going on; and in an adjoining cellar I was experimenting on magnetic strains. While Helmholtz and Kelvin were inspecting the arrangements and asking questions about the results a message came from the Secretaries of the Section demanding the presence of the three truants and especially of Lord Kelvin. Kelvin, however, was too eager over the problems of magnetic strains to pay immediate heed to the summons. Meanwhile Section A sat in silence like a Quaker's meeting. After a few minutes, a second and urgent message was sent to the effect that an important discussion in Section A could not be begun until Kelvin re-appeared on the platform. Reluctantly he tore himself away from the fascination of the research room, mounted the long stair, took his seat along with Helmholtz beside the President, and began almost immediately to occupy himself with a model on which he was to discourse an hour later.

In 1890 Tait tried his utmost to prevail upon Helmholtz to give the Gifford Lectures on Natural Theology in Edinburgh University. His letter of entreaty was as follows:

38 GEORGE SQUARE, EDINBURGH,
22/2/90.

My dear v. Helmholtz

I write to beg that you will give careful consideration to a formal document which will reach you in a day or two. It is to request that you will accept the post of Gifford Lecturer in the University of Edinburgh for the next two years.

The duties are not onerous, as they consist in giving 10 lectures in each year; and the remuneration is very handsome indeed. You would not require to spend more than a month, each year, in Scotland; and Glasgow is within such easy reach that you might spend part of the time there.

The terms of Lord Gifford's Will are such that the post may be held by *any one*; and we are particularly anxious that you should accept it, as a representative of so wide a range of thought. You have the inestimable advantage, over such men as Stokes and Thomson, of profound knowledge of Physiology. Besides, it is only a few years since Stokes occupied a somewhat similar (but more restricted) post in Aberdeen:—and we are of opinion that, at first at least, we should not appoint to the Gifford Lectureship a Professor (such as Thomson) in a *Scottish* University.

I can assure you of a most hearty welcome here; and we are sure to profit largely by your unfettered utterances.

Helmholtz, however, did not accept the offer; and Tait, who was anxious to have as Gifford Lecturer a man of recognised scientific reputation instead of the usual philosopher or theologian, prevailed upon Sir George Stokes to take up the burden. During the delivery of one of the second series in 1892 an amusing episode happened. It was a warm close afternoon, and Kelvin had come through from Glasgow to attend an evening meeting of the Royal Society. Wishing to honour his friend he accompanied Stokes to the platform along with Tait, Crum Brown, and other members of the Edinburgh University Senatus. Sir George had occasion to refer in his lecture to some of the views of Kelvin. When he came to the name he looked up with his beautiful smile and said "I little dreamed when I wrote those words some months ago that Lord Kelvin would be listening to me as I read them." The audience applauded heartily; and Kelvin who had been half dozing roused himself and joined in the applause!

Tait was invited by the Glasgow University Senatus to give the Gifford Lectures in that University; but he declined on the ground that so long as he had his Class Lectures to deliver he could not think of undertaking extra lecturing duties. When his last grave illness compelled him to resign he was no longer able for the task of preparing twenty lectures on natural theology. His own religious beliefs may easily be

5—2

inferred from the attitude of mind exhibited in the *Unseen Universe* or *Physical Speculations on a Future State* which Balfour Stewart and he wrote together. A more distinct utterance however is to be found in an article published in the *International Review* of November, 1878, and named "Does Humanity demand a New Revelation?" This article was largely polemical, being avowedly a reply to Froude who had communicated to the same Review some articles on "Science and Theology—Ancient and Modern." Towards the close of Tait's article these sentences occur:

"It would therefore appear, from the most absolutely common-sense view— independent of all philosophy and speculation—it would appear that the only religion which *can* have a rational claim on our belief *must* be one suited equally to the admitted necessities of the peasant and of the philosopher. And this is one specially distinguishing feature of Christianity. While almost all other religious creeds involve an outer sense for the uneducated masses and an inner sense for the more learned and therefore dominant priesthood, the system of Christianity appeals alike to the belief of all; requiring of all that, in presence of their common Father, they should sink their fancied superiority one over another, and frankly confessing the absolute unworthiness *which they can not but feel*, approach their Redeemer with the simplicity and confidence of little children.

* * * * * * * * *

All who approach the subject without bias can see from the New Testament records how some of the most essential features of Christianity were long in impressing themselves on the minds even of the Founder's immediate followers. And we could not reasonably have expected it to be otherwise. The revelation of Himself which the Creator has made by His works we are only, as it were, *beginning* to comprehend. Are we to wonder that Christianity, that second and complementary revelation, is also, as it were, only *beginning* to be understood; or that, in the struggle for light, much that is wholly monstrous has been gratuitously introduced, and requires a Reformation for its removal? What more likely than that, in the endeavour to frame a document for the stamping out of a particular heresy, over-zealous clergy should carry the process a little too far, and so introduce a new and opposite heresy? But this is no argument against Christianity; rather the reverse.

It might in fact be asserted, with very great reason, that a religion which, like any one of the dogmatic systems of particular Christian sects, should be stated to men in a form as precise and definite as was the mere ceremonial law, would be altogether an anomaly—inconsistent in character with all the other dealings of God with man—and altogether incompatible with that Free Will which every sane man feels and knows himself to possess."

Tait was indeed a close student of the sacred records. The Revised Version of the New Testament always lay conveniently to hand on his study table; and frequently alongside of it lay the Rev. Edward White's book on Conditional Immortality. I am not aware that he distinctly avowed himself a believer in this doctrine, as Stokes did, but he often expressed the high opinion he held of Edward White and his writings. His reverence for the undoubted essentials of the Christian Faith was deep and unmovable; and nothing pained him so much as a flippant use of a quotation from the Gospel writings. I have heard him reduce to astonished silence one guilty of this lack of good taste with the remark, "Come now, that won't do; that kind of thing is 'taboo'."

Tait's general outlook upon human affairs was fundamentally conservative. He had a deep distrust of Mr Gladstone as statesman and legislator. His strong political views did not however in any way interfere with his private friendships; and he refrained on principle from taking any public part in political discussions. He never failed to give his vote at an election; and was a consistent supporter throughout of the Conservative and latterly the Unionist Governments. When the South African War broke out he rejoiced to be able to send his son as a Lieutenant of the Black Watch to fight for his country and his Queen.

But swiftly came the stroke of sorrow as it came to many a family in the dark days of the South African War. Lieutenant F. G. Tait left this land with his regiment on October 24, 1899; on December 11 he was wounded at Magersfontein, where the Highland Brigade suffered so terribly; and after a few weeks in hospital he returned to the front only to meet his death on February 7, 1900, at Koodoosberg. The rumour of the tragic event came first through non-official channels and the uncertainty which hung over it for some days was harder to bear than if the worst had been immediately reported through the War Office. But there was no doubt of it; and all Scotland mourned the loss of her brilliant soldier golfer as she mourned few others of her warrior sons whose lives were cut short on the African veldt.

Tait's scientific work practically ended with his son's death. In December 1899 he communicated to the Royal Society a criticism on the "Claim recently made for Gauss to the Invention (not the Discovery) of Quaternions." It is a fitting finish to the publications of one whose controversies were always on behalf of others.

Meanwhile he was editing the second volume of his *Scientific Papers*, published by the Pitt Press, Cambridge. The Preface to the second volume is dated January 15, 1900. There is only one later printed statement by him—the preface to the seventh edition of Tait and Steele's *Dynamics of a Particle*.

The great physical and mental powers of the man were gradually beginning to fail. The vigour of his long stride was not what it had been. Yet in the keenness of ear and eye there was no abatement. Far beyond the years at which the great majority of normal-sighted men are forced to use spectacles or glasses, Tait was able to read his newspaper without artificial aid. Latterly, in reading an unfamiliar hand-writing he was occasionally compelled to hold it at the extreme stretch of his long arms; still he could read it—a very rare feat for a man of seventy.

During the spring and summer of 1900 he carried on his University work and his Secretarial duties at the Royal Society. He never failed to be present at the Council Meetings; but the general meetings of the Society saw less and less of him.

He and his family took their usual summer holiday at St Andrews, whose links in every hole and "hazard" were full of the memories of his son Freddie. But alas, the shadow of death had chilled these golden memories; and it was no surprise to his friends to learn that Tait returned to Edinburgh in the autumn none the better of his summer rest.

As he drew on his gown on the opening day of the session he confessed that for the first time in his experience he felt no desire to meet his new class. He was resolved in his own mind to complete the century at least in harness; but the task was too great for his waning strength. For nearly two months he carried on his lectures, to the great anxiety of all who knew and loved him best. On December 11, 1900, the anniversary of the Magersfontein disaster, he left the University, never again to pass within its portals.

He was indeed very ill: yet he himself never desponded, but spoke cheerily of looking in at College some day before the Christmas holidays, just to be able to say that he had completed the century. He was still able for mental work, and occupied himself forecasting his third volume of *Scientific Papers* and even criticising some of his own later papers published in the second volume. Once or twice in these days, when he was wholly confined to bed, he spoke to me of the linear vector function as something which still awaited development—there was a truth in it which had not yet been divined by the mind of man.

Tait formally retired from the duties of his chair on March 30, 1901. The Senatus expressed their appreciation of his long services in the following minute :—

"In taking regretful leave of their eminent and highly valued colleague, Professor P. G. Tait, the Senatus desire to place on record their warm appreciation of the ability and success with which, for the long period of forty-one years, he has discharged the duties and upheld the splendid traditions of the Chair of Natural Philosophy. They recognise with pride that his world-wide reputation as an original thinker and investigator in the domain of Mathematical and Physical Science has added lustre to this ancient university. A master in research, he is not less distinguished as an exponent of the Science with which his name will ever be associated. The zeal which inspired his Professorial work is well known to his colleagues, and has been keenly appreciated by successive generations of pupils, many of whom now risen to distinction have gratefully acknowledged their indebtedness to their teacher. In parting from their colleague, the Senatus would express the hope that he may speedily regain his wonted health and strength and be long spared to enjoy his well-earned leisure. He may be assured that he carries with him into his retirement their brotherly sympathy and affectionate regard."

On June 28, 1901, the Senatus resolved that the Honorary LL.D. degree be conferred on Emeritus Professor Tait. The formal intimation of this resolution was never seen by him.

Immediately after Tait's retirement a number of his former pupils resident in Edinburgh resolved to prepare an illuminated address, which would be signed by all former students who had made a specialty in laboratory work under his supervision. The address was illuminated by Mrs Traquair, who introduced round the margin illustrations of the various forms of apparatus which Tait had devised or used in carrying out his most important investigations. A portrait of Newton was placed at the top, and was flanked by scrolls, on which were inscribed certain Quaternion formulae and a few of the more characteristic lines of the Thermoelectric Diagram. Interwoven links and knots formed the foundation of the decorative design, and here and there appeared the names of Steele, Andrews, Thomson, Balfour Stewart, and Dewar, with whom he had collaborated in experimental and literary work. Immediately beneath the printed address was a group of curves taken from his papers : and then followed the sixty-three signatures in *facsimile* of the former students referred to above. Of these nearly thirty fill or have filled professorial appointments in universities and colleges both at home and abroad, while among the others we find eminent engineers and scientists, distinguished educationists, successful physicians, and vigorous self-denying clergymen.

Tait's constant companion through the weary months of illness was J. L. Low's *Record* of the life and golfing triumphs of Frederick Guthrie Tait. This finely written memoir gives a perfect picture of the generous hearted athletic Freddie, and traces with a genial literary touch his rise into the front ranks of golfers, among whom to this day his prowess is of undying interest. As Tait read and re-read the story of Freddie's peaceful victories he would live over again the happy rejoicings as medal was added to medal, or a new " record " was established, or another championship won.

As the summer of 1901 wore on there was no evidence of returning strength. In the hope that the change might be beneficial Sir John Murray offered his old Friend the use of his house and garden near Granton. Tait was greatly touched not only by the thoughtful care which prompted the act of kindness, but also by the loving solicitude with which Sir John gave all directions for his comfort and welfare. There in the secluded quiet of the garden of Challenger Lodge, carefully shielded from aught that might distract or weary, he passed through the last days of his pilgrimage. At first everything promised well.

On July 2 Tait felt able to return to his quaternion studies and covered a sheet of foolscap with brief notes of investigations in the theory of the linear vector function. This he handed to his eldest son, with the request to keep it carefully[1]. But it was the last effort of the keen vigorous mind.

Two days later on Thursday, July 4, 1901, the once strong life passed peacefully away.

There was cause for lamentation. Edinburgh had lost a son who had early brought fame to one of her oldest schools, and who had for forty years added to the renown of her University. Always strenuous, always devoted, always striving to extend our knowledge of the mysterious universe in which we live, full of interest in all that was best in humanity, and with a true reverence for the highest ideals of the Christ-like life, Peter Guthrie Tait had finished his appointed task.

On July 6 a large and representative company of Edinburgh citizens and University graduates assembled for the last sacred rites in St John's Episcopal Church, the Rev. Canon Cowley Brown and the Rev. H. S. Reid,

[1] The notes were afterwards published in *facsimile* by the Royal Society of Edinburgh, with a commentary in which I indicated their relation to his other papers on the same subject.

Professor Tait's son-in-law, officiating at the funeral service. The body was interred in the Church Yard immediately to the east of the church. The pall bearers were Professor Tait's three surviving sons, his two brothers-in-law (Professor Crum Brown and Mr J. S. Porter), Lord Kelvin, Sir Thomas R. Fraser, and Sir John Murray.

Among the many letters of sympathy which Mrs Tait and her family received during the sad days which followed Professor Tait's death, one may be given in full. It was from Sir George Stokes, to whom all through his life Tait looked as to a master, and from whom he had frequently taken advice and suggestions in his scientific work.

LENSFIELD, CAMBRIDGE.
9 *July*, 1901.

Dear Mrs Tait,

Now that the earth has closed over the remains of one most dear to you, permit me as a very old friend of your husband, and as one who not very long ago sustained a bereavement similar to that which you have just passed through, to express to you a feeling of sincere sympathy. When the last rites are over, and all is quiet again, the feeling of loneliness comes on all the more strongly. But we "sorrow not even as others which have no hope." Your husband was distinguished in the world of science. But it is more consolatory to you now to think of him who, with all that, looked "at the things which are not seen." We can think of him as one of those who in the beautiful language of the first reformed prayer book "are departed hence from us, with the sign of faith, and now do rest in the sleep of peace."

Pray do not trouble yourself to make any reply to this letter.

Yours very sincerely

G. G. STOKES.

The following extracts from letters written by former colleagues in Edinburgh University describe in appropriate language the real character of the man:

"To me he was a dear friend as well as a colleague, and in his loveable simplicity and warmth of heart one sometimes forgot his great gifts of intellect."

And again:

"No one could know him without being drawn to him by the warmest ties. My early recollections of him go back far into the past century. He was always so hearty and kindly, so ready to help and so pleased to have his friends around him. We all reverenced his gigantic intellectual power and were proud of all that he did for the advancement of science, but the charm of his buoyant and unselfish nature won our hearts from the very first."

Sympathetic letters were received not only from friends but from associations and corporations such as the Master and Fellows of Peterhouse, Cambridge, and the Students' Representative Council of Edinburgh University.

Full and appreciative notices of Tait's career and scientific work appeared in the leading newspapers, for the most part accurate, although here and there disfigured by some wild imaginings on the part of the writer. The able article in the *Glasgow Herald* is specially worthy of note. My own contribution to the *Scotsman* of July 5 was put together at a few hours' notice and was not of course seen by me in proof. I am not aware of anything inaccurate or misleading in the notice, although there were many points necessarily not touched upon. Professor Chrystal's article in *Nature* (July 25, 1901) gives an admirable sketch of his colleague's life and labours, with a sympathetic reference to the sincerity and honesty of purpose which were so characteristic of the man. Dr G. A. Gibson, who along with Sir Thomas R. Fraser attended him in the last illness, wrote a graceful biographical notice in the *Edinburgh Medical Journal* (1901).

Dr Alexander Macfarlane contributed to the pages of the *Physical Review* a sympathetic sketch of his old master; and Dr J. S. Mackay (mathematical master in the Edinburgh Academy) supplied a short biographical note to *l'Enseignement mathématique* (January 1905).

J. D. Hamilton Dickson's sketch in the Magazine of the Peterhouse Sexcentenary Club for the Michaelmas Term, 1902, gives, in addition to other matter, some interesting Peterhouse details as to Tait's undergraduate days.

Appropriate references were minuted by all the important organisations with which he was associated—the University, the Royal Society of Edinburgh, the Scottish Meteorological Society, the Cumming Club, the Scottish Provident Institution, etc.

After recording the main facts in connection with Professor Tait's labours as an official of the Royal Society of Edinburgh, the Council placed on record the following appreciation:

"This is not the occasion for an analysis of Professor Tait's work and influence. That will, no doubt, be given in due time by those specially qualified. What the

Council now feel is that a great man has been removed, a man great in intellect and in the power of using it, in clearness of vision and purity of purpose, and therefore great in his influence, always for good, on his fellowmen; they feel that they and many in the Society and beyond it have lost a strong and true friend."

The obituary notice in the *Proceedings* of the Royal Society (Vol. XXIII, p. 498) was prepared by Lord Kelvin. It contains, in addition to the customary biographical details, some interesting reminiscences of the days they worked together. Kelvin tells how they became acquainted in 1860 when Tait came to Edinburgh, and how they quickly resolved to join in writing a book on Natural Philosophy. He then continues:

"I found him full of reverence for Andrews and Hamilton, and enthusiasm for science. Nothing else worth living for, he said; with heart-felt sincerity I believe, though his life belied the saying, as no one ever was more thorough in public duty or more devoted to family and friends. His two years as 'don' of Peterhouse and six of professorial gravity in Belfast had not polished down the rough gaiety nor dulled in the slightest degree the cheerful humour of his student days; and this was a large factor in the success of our alliance for heavy work, in which we persevered for eighteen years. 'A merry heart goes all the day, Your sad, tires in a mile-a.' The making of the first part of 'T and T'' was treated as a perpetual joke, in respect to the irksome details of interchange of 'copy,' amendments in type, and final corrections of proofs. It was lightened by interchange of visits between Greenhill Gardens, or Drummond Place, or George Square, and Largs or Arran, or the old or new College of Glasgow; but of necessity it was largely carried on by post. Even the postman laughed when he delivered one of our missives, about the size of a postage stamp, out of a pocket handkerchief in which he had tied it, to make sure of not dropping it on the way.

One of Tait's humours was writing in charcoal on the bare plaster wall of his study in Greenhill Gardens a great table of living scientific worthies *in order of merit*. Hamilton, Faraday, Andrews, Stokes, and Joule headed the column, if I remember right. Clerk Maxwell, then a rising star of the first magnitude in our eyes, was too young to appear on the list....

After enjoying eighteen years' joint work with Tait on our book, twenty-three years without this tie have given me undiminished pleasure in all my intercourse with him. I cannot say that our meetings were never unruffled. We had keen differences (much more frequent agreements) on every conceivable subject,— quaternions, energy, the daily news, politics, *quicquid agunt homines*, etc., etc. We never agreed to differ, always fought it out. But it was almost as great a pleasure to fight with Tait as to agree with him. His death is a loss to me which cannot, as long as I live, be replaced.

The cheerful brightness which I found on our first acquaintance forty-one years ago remained fresh during all these years, till first clouded when news came of the death in battle of his son Freddie in South Africa, on the day of his return to duty after recovery from wounds received at Magersfontein. The cheerfulness never quite returned."

On opening his Divinity class the succeeding session Professor Flint uttered a beautiful tribute to the memory of his friend. This was published shortly afterwards in the *Student*, the Edinburgh University Magazine, and is now reproduced in full.

THE LATE PROFESSOR TAIT

AN APPRECIATION BY PROFESSOR FLINT

Since we last met here the University has lost through death the teacher who had been longest in her service, who was probably the most widely renowned member of her professorial staff. He was known to almost all of you not only by report but by personal contact and acquaintance, for almost all of you have come directly from his class room to the class rooms in the Divinity Hall. Undoubtedly it was a great advantage for our students here that they should have entered the Hall through that portal, and received the instruction and come under the influence of one universally recognised to have had not only a genius of the first order for research, but rare gifts as a teacher. He was not one whom his students were likely ever to forget, while many of them must have felt that they owed to him far more than they could estimate or express.

If you have not learned to be interested in the truths of Natural Philosophy, the fault cannot have been your teacher's, and unless altogether incapable of learning anything, you at least cannot have failed to learn the very important lesson that such a man's mind was immeasurably larger than your own.

Our deceased friend was a man of strong, self-consistent individuality. He was "himself like to himself alone." And he had about him the charm inseparable from such a character. He never lost the freshness of spirit which so soon disappears in the majority of men that it is apt to be deemed distinctive of youth. There was to the last a delightful boyishness of heart in him such as is assuredly a precious thing to possess. I am quite aware that great as he was, he had his own limitations, and sometimes looked at things and persons from one-sided and exaggerated points of view, but the consequent aberrations of judgment were of a kind which did no one much harm and only made himself the more interesting. His strong likes and dislikes, although generally in essentials just, were apt to be too strong. Although, like all great physicists, he was not really uninterested in metaphysics, yet he felt and professed the most supreme contempt for all that he called metaphysics. In connection with that I may mention an incident which once afforded much amusement to academic men in St Andrews, but is probably now forgotten even there. Shortly after Tait had delivered the remarkable lectures to which we owe the work entitled *Recent Advances in Physical Science*, he dined one evening at the house of the Professor of Mathematics in St Andrews, and among other guests present was a Glasgow Professor of Theology who had even less esteem for physical science than our dear departed friend had for metaphysics. Tait was very naturally drawn out to talk about the subjects which he had been lecturing on, and he did so largely and to the delight and edification of every one except the worthy and venerable Glasgow Professor, who, when he could stand it no longer, gravely put the question—"But, Mr Tait, do you really mean to say that there is much value in such inquiries as you have been speaking about?" After that the subject was changed, and during the rest of the evening the great physicist and great metaphysicist did little else than, as Tulloch expressed it, "glour at each other."

Tait was a genius, but a genius whose life was ruled by a sense of duty, and which was shown to be so by the vast amount of work he accomplished, and which is acknowledged by those who are ablest to judge of its worth, to be of the highest value. He was a genius with an immense capability of doing most difficult work, and he faithfully did it. His life was one of almost continuous labour. He faithfully obeyed the injunction, "Work while it is called to-day." And the work which he chose to do was always hard work, work which few could do, work which demands no scattering of one's energies, but the utmost concentration of them. He wasted no portion of his time in trying to keep himself *en evidence* before the world. He willingly left to others whatever he thought others could do as well or better than himself. But whatever he thought it his duty to undertake he did thoroughly. Thus for the last twenty years at least he was the leading spirit in an institution more closely connected, perhaps, than any other with the University of Edinburgh. I mean the Royal Society of Edinburgh.

It is natural for those of us who painfully feel that we shall not see his like again, natural for those who are most deeply deploring his loss, to wish that a longer life had been granted to him. Yet they may well doubt if he himself would have desired a mere prolongation of life. I cannot but think that he would not have cared for a life in which he could not labour.

While his friends must sorrow for his loss, they are bound also to acknowledge that God had been very good and gracious to him. He was favoured with many years of health and strength in which to work. His abilities were so conspicuous even in youth that they could not be hid. He could hardly have been earlier placed than he was in the very positions most favourable to the exercise of the gifts which had been bestowed on him. He was a Professor for forty-seven years, a Professor in Edinburgh for forty-one years. He was beloved by his students. His colleagues were proud of him. His country knew his worth. His many contributions are to be published in a suitable form at the cost of his English Alma Mater. He is among the rare few in a generation of whom the memories live through the centuries. Add thereto that his own worth and the value of his work were by none more fully appreciated than by those who were nearest and dearest to him, and that all distracting cares were spared him, and he was wisely left to follow the bent of his own genius. He had, so far as I know, only few great afflictions. The greatest which fell alike on him and his family was the loss of the generous, gallant, brilliant youth, who met a soldier's death near the Modder River, and in that loss a nation sympathised with him.

Our departed friend had no sympathy with theological dogmatism, and as little with anti-religious scepticism, and consequently held in contempt discussions on the so-called incompatibility of religion and science. At the same time he had a steady yet thoughtful faith in God, and in that universe which no mere eye of sense, aided by any material instrument, can see. That faith must have made his life richer, stronger, and happier than it would otherwise have been. And it must be a comfort to those who have the same faith, and to those who most deeply mourn his loss, to believe that he has entered into that universe which is so much vaster, and which may well have far greater possibilities of progress in truth and goodness in it than there are in the "seen" universe of us the passing creatures of a day. The things that are seen are temporal. The things that are unseen are eternal.

For none of his colleagues on the Senatus had Tait a greater esteem and affection than for Professor Flint. Sir Alexander Grant, who was Principal from 1868 to 1885, was regarded by Tait as the ideal tactful President, able to restrain the contending idiosyncrasies of the members of the Senatus and to guide their deliberations with unfailing courtesy. Professor Blackie, who ostentatiously scoffed at all things mathematical, used to ask Tait occasionally to give him some elementary instruction in analytical geometry. Tait drew the x and y axes and expounded their use with his accustomed clearness, and all went well until the teacher pointed out the need of the use of the negative sign, when the irrepressible Grecian broke away with the remark "Humbug, how can a quantity be less than nothing?" On one occasion in the Senate Hall shortly after Blackie had been uttering some strong patriotic sentiments Tait posed him with the conundrum, "What is the difference between an Englishman and a Scot?" The answer was, "Because the one is John Bull and the other is John (Kn)ox." Blackie replied to this chaff by throwing an ink bottle past the head of his tormentor.

In 1860, the Senatus numbered thirty, and in 1901 thirty-nine. During the forty years' tenure of his Chair, Tait had met in council with one hundred and seven colleagues, most of whom have left their mark in the history of theology, science, literature, or medicine. Of those who have passed away the following Principals and Professors may be mentioned, the latter in the order of their chairs as officially arranged in the University Calendar: Sir David Brewster, Sir Alexander Grant, Sir William Muir, Kelland, Pillans, Sellar, Goodhart, Blackie, Aytoun, Masson, Simon Laurie, Piazzi Smyth, Copeland, Fleeming Jenkin, Rev. Dr James Robertson, Sir Douglas Maclagan, Fraser Tytler, J. H. Balfour, A. Dickson, Hughes Bennett,

W. Rutherford, Laycock, Sir T. Grainger Stewart, Sir John Goodsir, Sir Lyon (Lord) Playfair, Sir J. Y. Simpson, Sir Robert Christison, Allman, Sir Wyville Thomson, Spence, Syme, Annandale, and Sanders. To them we may add the Chancellors, Lord Brougham and Lord Inglis.

Tait was awarded the Keith Prize twice (1867–9 and 1871–3) by the Royal Society of Edinburgh, and was the second holder (1887–90) of the Gunning Victoria Jubilee Prize. The Royal Society of London awarded him a Royal Medal in 1886 for his various mathematical and physical researches.

The following are the principal recognitions by Societies and Universities: Honorary Member of the Literary and Philosophical Society of Manchester, 1868; Honorary Doctor of Science of the University of Ireland, 1875; Honorary Fellow, Societas Regia Hauniensis (Copenhagen), 1876; Honorary Fellow, Société Batave de Philosophie Experimentale (Rotterdam), 1890; Honorary Fellow, Societas Regia Scientiarum (Upsala), 1894; Honorary Fellow of the Royal Irish Academy, 1900; Honorary Doctor of Laws, Glasgow University, 1901.

In 1882 some of Tait's many friends in Edinburgh commissioned George Reid (now Sir George) to paint a portrait of the Professor of Natural Philosophy in the act of lecturing. On the blackboard behind is the Curve of Vertices by means of which he elucidated the phenomena of mirage, with the Hamiltonian equation alongside. The general effect of the portrait is well described in the following contemporary criticism of the Exhibition of the Royal Scottish Academy.

"In portraits, George Reid's most characteristic effort is a portrait of Professor Tait. The grand domed cranium of the Professor of Physics, and his sagacious, solemnly comical face, seem to surmount a figure more likely to be met with in a Skye crofter's potato plot as a scarecrow, than among the amenities of a Scottish University. But the next look reassures you. The coat, as well as the noble head, is Professor Tait's veriest own—the coat, in fact, 'with which he divineth.' Even if the blackboard, and the high mathematical hieroglyphic thereon emblazoned, were silent, the Professor's 'office coat' is so redolent of chalk and experimental physics, that to old habitués of his class room, it would recount the tale of a hundred fights between the cutting mental gymnastic of the Professor and the mystic powers of mathematical abstraction. Altogether, this is a masterly portrait of a master, who knows no living rival in the sphere which he has made his own. As I stand and look on the characteristic picture I almost fancy that I can catch, on the solemn face of the grim mathematician demonstrator, some faint suspicion of a good-natured smile at the grotesqueness of the toggery in which he has chosen to be handed down to posterity."

This portrait was presented by the subscribers to Mrs Tait, who has now gifted it to the Natural Philosophy Department of Edinburgh University. It hangs in the library where the students gather to read the books of reference and study their notes.

Nearly ten years later Sir George Reid undertook a second portrait, which was subscribed to by Fellows of the Royal Society of Edinburgh. This portrait is the property of the Royal Society; but two replicas of it were made by Sir George Reid. One of these is hung in the National Portrait Gallery of Scotland, Queen Street, Edinburgh, and the other in the hall of Peterhouse, Cambridge. It is a three-quarter length portrait, and gives a faithful representation of Tait standing in a thoughtful attitude just in the act of elucidating some difficult point in mathematics or physics.

The Peterhouse portrait was unveiled on October 29, 1902, by Lord Kelvin, who gave some interesting reminiscences of how he and Tait worked together. The following report is from the *Cambridge Chronicle*:

"Lord Kelvin said he valued most highly the privilege of being allowed to ask the Master and Fellows of Peterhouse to accept for their College a portrait of Professor Tait. He felt especially grateful for this privilege as a forty-years' comrade, friend, and working ally of Tait. Their friendship began about 1860, when Tait came to Scotland to succeed Forbes as Professor of Natural Philosophy at Edinburgh. He remembered Tait once remarking that nothing but science was worth living for. It was sincerely said then, but Tait himself proved it to be not true later. Tait was a great reader. He would get Shakespeare, Dickens, and Thackeray off by heart. His memory was wonderful. What he once read sympathetically he ever after remembered. Thus he was always ready with delightful quotations, and these brightened their hours of work. For they did heavy mathematical work, stone breaking was not in it. *A propos*, perhaps, of the agonies (he did not mean pains, he meant struggles) of the mathematical problems which they had always with them, Tait once astonished him with Goethe's noble lines, showing sorrow as raising those who knew it to a higher level of spiritual life and more splendid views all round than it was fashionable to suppose fell to the lot of those who live a humdrum life of happiness. He did not know them, having never read 'Sorrows of Werther[1].'

> 'Who never ate bread in tears,
> Who never through long nights of sorrow
> Sat weeping on his bed,
> He knows you not, ye heavenly powers.'

But Tait gave it him in the original German, with just one word changed.

> 'Wer nie sein Brod mit Thränen ass,
> Wer nie die kummervolle Nächte
> An seinem Bette *rauschend* sass,
> Der kennt euch nicht, ihr himmlischen Mächte.'

[1] The passage is from the *Lehrjahre*, Book II, Chapter XIII

"Tait hated emotionalism almost as much as he hated evil, and he did hate evil with a deadly hatred. His devotion, not only to his comrades and fellow-workers, but also to older men—such as Andrews and Hamilton—was a remarkable feature of his life. Tait was a most attractive personality, and its attractiveness would be readily understood when he unveiled the portrait. It gave one the idea of a grand man, a man whom it was a privilege to know. His only fault was that he would not come out of his shell for the last twenty years, and that he never became a Fellow of the Royal Society of London."

Tait used to say that when he was young and would have liked to become a Fellow of the Royal Society, he could not afford it, and that later, when he could afford it, he had ceased to care about the distinction. It should be stated that from 1875 onwards Tait was never out of Scotland. His last visits to Cambridge were in 1874 and 1875, when he was Rede Lecturer and Additional Examiner in the Mathematical Tripos. Having, as it were, taken root in Edinburgh, he could have no very keen desire to become a Fellow of a Society whose meetings he would never have attended. For the last twenty-five years of his life he never left Edinburgh, except for a holiday at St Andrews of ten days in the spring and six weeks in the autumn[1]. Hence it came that he was personally unacquainted with most of the younger generation of scientific workers, and in this sense it is true that he did not come "out of his shell." But it must be repeated that the men of science who sought him out in his chosen haunts found the warmest of welcomes; and Mr Low's sketch will show how far Tait was from the crabbed recluse that the phrase suggests. About 1880 the President of the Royal Society suggested privately to Tait that he should allow his name to be submitted to the Council. Tait, who knew that the name of a valued friend whom he regarded as a genuine man of science had been recently rejected by the Council, replied that he had no pretensions to belong to a Society which was too good for his friend. This humorous excuse not only served its immediate purpose, but also, to Tait's delight, helped to procure for his friend soon afterwards the distinction he sought.

In "Quasi Cursores," the gallery of portraits of the Principal and Professors of Edinburgh at the time of the Tercentenary in 1884, the artist, William Hole, R.S.A., although very happy in most of his delineations, has not caught Tait quite satisfactorily. The attitude and figure generally are admirable, as are also the accessories of the Holtz machine,

---

[1] Tait's family can only recall one slight exception to this. In January, 1880, he delivered a popular lecture on Thunderstorms in Glasgow.

Leyden jars, and blackboard; but the expression of the face is not altogether suggestive to those who knew him well.

Of the likenesses reproduced in this volume one of the most striking is that from the photograph taken by the Rev. L. O. Critchley, when he was a student in the laboratory. He had been assisting Tait in some work requiring the camera; and, without the knowledge of the Professor he set the camera so as to photograph him in the act of writing a note. Tom Lindsay, the mechanical assistant, who is standing at the side ready to receive the note when finished, was in the secret. The portrait is admirable, giving not only a fine picture of the massive head, but also showing the usual condition of the writing table and general environment of what served as Tait's retiring room.

The establishment in 1903 of the "Tait Prize for Physics" at Peterhouse, Cambridge, was associated in an interesting manner with the execution of the portrait already referred to. Following up the proposal of Lord Kelvin and Sir James Dewar, the Master and Fellows of Peterhouse commissioned Sir George Reid to paint a replica of the portrait in the possession of the Royal Society of Edinburgh. The portrait was, however, more than a mere replica, for the painter worked into it reminiscences of his own long and intimate friendship with Professor Tait. Through the generosity of Sir George Reid a large portion of the funds contributed was left in the hands of the Treasurer, Mr J. D. Hamilton Dickson, who suggested that an effort should be made, by an appeal to a few other friends, to increase the fund until it should suffice for the establishment of a prize associated with Tait's name, to be given periodically for the best essay on a subject in Mathematical or Experimental Physics. In this way the fund for the foundation of the Prize was soon raised to two hundred pounds.

The idea of establishing a Tait Memorial in connection with the Natural Philosophy Department of the Edinburgh University occurred to many of Tait's pupils and friends. Considerations of general University policy prevented an authoritative appeal being made at the time. Nevertheless, quite unsolicited, a Tait Memorial Laboratory Fund took shape and began to grow. It has now reached the sum of nearly two thousand pounds.

On June 10, 1907, Sir John Jackson founded a Tait Memorial Fund, with the object of encouraging physical research in the University of Edinburgh on the lines of the work of the late Professor Tait. It is unnecessary here to give the whole Declaration of Trust, which may be found in the Edinburgh University Calendar; but as indicative of Sir John Jackson's personal feelings towards Tait the following quotation is of interest:

"LASTLY. I desire to place upon record that I have been induced to act in the premises as hereinbefore appearing from a deep sense of the advantages I as a student in the said University have derived from having been a pupil of the late Professor Tait and from a desire to assist instruction on similar lines to those followed by him for the benefit of future students in the said University."

In this closing sentence Sir John Jackson expresses the feelings of all who were serious students of Physical Science under Tait's guidance. One of the earliest of these was (Sir) James Dewar. There was then no Physical Laboratory, and I have heard Tait lament that he was unable to make use of Dewar's ability in those very early days. He was also in the habit of saying that one of the greatest services he did to experimental science was recommending him to Lyon Playfair as demonstrator and assistant. While still Professor of Chemistry in the Veterinary College in Edinburgh, Dewar frequently came to discuss physical problems with Tait at the laboratory; and in later years he never failed when he passed through Edinburgh to call on his old master and renew their fruitful intercourse. Shortly after Tait's death, Sir James was awarded the Gunning Victoria Jubilee Prize by the Council of the Royal Society of Edinburgh, and the sum received by him on this account he at once passed on to the Tait Memorial Fund as an expression of his regard for one to whom he owed so much.

Another frequent visitor at the Physical Laboratory was Dr Alexander Buchan, the well-known meteorologist, who could never rest satisfied with his own conclusions until he had sounded Tait on the physics of the problem. I have often heard Buchan express his great indebtedness to Tait for his valuable hints and criticisms.

But this feeling of indebtedness was not confined to those only who walked the pleasant paths of science. Many of his old pupils, who are now clergymen, physicians, teachers, lawyers, engineers, merchants, etc., retain not only a lively memory of the clear lecturer but a great deal of the principles of Natural Philosophy which he taught them. Some have even found his experimental illustrations useful in driving home spiritual and religious truths. Others, from their experience, have declared that what Tait taught them of the physical basis of things has been of more service in their pastoral work than most of the theology and church history they learned in their divinity course; and one maintains that a science degree in Mathematics and Natural Philosophy is probably more useful to a clergyman than a B.D. degree. This man, however, passed through Tait's laboratory, and was not an average specimen of the divinity student.

Before 1892 every Arts student was compelled to take Natural Philosophy as one of the seven sacred subjects; and even after 1892, although a certain amount of option was allowed to students, the majority who entered for the ordinary degree still passed under the spell of our great interpreter of Natural Law. Nevertheless, partly owing to the severity of the newly established preliminary examinations, partly to this introduction of option, the numbers of those attending the Natural Philosophy Class immediately fell off. From the outset Tait had little sympathy with the details of the New Regulations. In the diminished class which he met during the last eight years of his professoriate he saw one bad result of the University Commissioners' handling of the situation, and he never ceased to deplore that many students would hereafter pass out into the world with the degree of Master of Arts who had had no opportunity of learning the grand principles of Natural Philosophy. A great deal might be said in favour of this view of University study, more even now than formerly, when scientific developments bulk so largely in our modern civilisation. The difficulty mainly lies in the multiplicity of subjects now taught, all of them alike valuable as means of culture.

When Tait resigned his Chair in 1901 he was teaching the sons of men whom he had taught in the sixties and seventies; and it was with feelings of laudable satisfaction that he realised how he had served his University for two generations, and had impressed on the minds of fully nine thousand intelligent youths the great truths associated with the names of Archimedes, Newton, Carnot, Faraday, and Joule.

TAIT AT ST ANDREWS

By J. L. Low

It is the morning of a St Andrews' day in September; the early "haar" which had covered the Links like smoke has given place to sunshine and warmth, and the golfers are glad as they march in well matched parties, each player hopeful that he will make some notable performance. The last of the matches has left the first teeing ground, for it is nearing noon; the golfers are already in grips, and for every idle evening boast they are giving, as best they can, some sort of account. We enter the club-house and at once glance round the great smoking-room. It is deserted, save for the waiters who are gathering up the morning papers, which have had but a short perusal, and are placing them in order on the reading table. The scene is familiar to every golfer who remembers his September mornings, and there comes back quickly with this remembrance a figure which will not easily be severed from Golf and from its Fifeshire home. By the south fireplace on its right-hand side sits in the big arm-chair a venerable gentleman who was the oldest boy and the youngest old man we ever knew. The head is bent as the reader's eye glances quickly over the pages of the *Saturday* or the *Nineteenth*, his pipe is in his mouth, and by his side on a small table stands a tankard of small ale which he has ordered to make him not altogether forgetful of his Cambridge days. Here, alone in this big room, we would seem to have come across some recluse who would most strenuously oppose our interruption, and by his silence demand his peace. But in a moment the whole man changes; in a second he rebounds from sixty to sixteen, and by the mere raising of the head throws off the garment of his years. The head is the head of the scientist, and the brow, not without its furrows, tells of problems solved and yet to be solved. But the eye, though small, twinkles with an unquenchable boyishness which will not grow old, and the fullness which lies beneath it proclaims that sense, whether of measure, of words, or of music, which always accompanies this peculiarity of feature.

Before we can speak he is laughing; he greets us heartily, and demands, in order that we may laugh with him, that we read some passage he has just been enjoying. It is a dull passage on some subject we do not understand; but his eye twinkles when he marks that we detect in the writing some absurd incongruity of expression. "What do you think of that, my boy, from a professor of Philosophy?" he exclaims, and then, as if to be quit of the thing, he rises, shakes himself, knocks the tobacco ash off his waistcoat, and adds:—"Well, let's go out and meet Freddie, he will be past the turn by now." We, who were but golfers and fellow-sojourners in a city full of golf and professors, called this boy-man "The Professor," and we loved him.

I have been asked to add to the content of this biography as it were the side glance of the golfer to the all important view of Professor Tait as the great scientist. It was not far from the fitness of things that the Professor who was so full of Scottish character and was so well equipped as a mathematician and a philosopher should have found in the national Scottish game a field agreeable alike to his physical and mental recreation. Some recollection of him from the golfers' standpoint is therefore suggested, and is indeed the object of these reminiscences.

The Tait connection with golf is dual; for the Professor is known to many by the title he was wont, with his keen sense of humour, most to delight in, "The father of Freddie Tait." The Professor was a well-known figure at St Andrews from 1868 onwards to the end, and golf was his favourite recreation long before the prowess of his sons connected the name of Tait so closely with the national game; but it was not until his sons were beginning to show signs of great aptness for the sport that the father began those experiments which have not only been of importance to the student of natural philosophy but have intimated to the golfer the fact that he was playing a game which was a science as well as an art. It is reported of Freddie that, in reply to a question addressed to him by the Czar of Russia, he stated that he "took seriously to golf when he was eight years old"; of the Professor it may be said that he never took to the game seriously; by this I mean that his interest in the game was athletic and philosophical rather than competitive.

About 1860 the Professor made his beginning as a golfer in early morning rounds on Bruntsfield Links; golf is still played on the historic ground but the fair way is intersected by paths, and play is now allowed only at holes of a mean length. However, in these days of the early sixties the course was of sufficient importance to warrant its being the scene of an open tournament, to which came such heroes as the late Hugh and Pat Alexander.

For many years after this the Professor was in the habit of taking his pleasure at Musselburgh, and was a member of the Honourable Company of Edinburgh Golfers until that society removed to Muirfield. At Musselburgh he was in the habit of playing with Lord Inglis, with Mr A. D. Stewart, and with others who had long been accustomed to fight furiously with feather-stuffed balls in the depths of "Pandy." In 1868 the Professor began those visits to St Andrews which were continued without intermission until the year of his death. Being a regular glutton for play his daily rounds often amounted to five; and though his strength was equal to the task, he required, needless to say, several caddies to help him during his uncommon performance; these were at first chosen for him by the late Dr Blackwell, father of illustrious golfing sons. Of these five rounds the one he loved the best had its start at 6.30 a.m. and was not equally popular with the other members of his family. It is indicative of his boyish nature that not only did he play at an hour when birds alone should be playing but even sang about it in an ode which he appropriately signed "The Glutton."

THE MORNING ROUND (6—8 A.M.).

AIR—"BEAUTIFUL STAR."

1. Beautiful Round! Superbly played—
   Round where never mistake is made;
   Who with enchantment would not bound
   For the round of the morning, Beautiful Round?

2. Never a duffer is out of bed;
   None but the choicest bricks instead
   On the Links at six can ever be found;
   Round of the morning, Beautiful Round.

3. There they lie in a hideous doze
   Different quite from a golfer's repose—
   That from which he starts with a bound
   For the round of the morning, Beautiful Round.

4. Agile and light, each tendon strung,
   With healthy play of each active lung
   He strides along o'er the dewy ground
   In the round of the morning, Beautiful Round.

5. Beautiful Round! most cleverly won
   Under the gaze of the rising sun,
   And hailed with a pleasant chuckling sound
   Round of the morning, Beautiful Round.

6. Beautiful Round! vain duffers try.
   Thy manifold virtues to deny :—
   They ! ! ! mere specimens of a hound:
   Round of the morning, Beautiful Round.

7. Beautiful Round in thee is health,
   The choicest gem of earthly wealth :—
   Hands and face most thoroughly browned;
   Round of the morning, Beautiful Round.

8. Beautiful Round! to thee is due
    All the work I am fit to do:—
    Therefore in fancy stand thou crowned
    Queen of the morning, Beautiful Round.

9. Beautiful Round! I think of thee
    Through months of labour and misery:—
    Round thee the strings of my heart are wound,
    Round of the morning, Beautiful Round.

Among those who were his companions on his rounds early or late were Mr Tom Hodge and the Bethunes, Lord Borthwick, Lord Rutherford Clark, and Mr James Balfour; Professor John Chiene and Lord Kingsburgh were also at times his opponents; and as partners in foursome play he had Old Tom and Young Tom, and the Straths. He always stoutly maintained that given similar conditions Young Tom's play was equal to that of the best of the modern professionals. From the beginning of his golf the Professor used very upright clubs and played with the largest ball he could get—"a thirty"—as compared with the more common "twenty sevens" and "twenty eights."

In 1871 the meeting of the British Association was held in Edinburgh, the Professor being President of Section A. After the proceedings were finished some of the most distinguished members of this assembly accompanied him to St Andrews. Among these were Huxley, Helmholtz, Andrews, and Sylvester. Helmholtz took no interest in golf and "could see no fun in the leetle hole"; but Huxley played a round every afternoon during his stay of two months. He lived in the house known as Castlemount, hard by the Castle gate:—the house is now occupied by Dr Hunter Paton, whose family was at that time intimate with the Huxleys. In the afternoon round the Professor's eldest son Jack was in the habit of partnering Robertson Smith against Huxley and various people. Jack was only a small boy and no doubt too young to appreciate the excellence of the company in which he found himself; and indeed seems to have taken rather a high-handed position as regards these matters. At St Andrews there used to be an idea that the weaker player should make the easy drive at the first hole and Jack on one occasion was asked to perform this trivial task, but refused, declaring that he was "not the biggest duffer of the party"; this greatly amused Huxley, who willingly accepted the chastisement and topped his ball gently towards the road.

The St Andrews of those days was a city quite other than the fashionable watering place of to-day. The society, though small, was intellectual, and though intellectual yet devoted to the jests which are dictated by humour: the merry parties of the small colony were more than willing to enjoy at the seaside that freedom which is curtailed in the larger cities. The Professor was, from the nature of the man, the leader in everything which tended to humour and gaiety. It is difficult to imagine any man of years who day by day seemed so devoted to what, for lack of a more dignified term, must be called "fun"; one felt sure that he found jokes in his algebraical symbols, and jests even in his quaternions. It is the dinner hour and the Professor proposes to the company that a round may be played with phosphorescent balls. When proper arrangements have been made the party assemble at the first teeing ground. To this match come the Professor and his lady, Huxley, keen on the humour of the thing, Professor Crum Brown and another friend. The idea is a success; the balls glisten in the grass and advertise their situation; the players make strokes which surprise their opponents and apprise themselves of hitherto unknown powers. All goes well till the burn is passed, and Professor Crum Brown's hand is found to be aflame; with difficulty his burning glove is unbuttoned and the saddened group return to the Professor's rooms, where Huxley dresses the wounds. The pains of the phosphorescent hand having been mitigated by the tender care of the great scientist, it is not difficult to picture the fun which our Professor would derive from the night's adventure. In a nature so strong we cannot but expect to meet an accidental note which gives the theme originality. The Professor was a man of very strong, and as it seemed to some of us, almost unreasonable antipathies; endowed as he was with a humour which, had he given it vent, could have been magnificently satirical, he dealt by argument with those he did not favour, allowing the joy and humour of his nature to play only on his friends, and more particularly on his own family and his more intimate circle. Of a morning his opening words had relation to a small incident of home life; he would tell of something that had given him a chance of chaffing Freddie or Alec, or playing a practical joke on some member of the family:—one such story must suffice to exemplify. The Taits had a house in Gibson Place overlooking the Links on one side and the old station road on the other. The front door was generally open and an umbrella stand which stood by it seemed to the Professor to offer a too easy prey to the light-handed. Mrs Tait took another view and said that "no one stole in St Andrews"; but when the Fair day arrived, and she went for her parasol, she found that the stand had been pillaged. She immediately informed the police and went to the railway station to see if the thief was escaping by train. Returning she found the Professor and General Welsh finishing their round, and at once said, "Guthrie, you were quite right, the umbrellas are all gone." The Professor's eye sparkled as he asked, "What steps have you taken?" On being told he resumed his game; but when lunch was finished he pulled back the curtains and disclosed the umbrellas. Mrs Tait found herself in a position of some embarrassment as she had to tell the policeman that the affair had been a hoax; and this worthy, who afterwards became the well-known and respected Inspector, did not in any way relieve the situation by saying that he had suspected the truth from the first. For many years after the incident, Mrs Tait was in the habit of crossing the road rather than meet her late colleague in the cause of justice.

With the advancing years the exuberance of the Professor's golf decreased; the two round limit was never exceeded. In the later eighties he played but little, and after 1892 never a full round; but only the nine outward holes followed by a rapid walk home by way of the new course. This athletic decline on the part of the Professor synchronises with Freddie's advance as a golfer; it also marks the beginning of the transference of the former's interest to the philosophical side of the game. The Professor's famous experiments were begun in 1887 and reported in *Nature*, August, 1890, September, 1891; and his full theory was complete in 1893. He also wrote articles on "The Pace of a Golf Ball," *Golf*, Dec. 1890; "Hammering and Driving," *Golf*, Feb. 19, 1892; "Carry," *Golf*, August 25, 1893; "Carry and Run," *Golf*, Sept. 1893; "The Initial Pace of a Golf Ball," *Golf*, July 17, 1894; and he contributed an important summary of his work in a paper to the *Badminton Magazine*, March, 1896 (reprinted below).

One of the Professor's most interesting pieces of mathematical work deals with the subject of Rotating Spheres and Projectiles; but as this has been adequately discussed in another part of this biography, it will be sufficient if we glance at the general results as they appeared to the golfer. Prior to the Professor's investigations we imagined that speed of projection, elevation, and the resistance of the air, were the three things which determined the flight of a golf ball. The Professor indeed seems himself to have begun from this standpoint; and his first discovery was that we were all wrong. He told us that we imagined we knew all the laws under which a golf ball flew, but that these laws were in themselves insufficient to explain the duration of the flight, and that he proposed to find out what was lacking in the sum of our knowledge; he discovered in fact that there was a problem to solve. Mr H. B. Farnie, and afterwards Sir Walter Simpson had told us of the Art of Golf; the Professor detected that there was a Science of Golf, and afterwards worked out and communicated the problems which he had discovered and solved. There is a story that Freddie demolished his father's arguments by driving a ball further than the limit that had been set by the Professor. Freddie perhaps half believed that he had created this joke against "the Governor," for he never studied his father's articles very closely, as we can judge from the fact that it is not till the end of 1898 that we find him writing to Jack to announce that "the Governor's theory is underspin." The grain of truth that was in the story was made into a good jest by the facile pen of Mr Andrew Lang. The Professor indeed said in *Golf*, Dec. 1890, that from the *theoretical data* it appears that to gain ten per cent. of additional carry a long driver must apply nearly fifty per cent. more energy. But this statement must be read with his explanatory remark in *Nature*, "I shall consider the flight of a golf ball in a dead calm only, and when it has been driven fair and true *without any spin*." The essence of the Professor's discovery was that without spin a ball could not combat gravity greatly, but that with spin it could travel remarkable distances. In the first place the Professor found that a golf ball combated the attraction of gravity for a period nearly twice as long as he had expected. By floating marked golf balls in strong brine or mercury he found that they did not float truly, but wobbled, and that the marked spots ultimately came to certain fixed positions; from this he gathered that the centre of gravity of a ball seldom, if ever, coincided with its centre of figure. This fact, taken in conjunction with an assumed rotation, at once explained the violent wobbling in the air occasionally observed. Slicing and pulling proved the existence of spin about an axis not truly horizontal; and mathematical calculation showed that underspin, by introducing a lifting force, would increase the flight of the ball. The sufficiency of the omitted factor was made clear. This discovery has been of the utmost importance to the golfer, and is in fact the groundwork on which the modern school of scientific play has been built.

The Law which was known to Newton, and investigated by Magnus, viz. "That a sphere rotating and advancing in still air deviates from its straight path in the same direction as that in which the front side is being carried by the rotation," is the law which governs all slicing and pulling, topping and skying. We say that we slice if we stand in some particular position; but we may stand as we like and slice, if only we make the front side of the ball rotate from left to right during its progress. This knowledge of the power of spin having been placed in the hands of the golfer it became necessary for him to find out how he could make strokes which would cause the ball to turn from the right or left, or to rise in its flight and to stop without running, or to make but a short upward journey and then reach the ground with great power of run. Of the strokes indicated the last named is by those ignorant of the finer points in the game called the "common top"; but it is very far from this, and is a shot which was brought to great perfection by Freddie and by Mr J. E. Laidlay, and when well played from a suitable situation is a fine thing to see done. When the ball is topped it is struck above its centre and rolls in an irresponsible manner along the ground. In the proper stroke the ball is struck with a lofted club well below the belt, and is thus assured of a definite carry; but just as the head of the club reaches the ball an upward movement is given which imparts overspin and causes the ball to run after it touches the ground. This is the true overspin stroke, known to experts as the "rising club shot." Another stroke which has been understood through the Professor's discovery is the "long carry" over a hazard. The Professor showed that it was not necessary, or indeed advisable, to start the ball with a high trajectory, and that the low stroke which goes, because of underspin and in spite of gravity, concavely upwards produces the best result. These examples may be sufficient to show how deeply golfers were indebted both practically and intellectually to the increased interest he bequeathed to the game.

The Professor's experiments were of course conducted with the gutta ball and some of his conclusions have therefore been modified by the introduction of the more resilient rubber core ball. Speaking very roughly, he arrived at the conclusion that in the case of a full drive at the moment of impact the clubhead was travelling at the rate of 200 feet per second, and the initial velocity of the ball's projection was 300 feet; with the newer balls the initial velocity will no doubt be greater; and it is also possible that their greater carry may be influenced by their greater willingness to receive underspin, and as a consequence to allow of their being struck with a very low trajectory. The Professor, perhaps, laboured his theory of underspin too far, and his sons used to regard rather with amusement his famous underspin iron. This weapon was a very light upright cleek with ridges on the face running parallel to the base of the head. I remember the Professor asking me to have a shot with it and telling me that if I hit the ball fast enough I would drive from the "Sandy Road" over the burn. What he wished to impress upon us was that the speed at which the clubhead was travelling and the proper amount of underspin are the two chief factors in long driving; but he never looked to see us drive a great distance with this club, for he knew as well as we did that the head was too light to bring out the resilience of the ball, a most important practical factor.

The introduction of the "Bulger" of course interested him; but he was not in favour of the weapon, for it did not assist him in his theory of underspin, since it was intended to obviate the evils of rotation about a vertical not a horizontal axis. The *true Bulger*, he said, should have its vertical section convex. Over the initials G. H. there appeared in the *Scots Observer* some verses which the Professor afterwards acknowledged, describing them as "expressive at least, if not wholly elegant," which we reproduce as they have a ring of the author's humorous philosophy. The initials G. H., I believe, represented the name Guthrie Headstone, the play on the words Head and Tait and Peter and Stone being obvious.

THE BULGER.

1. From him that heeleth from the Heel,
   Or toeth from the Toe,
   The Bulger doth his vice conceal;
   His drive straight on doth go.

2. To him who from the Toe doth heel,
   Or from the Heel doth toe,
   The Bulger doth his faults reveal,
   And bringeth grief and woe.

3. And the poor slicer's awful fate,
   Who doth a-bulging go,
   Is sad indeed to contemplate;
   The Bulger is his foe.

4.   But whoso plays the proper game,
       His ball who striketh true,
     He findeth all clubs much the same;
       A goodly thing to do.

MORAL.

Bulgers, and Mashies, Presidents,
  Are for weak players made;
As spectacles and crutches be
  For eyes and limbs decayed.   G. H.

Returning with the Professor to the club-house, we notice that the golfers freely greet him, as he quietly retires to his accustomed seat, or finds a companion for an afternoon game at billiards. In this community, full of cosmopolitan elements, the great man walked humbly and was accessible to everyone. On a doubtful morning no one started for a round without asking him if an umbrella should form part of the caddie's burden; and his opinion was always backed against the barometer. The Professor seldom addressed anyone, but of all the notables he was the most easy of approach. No topic of conversation was foreign to his interest; and the more remote the subject from the beat of his scientific enquiries the more were we astonished by the intimate manner in which he threw himself into the discussion. On politics he held tremendous views; and his eye glistened as he read a slasher in the *Saturday Review*. In his Edinburgh home he was not a club man, and I believe he refused to join in any way in club life; but in his holiday time he loved to mingle with the golfers, and enjoyed greatly his billiards. Although not a great player, his intimate knowledge of angles gave him a fine field for amusement and experiment as he tried almost impossible cannons. To an opponent who had indulged in a very forceful game, I remember him remarking that the play had seemed to be a combination of bagatelle and racquets. But these hours in the billiard room were for him, especially in later days, sources of splendid recreation.

Many great men have been drawn to St Andrews, and have gone in and out of the Royal and Ancient Club; but probably no man so great has ever come so closely in touch with its members. We knew that he knew the mysteries which our minds could not grasp; but the man as he walked among us put himself, almost with diffidence, on our level and invited our opinion. We, who had not been his pupils, were thus able to guess the cause of that power and fascination which he had exercised over generations of Edinburgh students.

The Professor never seemed to be far from any one of us; he disguised the fact that he was in touch with the immortals.

Mixing with all, and always friendly with all, his heart was nevertheless fixed within the circle of his own home; and, as we write of him, more particularly on Freddie and his doings on the links. What Freddie had done, what match he was playing, what chance he had at the next Championship, or medal, these were the thoughts always near to him.

Freddie was his companion in his experiments, making herculean drives against the apparatus prepared by the Professor. Freddie chaffing "the Governor," is still the better loved Freddie. Freddie fighting in South Africa, wounded, but making a good recovery, remains the father's idol. It was little wonder then that in that dark February of 1900, when the bad news came, the Professor, the man of rock, was rent.

A few months later, when on my way from St Andrews to Sandwich for the Championship meeting, I dined with the Taits in Edinburgh before starting on the night train. Through dinner the Professor seemed very depressed as though afraid to enter into any conversation which might become reminiscent of the golf which had Freddie for its central figure. I tried to draw him on to subjects which involved no risk; but a most unnatural heaviness seemed to hang over him. After dinner, in his study at the back of the house, he showed some return of his old boyish nature, and made some pithy remarks about the players who were likely to be at Sandwich. I was looking at some shelves full of old text books while he was attending to some small note he had to answer; suddenly he turned round and called out, "We have new editions of all these." This pregnant remark was followed by his old laugh; and until I left his conversation was as bright as in former days. Yet I do not think that he ever got back into his true gait after Freddie's death; the light seemed to have left the eyes which in repose often wore an expression of weariness.

The passings of Father and Son were in striking contrast; Freddie died before his life was fulfilled: the Professor died after he had searched the philosophies and completed his investigations. The Professor's favourite theme was the Law of Continuity. It has been well said that every ultimate fact is but the first in a new series; the Professor was still a boy when he left us.

CHAPTER II

EXPERIMENTAL WORK

CLEAR indications have already been given that from his early student days Tait's main interest was in physical rather than in pure mathematical science. His first experimental work was done in Belfast under the guidance of Andrews, whom he assisted in the preparation of three papers on Ozone. These appeared in the *Proceedings* of the Royal Society of London between 1856 and 1857. Already in 1855 he had visited the Paris Exposition, one of his chief objects being the study of scientific apparatus. This we learn from the following letter written from Cambridge:

ST PETER'S COLLEGE,
CAMBRIDGE, *Sept.* 21/55.

My dear Dr Andrews,

I have just received your note. I am sorry it will be impossible for me to revisit Paris this vacation. Everything has been going on so wretchedly here during my absence, so far as regards printing[1], that even with a month's hard work from this date, I fear not more than $\frac{3}{4}$ of the work will be ready....

I have made attempts to see Ruhmkorff, Soleil, and Tyndall. The former was out of the way, Soleil was in Glasgow, and I believe so was Tyndall. I extracted from the woman in Soleil's shop all the information they could give about the Saccharimeter. I saw the instrument, pr. 260 fr., and bought a description of it and its use by Moigno.

I found and examined all the electromagnetic apparatus in the Exposition, and it was my decided opinion that an instrument in Ruhmkorff's stall called "Appareil de Faraday" was the very thing for us....

I hope you agree with me in the matter of the apparatus for Faraday's experiments. The only objection that I could see to it is that possibly it might not be powerful enough; but of that you will be a much better judge.

Not far from Ruhmkorff's there is a collection of clockwork, and along with it a small machine for exhibiting the permanence of the plane of rotation. I have not seen the gyroscope itself—this machine seemed to me not only comparatively useless, but even dangerous.

---

[1] The printing of Tait and Steele's *Dynamics of a Particle*.

SOMERSET COTTAGE,
COMELY BANK,
EDINBURGH, 21/7/59.

My dear Dr Andrews,

I was very glad to find from your letter that you had been successful in procuring apparatus in London....

I did not expect more from Faraday than you seem to have obtained, for I thought it scarcely possible that he could suggest at an hour's notice anything that we might have missed for three years.

My paper on the Wave-surface has reached me in separate form—and I have been asked by several men of note, to whom I have sent copies, to publish an elementary work on Quaternions. Todhunter of Cambridge, about the best authority on matters of that sort, is one of them—and I have written to Macmillan (the publisher) to enquire about terms etc....

Sir W. Hamilton has expressed his satisfaction with the project—and has only asked me to refrain from laying, or trying to lay, new metaphysical or other foundations for the Theory, wishing to reserve such for himself; and I am quite sure that I shall not feel this in any way a restraint....

I have ordered the addition to the small electrical machine....There is only one novelty here, so far as I can see, and as it is extremely interesting, I have given an order for one. Its object is the compounding of colours by rapid rotation, and so far it is simple—but when used in combination with a looking glass (like the Thaumatrope) it gives some most startling but easily explained and instructive effects....

SOMERSET COTTAGE,
COMELY BANK,
EDINBURGH, 18/6/60.

My dear Andrews,

I shall probably leave this for Cambridge on Monday next, and it will not be possible for me to be in Oxford as Hopkins and I are to be engaged in getting up our Ex^n Papers just at the time of the Ass^n Meeting....

Dr Bennett showed me on Saturday the whole series of frog experiments with a splendid galvanometer from Berlin and *German Frogs* which he had imported! But what interested me most was the perfect success of the experiment showing the muscular current in the operator himself, that you remember which we could not repeat and had begun to doubt. Mr Pettigrew, his assistant, produced by contracting his right arm a deflection of 15° E., then by contracting his left arm, one of 35° W. 50° in all. Neither Dr B. nor a Russian who was present could produce more than very uncertain results. I no longer entertain any doubt as to the reality of the phenomenon. The explanation, however, does not seem quite satisfactory. Dr B. told me that Humboldt had skinned his forefinger by raising blisters in order to get rid of the great resistance of the skin, and that then he produced extraordinarily great deflections....

T.

Towards the close of Tait's sojourn in Belfast, Andrews was preparing to attack the problem of the compressibility of gases. In this research Tait was to join him; but his election to the Chair of Natural Philosophy in Edinburgh altered all these plans.

The duties of his new Chair compelled him to give still more attention to the experimental than to the mathematical side of Natural Philosophy. In the early years he devoted much time to the preparation of his lectures and lecture experiments. In arranging the experimental illustrations he had the able help of James Lindsay who had served both Sir John Leslie and Professor Forbes as mechanical assistant. His scientific activities are clearly displayed in his letters to Andrews; and from these a few quotations will show how this kind of work grew upon his hands. A long extract referring to his first lecture has already been given (page 22). On December 1, 1860, Tait wrote:

My dear Andrews,

I am very much obliged to you for your note to Faraday. I enclosed it in a letter to him, telling him that I wished to ask his opinion on a point in the optical effects of magnetism; and as I sent him a copy of my lecture[1] I ventured to ask him to inform me at his leisure whether I had in it *fairly* stated the case at issue between him and the pure mathematicians about conservation of force. I got a very kind answer yesterday. He requests me to postpone my question (if a difficult one, and it is so) till after Christmas—but about the other matter he says "I thank you for the way in which you have put the Gravitation case. It is just what I mean." He says he has been working at it all summer, but still with negative results—and that he had drawn up a new paper for the Royal Society, but that Stokes had advised him not to present it....

COLLEGE, EDINBURGH,
Jan. 29, 1861.

My dear Andrews,

I would have written to you sooner, had not my hands been full of the January Examinations, and some experiments which Principal Forbes asked me to make.... In a paper which is I believe to appear in the Phil. Mag. for February, and which was read some weeks ago at the R. S. E., he states that few people living have ever seen Ampère's experiments for the repulsion of a current on itself—and that he had never succeeded in getting it. At his request I tried it, and succeeded with a single cell of Grove's battery. With twelve cells the floating wire almost jumped out of the trough! As there is some slight objection to this form of the experiment on account of the thermoelectric effects which occur at every change of metal in the circuit,

[1] This refers to Tait's inaugural lecture, in which he discussed Faraday's attempts to demonstrate the Conservation of Force in the sense of attraction.

I devised a floating conductor of *glass tube* full of mercury to replace the copper wire. The mercury is so much worse a conductor than copper, that it required four cells to give a good effect.

6 GREENHILL GARDENS,
EDINBURGH, 18/12/61.

My dear Andrews,

I find that I cannot manage to visit Belfast at present—my simple reason is that I am to bring home from Glasgow (where I am going to stay a day with Thomson) two galvanometers and an electrometer on Saturday next—and I must have one galvanometer and electrometer fitted up during the holidays, as I shall just have reached the critical point of *Radiant* heat when we stop. The new galvanometer works by reflexion, and can therefore be easily shown to a large class, which was impossible with the needle ones—besides it is delicate enough to show an effect even by frog-currents.

The electrometer also works by reflexion, and gives a deflection of some inches on a scale for $\frac{1}{100}$ th of the electromotive of one cell (Daniell). Of course the gold-leaf electroscope must now remain unused on the shelf, or at most be brought out to show what we used to be content with....

This prophecy of Tait's was not fulfilled even by himself during the succeeding forty years of lecturing. There is a simplicity about the gold-leaf electroscope which will ever keep it a prime favourite for purposes of demonstration, especially now when it is so easy to project the moving and divergent leaves magnified upon a distant screen.

GREENHILL GARDENS,
EDINBURGH, *Jan.* 15, 1862.

My dear Andrews,

Three reasons especially urge me to write to you to-night—the first and most pressing I shall detail at once.

I wish to know (by return of post if possible) what is the nature of the new ammonia process for procuring cold, and from whom, and at what price, it can be procured. This urgent business having been got over, I can be more easy in my future remarks.

You should *at once* get William Thomson's galvanometers—acting by reflexion. I have been lecturing on heat for some 4 weeks back; and I have shown, to my *whole* class, not only Melloni's experiments about diathermancy &c., but on a large scale the polarization of dark and bright heat....

Next I wish to know where *your* (and others') results as to Heat of Combination are to be found.

As to myself I may say that I have done nothing experimentally for a long time except with a view to familiarising myself with new apparatus....The beauty of the new galvanometers is such that today I arranged to show in a future lecture the Inductive Effects of the Earth's magnetism on a coil of wire about 30 feet long, coiled in a circle of about eight inches diameter. Turning that through 90° from a position perpendicular to the dipping needle, I got sufficient deflections of the galvanometer to throw the light off the scale. My own peculiar experiments on light, which you assisted at two years ago, I have arranged to try the very first fine day, and now with some hope of success, although Thomson is not at all sanguine about the idea.

I intend to repeat (if true) Tyndall's observation on the Adiathermancy of Ozone with an instrument far superior to his. Perhaps something may come of it.

The invention of the Divided Ring Electrometer indeed opened up many new lines of research; and in 1862 Tait and Wanklyn[1] published a joint paper on the electricity developed during evaporation and during effervescence from chemical action (*Proc. R. S. E.*), in which attention was called to the large charges produced by the evaporation of a drop of bromine and especially a drop of aqueous solution of sulphate of copper, from a hot platinum dish.

On January 23, 1862, in a letter mainly taken up with the projected treatise on Natural Philosophy, Tait again got into ecstasy over Thomson's galvanometers and electrometer.

"They are splendid instruments. If you are in no hurry I will be over in Belfast in April or May and will set them up for you. It requires some practice, but the gain in visibility to the class is ENORMOUS. I showed by his electrometer today to my whole class (150) in lecture the tension of a cell without condenser or anything of the sort."

On July 7 of the same year Tait mentioned the visit of Stas of Brussels to Edinburgh and referred to experiments which he was doing along with Wanklyn. With the preparation of the great treatise on hand, and the consideration of the experiments on the rotation of a disk *in vacuo* which Balfour Stewart and he had begun upon, there was not much time for undertaking any other experimental work on his own account. Tait was moreover at this time working hard at quaternions. One very fruitful piece of experimental illustration we owe, however, to this period.

As will be more clearly brought out in the chapter on quaternions, Tait was greatly impressed with Helmholtz's famous paper on vortex motion, so much so that for his own private use he took the trouble of making a good English translation of it. Early in 1867 he devised a simple but effective method of producing vortex smoke rings; and it was when viewing the behaviour of these in Tait's Class Room that Thomson was led to the conception of the vortex atom. In his first paper on vortex atoms presented

[1] Dr J. A. Wanklyn was assistant to Lyon Playfair the Professor of Chemistry. He was a well-trained chemist, ingenious and resourceful.

to the Royal Society of Edinburgh on February 18, 1867, Sir William Thomson refers as follows to the genesis of the conception:

"A magnificent display of smoke-rings, which he recently had the pleasure of witnessing in Professor Tait's lecture-room, diminished by one the number of assumptions required to explain the properties of matter, on the hypothesis that all bodies are composed of vortex atoms in a perfect homogeneous liquid. Two smoke-rings were frequently seen to bound obliquely from one another, shaking violently from the effects of the shock....The elasticity of each smoke-ring seemed no further from perfection than might be expected in a solid india-rubber ring of the same shape....

"Professor Tait's plan of exhibiting smoke-rings is as follows:—A large rectangular box open at one side, has a circular hole of six or eight inches diameter cut in the opposite side....The open side of the box is closed by a stout towel or piece of cloth, or by a sheet of India-rubber stretched across it. A blow on this flexible side causes a circular vortex to shoot out from the hole on the other side. The vortex rings thus generated are visible if the box is filled with smoke."

Then follows a description of one way of producing a cloud of sal-ammoniac, not the way however as generally practised by Tait; and the paper ends with a description of the effects of collision between vortex rings produced from two boxes. This seems to be the earliest printed account of Tait's experiments on vortex rings which gave the start to Thomson's famous theory of vortex atoms.

From 1859 till his death in 1868 Sir David Brewster was Principal of Edinburgh University. In spite of his eighty winters the famous experimenter still continued his researches, and Tom Lindsay, then a youth training as mechanical assistant under his father, James Lindsay, tells how Brewster made considerable use of the optical facilities of the Natural Philosophy Class Room, and discussed many optical phenomena with the young Professor. Sir David had made his residence at Allerly near Melrose and travelled to and from Edinburgh by train whenever his University or Royal Society duties demanded his presence. Had he lived in Edinburgh, he would no doubt have spent a large part of his time in the Natural Philosophy Department; for Tait, then as ever, cordially welcomed any one who had a physical problem to investigate. Among the subjects which specially occupied Brewster's attention during the later years of his life were the colours of soap films and the phenomenon which he had discovered in 1814 and had described under the name of the Radiant Spectrum. When a bright small image of the sun, such as may be obtained by reflexion from a convex mirror, is viewed through a prism, there appears in addition to the usual spectrum a bright radiant spot beyond the

violet.   Brewster described his latest experiments in a short communication to
the Royal Society of Edinburgh on April 15, 1867, but gave no explanation.
At the next meeting, on April 29, when Sir David, as President, was again in
the chair, Tait read a very brief communication on the same subject, tracing
the phenomenon to the peculiar texture of the membrane covering the cornea
and to the effect of parallax.   There can be no doubt that the experiments on
which Tait based his conclusions were made in conjunction with Brewster, who
probably agreed with the explanation brought forward by his colleague.

It was just at this time (April, 1867) that Tait's efforts to establish a
physical laboratory, in which doubtless he was strongly backed by Sir David
Brewster, received formal recognition by a grant of money from the Senatus.
The minutes simply record the fact, but give no indication of how long a time
was required by Tait to educate his colleagues up to the point of admitting that
such a new departure was desirable.   But to vote the money was one thing, to
find accommodation even for a small laboratory was another.   Six months
seem to have elapsed before the next step was taken; and then in a letter of
date December 20, 1867, Tait wrote to Andrews:

"I am about to get a Laboratory for practical students.   The money has been voted.
Henderson[1] has been induced to give up his class room (which is situated just over
my apparatus room), and during the holidays it will be put in order for work....
I want to ask if you can give me hints as to good subjects of experimental work for
practical physical students, not subjects that require a Faraday, still less such as
require a Regnault."

In his opening lecture of the session 1868–9 Tait was able to make a
definite announcement regarding the Physical Laboratory.   The following
report of part of the lecture is taken from the *Scotsman* of November 3, 1868.

"In several respects the present session may be expected to differ for the better,
as regards the class of Natural Philosophy, from at least the last eight during which
I have been connected with this University....From the miserable resources of the
University enough has been granted me to make at least a beginning of what will
I hope, at no very distant time, form one of the most important features in our
physical education.   A room has been fitted up as a practical laboratory, where a
student may not only repeat and examine from any point of view the ordinary lecture
experiments, thereby acquiring for himself an amount of practical information which
no mere lecturer can pretend to teach him; but where he may also attempt original
work, and possibly even in his student days make some real addition to scientific
knowledge.   That this is no delusive expectation is proved by the fact that in Glasgow,

---

[1] The Professor of Pathology at the time, the predecessor of the well-known Professor
Sanders.

under circumstances as to accommodation and convenience far more unfavourable than I can now offer, Sir W. Thomson's students have for years been doing excellent work, and have furnished their distinguished teacher with the experimental bases of more than one very remarkable investigation. What has been done under great difficulties in the dingy old buildings in Glasgow, ought to be possible in so much more suitable a place as this."

The most complete account given by Tait himself of his method of running a physical laboratory is to be found in his evidence before the University Commission of 1872, which consisted of Professor William Sharpey, Professor G. G. Stokes, and Professor H. J. S. Smith. The following successive answers to questions form a concise statement of Tait's views.

"I have made the laboratory open to all comers, limited of course by the number of students which my assistant and I can look after, and which my space can accommodate....They (the students) are free to spend their whole time in the laboratory when it is open each day, and thoroughly to devote themselves to their work....

"There is a small fee of two guineas for each student, but...that does not pay for the mere chemicals and other materials used by each student....With the help of my assistant I put each student as he enters the laboratory through an elementary course of the application of the various physical instruments, the primary ones. For instance, I begin by practising them in measuring time, estimating small intervals of time, then measuring very carefully length, angle, temperature, electric current, electric potential, and so on....

"When I find that they have sufficiently mastered those elementary parts of the subject I allow them to choose the particular branch of natural philosophy to which they wish to devote themselves, and when they have told me that, it is not by any means difficult to assign to them, if they carry it out properly, what may be excessively useful and valuable work."

The assistant under whose care the Laboratory first took shape was William Robertson Smith[1], M.A., afterwards well known as a theologian and Semitic scholar, the final editor of the ninth edition of the *Encyclopaedia Britannica*, and Librarian of the University of Cambridge. Smith was an Aberdeen graduate who shortly before had gained the Ferguson Scholarship in Mathematics open to the four Scottish Universities. Tait was examiner that year; and, impressed with the brilliant though untrained, indeed "almost uncouth," powers of the young student, he invited him to become his assistant. When Robertson Smith saw that he could combine the duties of the post with his theological studies at the Free Church College, he accepted Tait's offer; and after training himself in physical manipulation

[1] A biographical note communicated by Tait to *Nature* is reprinted below.

during the summer months of 1868 undertook, the next winter session, the systematic teaching of students in practical physics.

In this small upper room stripped of its benches, but with the terraced floor left intact, the men were put through a short course of physical measurements, such as specific gravities, specific heats, electrical resistance, and the like. Any who showed talent were soon utilised by Tait in carrying out original research ; and, to facilitate this kind of work, every possible corner of the old suite of rooms of the Natural Philosophy Department was adapted by means of slate slabs built into the thick steady walls for the installation of galvanometers and electrometers. The small room which Professor Forbes had used as his sanctum became the centre of experimental work. In this room Forbes had made his classical researches in polarisation of heat ; and here also Tait, with the help of successive sets of students, made his novel discoveries in thermoelectricity.

The large class room was also used as a research room, especially during the summer session when (at least until well on in the seventies) no class met. Two slate slabs were built into the wall, one on each side of the blackboard ; and on these were placed the mirror galvanometers and electrometers necessary for delicate electrical investigations.

Robertson Smith remained with Tait till 1870, and found time to carry through an interesting piece of experimental work on the flow of electricity in conducting sheets. In the paper giving an account of these experiments he considerably simplified the mathematical treatment, which had already engaged the attention of Maxwell and Kirchhoff. Among the students who passed through the Laboratory during the first and second years of its existence were Sir John Murray, Sir John Jackson, and Robert Louis Stevenson. Stevenson was paired off to work with D. H. Marshall, who succeeded Smith as assistant in 1870 and is now Emeritus Professor of Physics of Queen's University, Kingston, Ontario. Marshall of course was keen in all things physical, while Stevenson's preference was for a lively interchange of thought on every thing of human interest except science. When, as frequently happened, Stevenson got weary of reading thermometers or watching the galvanometer light-spot, he easily found some excuse to bring Robertson Smith within hearing and set him and John Murray arguing on the age of the earth and the foundations of Christianity. In some idle moments these lively students broke Tait's walking-stick. In haste and trepidation they commissioned two of their number to buy another as like the shattered one as possible. Tait who had been attending some Committee meeting returned ere long, and went to the usual corner to take possession of the stick. He paused doubtfully for a moment, then advanced, took the stick in his hand, and felt its weight and surface with considerable uncertainty. He looked at it again, glanced round the room, and then walked off towards the door. Back he came again almost immediately, glanced more carefully into various corners, swung the unfamiliar weapon to and fro, and at length, deciding that it was not what it seemed to be, put it back in the corner, and walked briskly home. Nothing was possible now save a full confession; and Tait accepted the gift in token of forgiveness.

Stevenson's father was Thomas Stevenson, the well-known lighthouse engineer. He hoped that his son would carry on the family traditions, and expressly desired Tait to let him work with optical apparatus. But the future essayist and writer of romances had not the smallest elementary knowledge of the laws of reflexion and refraction. The immediate purposes of the Physical Laboratory were lost on him; although no doubt what little training he allowed himself to undergo bore some fruit when a few years later he read a paper before the Royal Society of Edinburgh comparing rainfall and temperatures of the air within and without a wood. It was published in the *Proceedings*: literary critics have, however, left it severely alone.

Nevertheless, Stevenson's familiarity with the Physical Department led in after years to the writing of a charming picture of James Lindsay, the mechanical assistant already referred to. In 1886 when the University students held their great Union Bazaar, Stevenson contributed "Some College Memories" to the *New Amphion*, a beautiful volume (32mo.) printed in exquisite old-fashioned style by T. and A. Constable after designs and plans by W. B. Blaikie of that firm. After giving a quaint picture of himself in the third person, Stevenson continues,

"But while he is (in more senses than one) the first person, he is by no means the only one I regret, or whom the students of to-day, if they knew what they had lost, would regret also. They have still Tait to be sure—long may they have him !—and they have Tait's class-room, cupola and all; but think of what a different place it was when this youth of mine (at least on roll days) would be present on the benches, and at the near end of the platform, Lindsay senior was airing his robust old age. It is possible my successors may have never even heard of Old Lindsay; but when he went, a link snapped with the last century. He had something of a rustic air, sturdy and fresh and plain; he spoke with a ripe east-country accent, which I used to admire; his reminiscences were all of journeys on foot or highways busy with post-chaises—a Scotland before steam; he had seen the coal fire on the Isle of

May, and he regaled me with tales of my own grandfather. Thus he was for me a mirror of things perished; it was only in his memory that I could see the huge shock of flames of the May beacon stream to leeward, and the watchers, as they fed the fire, lay hold unscorched of the windward bars of the furnace; it was only thus that I could see my grandfather driving swiftly in a gig along the seaboard road from Pittenweem to Crail, and for all his business hurry drawing up to speak good-humouredly with those he met. And now, in his turn, Lindsay is gone also; inhabits only the memory of other men, till these shall follow him; and figures in my reminiscences as my grandfather did in his."

James Lindsay retired from his College duties in 1872, after having acted as mechanical assistant since 1819 when Sir John Leslie became Professor of Natural Philosophy. He had for the five previous years acted as Leslie's door-keeper at the mathematical class room. He had thus been connected officially with the University for fifty-seven years; and his memory went back to the days when Carlyle was still a student. He was a native of Anstruther; and—to quote from an obituary notice which Tait himself supplied to the *Scotsman* of January 5, 1877—"during the summer months, for at least the half of his life, he pursued the arduous occupation of a fisherman, in order to eke out his scanty income; and even in later years, when unable to go to sea, the position he had deservedly acquired among the fishing population of the district, led to his being employed during the herring season as an agent in the interests of some of the great fish curers. In this position his punctuality and rectitude were as much displayed at the pier head as in the Natural Philosophy class room." Under Leslie he became wonderfully dexterous in many difficult experimental processes, especially excelling in glass-blowing; and he rendered most efficient and indeed valuable aid both to Leslie and to Forbes in their experimental investigations. For twelve years he continued to assist Tait in the lecture experiments; and after he had trained his son Thomas to all the duties of the post, he retired to spend his last days in his native village. After his retirement he used occasionally to pay a visit to the scenes of his scientific labours, and I remember him on one such visit expressing great indignation at the careless way in which a box-full of small differential thermometers had been allowed to gather dust in a dark corner. These he had made with his own hand; and he had not realised that the thermopile and galvanometer had completely displaced the differential thermometer as a delicate instrument of research.

The following letter to Thomson touches on several pieces of experimental work which were engaging Tait's mind in the early years of the Laboratory.

17 D. P. E. 5/7/69.

Dear T.,

I have just heard from T″ [i.e. Tyndall] that you are in Largs. I feared you would be in a state of suspense and uselessness at Brest.

Do you mean by multiple-arc coils the set which has a *separate* frame for plugs—one in fact *into which* plugs are to be put, not *out of which* they must be taken, in order to work them? If so I shall send them off at once on hearing from you, for I have not even attempted to work with that set.

The other set works capitally and I have almost finished my copper wire determinations by its help—besides having carefully got the values of the coils of my own set; the unit in which is curiously (purposely?) 1·5 B.A. units very nearly.

You did not answer my query about the equation for heat in a bar. Do so now.

$$\frac{d}{dx}\left(k\frac{dt}{dx}\right) - h\phi t = 0$$

for two similar bars which when heated and left to cool work exactly together—Is not $k \propto (\Delta x)^2$? h is as nearly as possible the same in both.

I am working now with a platinum spiral heated by a current. I measure its radiation by a pile and galvanometer, then suddenly[1] for an instant shunt it into the bridge and find its resistance. I am getting very steady results with different battery power.

One of my students has attained great skill in finding specific heats; and has found that of best conducting copper to be slightly above that of bad, but to rise more slowly with increase of temperature.

I have asked Tyndall whether he couldn't induce the Shoeburyness people to fire a few stone bullets at a stone wall and get a party with spectroscopes to examine the resulting flash. I think comets might be thus elucidated.

I sent a copy of my article to Lady Thomson last week.

Yours T′.

PS. Are you remembering poor Balfour and the Vortices?

PS. [*Written across the top of the first page of the letter.*] Your sets of tenths of a unit not o.k. I get different values when I use 100 and 1000 as the next sides of the quadrilateral. For instance I find 1775 to 1000 and 179 to 100 for the same pair of wires.

In 1870 Tait began to communicate to the Royal Society his brief Notes from the Physical Laboratory, the first set including J. W. Nichol's[2] experi-

---

[1] A marginal note by Thomson reads "March 28/71 Why suddenly? Rather keep it always in the bridge under a constant El. M. F."

[2] J. W. Nichol, F.R.A.S., accompanied the Transit of Venus Expedition to the Hawaian Islands, and published in the *Proc. R. S. E.* (Vol. IX, 1875) a graphic account of a visit to Mauna Loa and Killauea, the remarkable volcanos with their lava lakes only 15 miles apart but differing in level by 10,000 feet. He died young; and his mother founded in his memory the Nichol Foundation in the Physical Laboratory of Edinburgh University.

ments on Radiation at various pressures of the surrounding gas, Brebner's work on electrolysis, and Meik and Murray's investigations on the effect of load on the resistance of copper wires.

Robertson Smith also found time for an exposure of Hegel's attack upon the principles of the calculus as laid down in Newton's *Principia*, a kind of criticism for which Smith, by virtue of his profound knowledge of both mathematics and metaphysics, was singularly well equipped.

During the early years of his professoriate, Tait was on intimate terms of friendship with W. H. Fox Talbot, best known for his discoveries in photography and his deciphering of the cuneiform inscriptions. Fox Talbot was a mathematician of distinct originality and was keenly interested in experimental physics. He lived a good deal in Edinburgh during the sixties and early seventies; and on Saturday forenoons he often paid Tait a visit at the College to experiment in light and magnetism. On May 15, 1871, Fox Talbot communicated three short papers to the Royal Society of Edinburgh, the first of which, "Note on the early History of Spectrum Analysis," was probably suggested by Tait's address on that subject delivered the same evening before the Society. The second, "On a New Mode of observing certain Spectra," ends with the remark that "all these experiments were made in the Physical Laboratory of the University of Edinburgh by the kind permission and assistance of Professor Tait." The third, "On the Nicol Prism," recalls some of his earlier investigations and contains the description of a modified form of polarising prism, which is made half of calc spar and half of glass. I have often heard Tait express the very high opinion he held in regard to Fox Talbot, whose discovery of anomalous dispersion was kept back from the world by his own modesty and the too great caution of Sir David Brewster, and had to be rediscovered many years afterwards by Le Roux and Christiansen.

The following letter touches on several points of interest.

17 DRUMMOND PLACE,
EDINBURGH, 11/1/71.

My dear Andrews,

We all heartily join in wishing you and yours many happy new years. We are all well, but *very* busy—I at Physics, the rest at *skating*! Even my wife has become an enthusiast. 23 years ago I was wild about it, but I feel no inclination to waste time on it now....

I am delighted to hear that you are getting on so well with your high pressures. I often wish I were back again in Belfast. True I had more lecturing to do, and less pay, but I had a great deal more leisure for private work. In fact I have barely time for any private work during the winter session now-a-days.

However, I have got some students who are able and willing to work and I have handed over my apparatus to them to make the best of it. At present I am entirely engaged with "l'effet Thomson" if you know what that is—the so-called specific heat of electricity in different conductors, which I think I have proved both experimentally and theoretically to be proportional to the absolute temperature. This has led me to construct a thermometer depending on two separate thermoelectric circuits working against one another, so as to give galvanometric deflections *rigorously* proportional to differences of absolute temperature through all ranges till the wires melt. I hope to get the specific heats and melting points of various igneous rocks, &c., &c., true to a very few degrees.

My Holtz machine—perhaps about the last thing that Ruhmkorff sent out of Paris[1]—is a splendid success; 2-inch sparks from a jar with $\frac{1}{4}$ square yard of coated surface at intervals of 4 seconds.

Tait was now in the heart of his thermoelectric investigations, which for several years dominated the work of the Physical Laboratory. The difficulties encountered and the methods by which they were overcome are discussed in a series of short papers communicated to the Royal Society of Edinburgh, afterwards worked up into the great *Transactions* paper of 1873. In the earlier pioneer work Tait was helped by May and Straker, and a little later by John Murray and R. M. Morrison. In the summer of 1873 he instructed C. E. Greig and myself, who had spent one winter in the Laboratory, to investigate by one and the same method the thermoelectric properties of some twenty different metals paired in a sufficient number of ways; and these experiments which were made in the Natural Philosophy class room formed the basis of the "First Approximation to the Thermoelectric Diagram." The hot junctions were heated in oil up to a temperature of nearly 300° C. Meanwhile Tait himself had been working with iron at still higher temperatures, and making the first of what proved to be the most novel of his discoveries in thermoelectricity, namely, the remarkable changes at certain temperatures in the thermoelectric properties of iron and nickel.

Nearly all pairs of metals up to the temperatures of their melting points have the thermoelectromotive force a parabolic function of the difference of the temperatures of the junctions. When, however, iron or nickel is one of the metals forming the thermoelectric couple this rule breaks down. Nevertheless between particular limits of temperature the parabolic law is satisfied, so that the relation between electromotive force and temperature can be fairly well

---

[1] That is, before its investment by the German troops.

represented by a succession of three parabolas with quite different parameters. In the case of iron these peculiarities occur at high temperatures, which Tait was able to measure by means of two alloys of platinum and iridium whose thermoelectromotive force was very approximately proportional to the temperature difference. These were known as M and N. Tait hoped to get a series of such alloys having the same properties; but though many specimens of various percentage compositions were supplied him by Johnson and Matthey, never again did he obtain a pair possessing the same simple proportionality. The final experiments on iron at high temperatures were entrusted to C. Michie Smith and myself in the winter of 1873. The three wires M, N, and the particular specimen of iron under investigation had their ends bound together to form one triple junction, while the other ends were arranged so that the circuit M-N or the circuit N-Iron could be alternately thrown into the galvanometer circuit. The triple junction was then inserted within the hollow of a white-hot iron cylinder; and as this cylinder cooled to lower temperatures, the two circuits were thrown in rapid alternation into the galvanometer circuit, and practically simultaneous measurements were obtained of the N-Iron and M-N currents.

Nickel and cobalt were not easily obtained in the early seventies; and the first piece of nickel experimented with was a narrow ribbon not more than two feet long, supplied by F. Lecoq de Boisbaudran. The following letter to Andrews touches upon the work with these magnetic metals.

38 George Square,
Edinburgh, 13/12/75.

My dear Andrews,

Many thanks for your letter. I have been extremely remiss in not long ago thanking you for the Nickel and Cobalt you kindly sent me. I know you will be glad to learn what they have told me. Here it is:—

1. The new specimen of nickel gives almost exactly the same results as those in my Thermoelectric Diagram. So *that* very curious result is verified.

2. The Cobalt specimen was not coherent enough for any but qualitative results:—but it has shown me that cobalt lies (in the diagram) *between* Iron and Nickel (at moderate temperatures), cutting copper, platinum, lead, zinc, cadmium, &c., so that the observations of a few neutral points will tell me all about it—except (of course) the sinuosities which I have reason to think its line will show somewhere about a white heat. But I may be altogether wrong in this. Meanwhile with Crum Brown's assistance I am preparing to deposit electrolytically films or foil of pure cobalt.

The cobalt supplied by Andrews was probably far from pure; for with the rod of pure cobalt obtained by electrolytic deposition on aluminium, the aluminium being afterwards dissolved away, J. G. MacGregor and C. M. Smith found that the cobalt thermoelectric line lay below the nickel line and therefore further away from the iron line[1].

Some of the difficulties encountered in these early days are not described either in Tait's *Transactions* paper, or in the short laboratory notes which Tait communicated from time to time to the Royal Society of Edinburgh.

Particularly interesting were the experiments on sodium and potassium, the carrying out of which was entrusted to C. Michie Smith and myself. The metals were prepared for Tait by (Sir) James Dewar, who sucked them in the molten state up glass tubes under the surface of melted paraffin and then allowed the whole to solidify. Each of the sodium and potassium bars was thus enclosed in a glass tube, with solid paraffin ends protecting it from the air. The ends were then slightly melted and platinum wires pushed through the paraffin into the sodium or potassium. Sodium-platinum and potassium-platinum circuits were thus constructed. Each bar was only a few inches long, and as the one end had to be kept cool in running water while the other was gently heated in an oil bath, the manipulation of the experiments was not easy. There was moreover some risk of accident to the eye of the operator who attended to the warmer junction.

Tait seems to have been led into his thermoelectric work in the hope of testing a theoretic result he had obtained with reference to the "Thomson Effect." Experimentally the work was a following up of much earlier investigations made by Thomson himself, to whom indeed the idea of the thermoelectric diagram was due. What Tait did was (1) to establish for most metals and through a considerable range of temperature the parabolic law for electromotive force, or the linear law for thermoelectric power, in virtue of which each metal was represented by a straight line on the diagram; (2) to show how the "specific heat of electricity" was indicated by the inclination of the thermoelectric line and how the Peltier Effect and the Thomson Effect were represented by areas on the diagram; and (3) to discover the remarkable changes of sign in the Thomson Effect for iron and nickel. His attempts to measure the Thomson Effect directly were not successful, although he made repeated attacks on the problem. For example, by passing a current first in

---

[1] Working with a fairly pure specimen of rolled cobalt in 1891, I found that its thermo-electric line lay above the nickel line at temperatures below 100° C. but below it at higher temperatures.

one direction and then in the other along a piece of thin platinum foil which was cut away towards the centre until it became very narrow, he hoped to be able to witness the shift of the glow at this narrowest part. When he got the Gramme Dynamo about 1877, one of the first experiments he tried was to pass the current from the Gramme machine along an iron bar when it had been brought to a steady gradient of temperature along its length, after the manner of Forbes' experiment in thermal conductivity. He hoped to detect a change in the gradient of temperature; but here again there was no success, the current density not being great enough.

Another line of experiments on related effects, at which A. Macfarlane, C. M. Smith, and I worked, was the coordination of the striking phenomena which occur in iron about the dull red heat, namely, the loss of magnetic susceptibility, the reglow as the iron wire cooled, the change of sign of the Thomson Effect, and the change in the law of alteration of electrical resistance with temperature, all of which Tait proved to be in the neighbourhood of the same temperature. In one of these experiments iron and platinum wires were led through a white-hot iron cylinder side by side, while to the middle of the iron were attached the M and N platinum-iridium wires. As the whole gradually cooled, observations were taken in rapid succession of the resistances of the iron and platinum wires and the thermoelectric currents in the N-iron and N-M circuits. The method was no doubt rough and ready and not susceptible of great accuracy, but it was effective enough to establish conclusions which more carefully designed experiments of later date have fully corroborated.

Among Maxwell's letters to Tait about this time the following quaint remark was found written in three lines on a long strip of paper.

"If your straight lines, parabolas, &c. have no resemblance at all to those things which men call by those names, I would as soon be J. Stuart Mill as call them so. But if they differ very slightly, then T' is enrolled among the Boyle and Charles of ΘH[1] who remain unhurt by Regnault, &c. But in Physics we must equally avoid confounding the properties and dividing the substance. In the one case we fall into the sin of rectification (Eccl. i. 15) and in the other we see in every zigzag a proof of transubstantiation."

Although himself greatly taken up with the thermoelectric experiments, Tait never lost sight of the investigation into the thermal conductivity of metal bars, which was the first serious piece of experimental work he tackled in Edinburgh. This following up of Forbes' important researches was begun under the

[1] The Greek initials of Thermo-Electricity.

auspices of the British Association; and Tait sent in two short Reports in 1869 and 1871. Most of the 'veteran' students had a turn at the bars during the seventies and eighties; and Tait's paper on the application of Ångström's method of sending waves of heat along the bar (*Proc. R. S. E.* Vol. VIII), was based on observations made by A. L. MacLeish and C. E. Greig[1] in the early part of the year 1873. The harmonic analysis is fully worked out so as to give the amplitudes and phases of the temperature oscillations at each chosen point; but the final calculation of the conductivities is not given. In fact the simple and solvable form of the equation of conduction did not apply even to a rough approximation. Tait therefore fell back upon Forbes' method, and in 1878 he published a detailed account of his investigations, the main purpose of which was to extend to other metals what Forbes had done for iron. An important supplement to this memoir appeared in 1887 by (Professor) Crichton Mitchell, who as an advanced student went over the whole ground again, the one difference being that all the bars were now nickel plated. Their surface conditions were thus rendered more nearly identical than in the first set of experiments. One of the final conclusions come to was that

"We cannot yet state positively that there is any metal whose conductivity becomes less as its temperature rises; and thus the long sought analogy between thermal and electric conductivity is not likely to be realised."

Early in 1875 Tait and Dewar made together a series of well planned experiments on the phenomena of Crookes' radiometer. They gave a demonstration of these before the Royal Society of Edinburgh on July 5, 1875; but unfortunately no authoritative account of them was ever published. In *Nature* of July 15, 1875, a report of the communication was given under the title "Charcoal Vacua" which does not bring out clearly the real significance of certain parts of Tait and Dewar's investigations. The following quotation from Lord Kelvin's obituary notice read before the Royal Society of Edinburgh puts the question in a clearer light:

"In a communication on 'Charcoal Vacua' to the Royal Society of Edinburgh of July 5, 1875, imperfectly reported in *Nature* of July 15 of that year, the true dynamical explanation of one of the most interesting and suggestive of all the scientific wonders of the nineteenth century, Crookes' radiometer, was clearly given. The phenomenon to be explained is that in highly rarefied air a disc of pith or cork or other substance of small thermal conductivity, blackened on one side, and illuminated by light on all sides, even the cool light of a wholly clouded sky,

[1] Dr A. L. MacLeish is now a physician resident in Los Angeles: the Rev. C. E. Greig is a pastor in Paris.

experiences a steady measurable pressure on the blackened side. Many naturalists, I believe, had truly attributed this fact to the blackened side being rendered somewhat warmer by the light; but none before Tait and Dewar had ever imagined the dynamical cause—the largeness of the free path of the molecule of the highly rarefied air, and the greater average velocity of rebound of the molecules from the warmer side. *Long free path* was the open sesame to the mystery."

I had the good fortune to be present in the Laboratory when some of the experiments were being made. One especially struck me as being of peculiar significance. I cannot remember if this was shown before the Royal Society of Edinburgh; but it is not referred to in the published report. A transparent light vane of rock salt was suspended under an ordinary air-pump receiver and placed in front of and fairly close to a fixed blackened surface. The energy rays were directed through the transparent vane on to the blackened surface. At very moderate exhaustions repulsion was set up, whereas for the ordinary form of Crookes' radiometer a very high vacuum is needed. The whole question was thus proved to be one of the relation between the free path and the distance between the repelling surfaces.

The following among other experiments are described in the *Nature* Report. Two equal disks, one of glass and the other of rock salt, were attached to the ends of a delicately suspended glass fibre. When the radiation fell on the glass disk there was repulsion due to the heating of the disk; but when the radiation fell on the diathermanous rock salt there was no repulsion—the heat was not absorbed sufficiently to produce the necessary rise of temperature. The back of the rock salt disk was next coated with lamp black, and after sufficient exhaustion was produced in the enclosing vessel, the radiation was thrown through the rock salt on to the blackened surface. At first one might expect an apparent attraction due to the repulsive action on the far-away side; but the disk was repelled exactly like the glass disk. This was due to the bad conducting power of the lamp black, so that the rock salt on the near side became heated by conduction more quickly than the outside parts of the lamp black layer on the further side. In these experiments it was necessary to use a very thin-walled enclosing vessel within which the vacuum was formed, otherwise the glass vessel would itself absorb so much of the low heat rays that the differential action of the glass and rock salt disks would not be great enough to make itself apparent.

The next engrossing piece of experimental work was in connection with the "Challenger" Reports. On the return of the "Challenger" Expedition in 1876, Sir Wyville Thomson consulted Tait as to the corrections to be applied to the readings of the deep sea temperatures given by the self-recording thermometers which had been used. Experiments made by Captain J. E. Davis, R.N., before the Expedition started on its four years' voyage, indicated that a correction of at least half a degree Fahrenheit for every mile depth under the sea had to be applied. A careful examination of the thermometers with their protected bulbs convinced Tait that only very slight corrections would be required; and the necessity arose for retesting the thermometers. In these laboratory experiments, as conducted first by Captain Davis and then by Professor Tait, the conditions are very different from those under which the thermometers record the temperatures of the ocean deeps. For example, under the increasing pressure in the hydraulic press the temperature of the surrounding water will be raised. Captain Davis and Professor Miller tried to determine this rise of temperature by direct experiment; and after taking it into account they found a correction still to be applied, and this they referred to the direct effect of pressure. This pressure correction accordingly was to be applied to the readings obtained in deep sea observations. Tait's acuter physical instinct saw no necessity for such a correction; and after a prolonged investigation into all the possible causes of temperature change he found that the vulcanite mounting of the thermometers was the principal source of the change which Davis and Miller failed to account for. The heating of the vulcanite mounting due to compression would be of no consequence in the deep sea experiments; consequently no correction was needed. Or, to put it quite accurately, the correction due to pressure was of an order distinctly smaller than the errors of observation and therefore negligible. See Tait's Report, "*Challenger*" *Narrative*, Vol. II, Appendix A; *Sci. Pap.* Vol. I, p. 457.

The beautiful hydraulic apparatus designed by Professor Tait and supplied by the Admiralty for making these tests was utilised by him in making further investigations in the realm of high pressures. Some of these investigations form the substance of a second "Challenger" Report (*Physics and Chemistry*, Vol. II, Part IV, 1888), bearing upon certain physical properties of fresh water and of sea-water (*Sci. Pap.* Vol. II, p. 1). The wide scope of this enquiry may be best indicated by a few quotations from his own summary of results. The compressibility of the glass of the piezometers was measured by means of J. Y. Buchanan's apparatus, and found to be 0·0000026 per atmosphere. By a modified form of piezometer the compressibility of mercury was determined, the value being 0·0000036 per atmosphere. These data were necessary for the accurate determination of the compressibilities of the various kinds of water and solutions. Within a range of temperature 0° to 15° C. and a range of pressures from 150 to nearly 460 atmospheres, the compressibility of fresh water was approximately represented by the empirical formula

$$\frac{0\cdot00186}{36+p}\left(1-\frac{3t}{400}+\frac{t^2}{10,000}\right).$$

The corresponding formula for sea-water was

$$\frac{0\cdot00179}{38+p}\left(1-\frac{t}{150}+\frac{t^2}{10,000}\right).$$

In these t is Centigrade temperature and p is pressure in tons weight per sq. inch. The point of minimum compressibility of fresh water is about 60° C. at atmospheric pressure, and that of sea-water at about 56° C.; both are lowered by increase of pressure.

The average compressibility of solutions of NaCl for the first p tons of additional pressure, at 0° C., s being the amount of NaCl in 100 parts by weight of water, could be represented very accurately by the formula $0\cdot00186/(36+p+s)$.

The maximum density point of water was found to be lowered about 3° C. by 150 atmospheres of pressure; and from the heat developed by compression of water Tait calculated that this lowering of the maximum density point should be 3° per ton weight per square inch. (1 ton weight per sq. inch = 152·3 atmospheres.)

In most of his experimental work Tait did not apply his mind specially to the invention of elaborate apparatus; but that he could when the necessity arose devise useful and ingenious forms appears very clearly in his compression work.

Consider for example his high pressure gauge, constructed of a steel cylinder, the measured change of volume of which under hydrostatic pressure gives by a simple elastic formula the value of the pressure. In this instrument the pressure is applied to the outside of the cylinder, and the change of volume is measured by the alteration in level of mercury which fills the cylinder and the narrow glass tube fitted to it above. This glass tube which is in continuous connection with the interior of the steel cylinder is open above and is not itself exposed to pressure. It projects through the top of the outer vessel which surrounds the steel cylinder and within which the pressure is applied. By a very simple but ingenious device Tait practically got rid of the disturbing effects of temperature changes in the mercury filling the steel cylinder. He placed within the cylinder a glass tube closed at both ends which all but filled the cylinder. This left the action perfect as a pressure gauge, and rendered negligible its action as a thermometer. Professor Carl Barus in his memoir on the volume Thermodynamics of Liquids[1] found his modified form of the Tait Gauge highly efficient. One great merit was the complete absence of cyclic quality so that the same pressure readings were obtained whether they formed a series of ascending or descending pressures.

Another example of Tait's ingenuity is his electric contact device for indicating when a definite compression has been produced in a piezometer which is enclosed in an opaque hydraulic press and cannot therefore be seen by the eye. His own description is in these words (second "Challenger" Report, Appendix A).

"We have, therefore, only to fuse a number of platinum wires, at intervals, into the compression tube, and very carefully calibrate it with a column of mercury which is brought into contact with each of the wires successively. Then if thin wires, each resisting say about one ohm, be interposed between the pairs of successive platinum wires, we have a series whose resistance is diminished by one ohm each time the mercury, forced in by the pump, comes in contact with another of the wires. Connect the mercury with one pole of a cell, the highest of the platinum wires with the other, leading the wires out between two stout leather washers; interpose a galvanometer in the circuit, and the arrangement is complete. The observer himself works the pump, keeping an eye on the pressure gauge, and on the spot of light reflected by the galvanometer. The moment he sees a change of deflection he reads the gauge...."

Amagat, between whom and Tait much correspondence passed at one time with reference to pressure measurements, adopted this method with great success in his later experiments. Regarding its efficacy he writes

"Sur la recommandation de l'eminent physicien, je l'ai essayé tout de suite et n'en ai plus employé d'autre, non seulement pour les liquides, mais encore pour le gaz, dans les series allant jusqu'aux plus fortes pressions et pour les températures ne depassant pas 50."

One general conclusion of great interest in these experiments is the representation of the compressibility by an expression of the form $A/(B+p)$, where p is the pressure and A and B depend only on the temperature. In several subsequent papers Tait tested the applicability of this empirical formula to experimental results obtained by other experimenters, notably

[1] *Bulletin of the United States Geological Survey*, No. 96 (1892).

Amagat. He also projected a series of investigations upon solutions of varying concentration, so as to test the applicability of the formula $A/(B+s+p)$, where s represents the percentage amount by weight of the solute. In 1893 he published results of a preliminary character on three solutions of the substances Potassium Iodide, Potassium Ferrocyanide, Ammonium Sulphate, Magnesium Sulphate and Barium Chloride, and found that the formula applied fairly well. Five years later he published a preliminary note on the compressibility of solutions of sugar based upon experiments which were carried out by A. Shand (Nichol Foundationer). The results were not very concordant; but they indicated that the effect of sugar was, weight for weight, barely one-third that of common salt in reducing the compressibility. Mr Shand was planning a continuation of the experiments, when his early death deprived the Edinburgh University of an experimenter of real ability and resourcefulness.

The new compression apparatus, familiarly known as the " Big Gun," was not received till 1879; and it was first set up in a small cellar on the basement of the north side of the College. Here all the experiments dealing with the testing of the " Challenger" thermometers were carried out. The accommodation was very limited, and the light was poor; but in a few years the apparatus was transferred to a much larger basement cellar, in the north-west corner; and here all the later experiments on compression were made.

This change was part of a general expansion of the Physical Laboratory consequent on the removal of the Anatomical Department in 1880 to the New Buildings which were to be wholly devoted to medical studies. Till that date the Dissecting Rooms occupied the top story of the north side of the College with the exception of the small room which had served for a physical laboratory under the care of Tait's successive assistants, W. Robertson Smith, D. H. Marshall, and P. R. Scott Lang. During my first year of the assistantship (1879–80) the whole suite of four rooms became transformed into the physical laboratory. There was ample accommodation, so far as mere area of floor space was concerned; and it was possible to arrange a junior laboratory and rooms for special magnetic and optical work. On the basement Tait secured the large cellar already mentioned, in which were installed the compression apparatus, the dynamo, the gas engine for driving the dynamo and for working up to high pressures, and latterly the " guillotine" for the impact experiments to be afterwards described. In a neighbouring cellar fifty secondary cells were in due course installed; and there was also a third cellar which was used originally as a store room but which after 1892 was equipped as a research room for my work on magnetic strains.

The old anatomical theatre was adapted to the purposes of the Mathematical Department under Professor Chrystal, who began his Edinburgh professoriate in the same transition year 1879–80. These changes brought the Mathematical and Natural Philosophy Departments into closer contiguity; but, what was of still greater importance, Professor Tait found in his new colleague an enthusiastic experimentalist, who from 1880 to about 1886 passed the summer sessions in the Physical Laboratory, exercising a stimulating influence upon many of the students who were devoting themselves to practical physics. Chrystal had just written the articles on Electricity and Magnetism for the *Encyclopaedia Britannica* and was thoroughly posted on all the recent work in these rapidly developing branches of physics. In carrying out his important researches on the differential telephone and the measurements of inductances, he had all the facilities of the laboratory placed at his disposal; and both directly and indirectly he gave many a hint to the students who were able to take advantage of their opportunities. When I left for Japan in 1883 Chrystal was almost as strong an influence in the Laboratory as Tait himself; but after a few years the increasing duties of his own chair and the fact that he found himself to be appropriating more and more of the really serviceable apparatus for his own experiments obliged him to relinquish experimenting for some time.

When, mainly through the exertions of Dr Buchan, the Ben Nevis Observatory was started in 1883, attention was drawn to the difficulty of measuring humidities of the atmosphere under the conditions which frequently existed on the top of the mountain. Both Chrystal and Tait suggested forms of instrument for the purpose. Chrystal's was on the principle of Dine's hygrometer, the nickel plated copper box, into which the thermometer bulb was inserted, being supplied by means of a double tap arrangement with warm or cold water at will. The temperature was adjusted until a film began to form on the box. Tait's instrument was constructed on a totally different principle, that of the atmometer. The following is Tait's description from his paper of February 16, 1885 (*Proc. R. S. E.* Vol. XIII, p. 116).

"The atmometer is merely a hollow ball of unglazed clay, to which a glass tube is luted. The whole is filled with boiled water and inverted so that the open end of the tube stands in a dish of mercury. The water evaporates from the outer surface of the clay (at a rate depending partly on the temperature, partly on the dryness of the air) and in consequence the mercury rises in the tube. In recent experiments this rise of mercury has been carried to nearly 25 inches during dry weather. But it can be carried much farther by artificially drying the air round the bulb...I found, by inverting over the bulb of the instrument a large beaker lined with moist filter paper, that the arrangement can be made extremely sensitive. The mercury surface is seen to become flattened the moment the beaker is applied, and a few minutes suffice to give a large descent, provided the section of the tube is small compared with the surface of the ball.

"I propose to employ the instrument in this peculiarly sensitive state for the purpose of estimating the amount of moisture in the air, when there is considerable humidity; but in its old form when the air is dry. For this purpose the end of the tube of the atmometer is to be connected, by a flexible tube, with a cylindrical glass vessel, both containing mercury. When a determination is to be made in moist air the cylindrical vessel is to be lowered till the difference of levels of the mercury amounts to (say) 25 inches, and the diminution of this difference in a definite time is to be carefully measured, the atmospheric temperature being observed. On the other hand, if the air be dry, the difference of levels is to be made nil, or even negative at starting, in order to promote evaporation."

Experiments were made to test the applicability of the method; but the manipulation demanded more care and attention than could be expected from a busy observer at a meteorological station.

In 1887 Tait who had been for many years a keen devotee of the game of golf was led to consider various physical problems suggested by the flight of a golf ball, from the moment of impact of the club to its final fall to earth. The first consideration is the manner in which the momentum of the club is communicated to the ball. Given the club moving with a certain speed, with what speed will the ball be projected? This is the one stage over which the player has any control. After the ball has left the club its further progress is conditioned by the initial conditions of flight and the continuous subsequent interplay between the moving ball and the surrounding air.

Accordingly to study the laws of impact of various materials Tait set up a simple but very effective form of apparatus which he humorously called the "guillotine[1]." The name occurs early in the third paragraph of the first paper on Impact (*Sci. Pap.* No. LXXXVIII). The block whose impact on the material was to be studied "slid freely between guide rails, precisely like the axe of a guillotine." As this block fell and rose again after several rebounds, a pointer attached to it bore with sufficient pressure upon the blackened surface of a revolving plate-glass wheel. The curve traced out in this manner

[1] The name is preserved historically in the new Physical Laboratory of Edinburgh University, the room in which the apparatus is now installed being called the Guillotine Room·

contained a complete record of the whole motion of the impinging block; and from this record all the numerical data of the experiment could be obtained, such as the successive heights of rebound, the time of duration of impact, and the amount of compression of the substance on which the block fell.

To be able to measure time intervals it was necessary to have a definite time record impressed on the revolving plate. This was simply effected by means of a tuning fork delicately adjusted so that a tracing point attached to one end traced out a sinuous curve on the plate concentric with the curves traced out by the impinging block.

Although originally undertaken with the aim of determining the resilience of rubber and guttapercha, the experiments were not confined to these golf ball materials. The first series of experiments dealt with the impact properties of plane tree, cork, vulcanite, and vulcanised indiarubber; and in the second series lead, steel, glass, new native indiarubber, and various kinds of golf balls were added to the list of substances experimented with.

The results embodied in these papers are of the highest physical interest. Among the practical applications we may mention Tait's estimates of the duration of impact of a hammer and nail (0·0002 sec.) and of the time-average force (300 lb.-wt.). As regards the golf ball problem which suggested the experiments, the very much smaller speeds of approach attainable in the experiments render the data not very directly applicable. But it was estimated that the time-average of the force during the collision (which may have lasted about 0·005 sec.) of the golf club and ball must be reckoned in tons' weight.

Closely connected with the golf ball enquiry were the ballistic pendulum experiments, described in the second paper on the path of a spherical projectile (1896, *Sci. Pap.* Vol. II, p. 371). The final type of pendulum used was a bifilar suspension with a bob formed by a long horizontal bar oscillating in the plane of the bifilar. The one end of the bar was faced with clay, and into this soft material a golf ball was driven from a "tee" a few feet away. The momentum of the ball at impact was transferred to the pendulum and ball together, and could be easily estimated in terms of the distance through which the bob was driven to the extremity of its range. This was observed directly by an observer who was protected from being hit by the ball (should that by any chance miss the clay) by the half-closed door past which the pendulum swung. Tait's son Freddie and other powerful players visited the laboratory and experimented with this form of apparatus. The general conclusion was that under ordinary conditions a well driven golf ball left the "tee" with a speed of not more than 300 feet per second. Assuming 2/3 as the coefficient of restitution, Tait found that the head of the club must have been travelling at the rate of about 200 feet per second at the instant of impact. Other questions relating to the flight of the golf ball have been already discussed in Chapter I.; and Tait's own final views will be found below in his article on Long Driving, which appeared in the *Badminton Magazine*.

With a mind always on the alert for scientific problems, it is not surprising that Tait occasionally failed to find what he was in search of. His attempts to obtain distinct evidence of the Thomson Effect in thermo-electricity have already been noted; and I remember him spending the better part of a summer session in the experimental study of electrification due to sudden evaporation or condensation. Morning after morning he would come with a new arrangement to try, meeting my enquiry with the remark, " Now, at last, I have got the crucial experiment." He devised for this research large flat metallic dishes, which we facetiously dubbed "frying pans"; but nothing came of it. Tait's conclusion was that his surfaces were not big enough.

Another enquiry which occupied his mind at intervals from his Belfast days was the possibility of doubling an absorption line through the influence of magnetism. That such an effect should take place was an inference he made from Faraday's discovery of the rotation of the plane of polarization in a strong magnetic field. In a short paper read before the Royal Society of Edinburgh in 1876 he gave briefly the grounds for his belief (*Sci. Pap.* Vol. 1, p. 255). With more powerful magnetic fields than were at that time available there is little doubt that he would have observed an effect of the kind looked for, and thus anticipated Zeeman's closely allied discovery of 1896.

"In consequence of the severe lightning stroke with which Skerryvore Lighthouse was visited on 2nd February 1876, occasioning considerable disturbance to the internal fittings of the lighthouse and the destruction of the entrance door," D. and T. Stevenson, Engineers to the Board of Commissioners of the Northern Lights, suggested "the propriety of consulting Professor Tait on the general question of protecting the lighthouse towers against the effects of lightning." Professor Tait accordingly accompanied David Stevenson and others of the Commissioners on their annual visitation during the ensuing summer, and his opinions and advice are given in the Report, of which the opening sentences have just been quoted. During this trip of inspection Tait's attention was drawn to the methods of producing fog signals, and experiments were afterwards made in Edinburgh to test the applicability of a method of alternating out-blast and suction for producing economically sounds of high intensity. The original idea did not develop satisfactorily; but some experiments in conjunction with Crum Brown led him in 1878 to the construction of a new form of siren suitable for fog-signalling. In these experiments an organ note was made discontinuous by being sounded through a partition and a revolving disk cut into separate sectors. Unfortunately the siren effect superposed on the effects which were being studied disturbed somewhat the quality of tone. Tait found that when there was no organ note being sounded the mere rotation of the perforated disk produced a sound whose intensity could be greatly increased by soldering plates perpendicularly to the revolving disk so as to increase the thickness of the back edges of the apertures. When rapid rotation was set up the sound emitted was almost terrifying in its intensity. It shrieked out through the open window of the Natural Philosophy Class Room into the quadrangle to such a degree as to interfere with the lectures in the neighbouring class rooms. Tait was accordingly obliged to conduct his experiments at hours when no classes met.

In February, 1880, Tait communicated a short note to the Royal Society of Edinburgh describing his unsuccessful attempt to measure the velocity of the particles which constitute the cathode rays in a Crookes' tube, by means of observations of the spectrum made in directions perpendicular and parallel to the lines of motion of the charged particles. One cause of the failure was the loss of light by multiplied reflections when a powerful spectroscope was used. This led him to construct a rotatory polariscope whose principle depended upon the rotation by quartz of the plane of polarisation, combined with sufficient prism dispersion just to separate the various bright lines of the source from one another. The final form of the apparatus was described in *Nature*, Vol. XXII (*Sci. Pap.* Vol. I, p. 423). When a plane polarised ray of light is subjected to rotatory dispersion by transmission through quartz, and, after further transmission through a double image prism, is dispersed prismatically by means of a direct vision spectroscope, there appear side by side two spectra of the original ray crossed by one or more dark bands according to the thickness of the quartz plate used. The dark bands in the one spectrum correspond in position with the bright bands on the other. When the polarising nicol is rotated the bands move along the spectra. Tait's idea was to use this form of apparatus for studying the bright line spectra of faintly luminous objects, such as nebulae and comets. By employing first a thin piece of quartz and then a much thicker piece, he showed how the wave length of the light examined could be determined. There were however several practical difficulties in the way, and the method on trial did not turn out to be so sensitive as Tait had expected.

The following brief note to Thomson shows that Tait was thinking over the still debated question of the relative motion of the earth and the aether.

THE CLUB HOUSE,
ST ANDREWS,
26/4/82.

Dr T.

Στωξ says No! He says that in such a case *period* is everything.

But I have set Piazzi[1] on to try his magnificent Gitter. If the ether be in motion relatively to the earth, the absolute deviations of lines in the diffraction spectrum should be different in different azimuths: unless (of course) the relative motion of earth and ether be *vertical*. Anser. Yrs.

I am not aware however that either Piazzi Smyth or Tait ever tried experiments of the kind indicated. Tait had clearly taken Stokes' opinion, but was not convinced.

Another problem to which Tait again and again recurred was the question of the diathermancy of water vapour. He strongly doubted the accuracy of Tyndall's well-known experiments on this subject—see, for example, his letter to Andrews given above, p. 68. In 1882 he described in a letter to Thomson, who communicated it to the B. A. Meeting at Southampton, a new form of apparatus for investigating absorption of radiant heat by gases. The letter was published in *Nature*, Oct. 26, 1882. (See *Sci. Pap.* Vol. II, p. 71.) The general idea was to measure the absorption by the increase of pressure in the gas due to the heating. The apparatus was simply a double walled cylinder. While cold water was kept circulating in the jacket, steam could be blown into the double top. The changes of pressure in the gaseous contents were measured by a manometer U tube placed at the bottom. Several series of experiments were carried out by J. G. MacGregor and T. Lindsay (*Proc. R. S. E.* Vol. XII, 1882, p. 24), the conclusion being "that the absorption of air containing 1·3 per cent. of water vapour is between that of

[1] i.e., Piazzi Smyth, Astronomer Royal for Scotland and Professor of Astronomy in Edinburgh University, a well-known worker in spectroscopy. The signature is a compact monogram giving all three initials P. G. T. The phonetic spelling of "answer" is of course intentional, just as in the case of "Stokes"; these contractions were frequent, especially between Tait and Maxwell.

air containing 0·06 per cent. and that of air containing 0·2 per cent. of olefiant gas." This result was in agreement with what Tait himself obtained from a first rough experiment, and indicated that damp air was less absorbent of low heat rays than air mixed with a small quantity of olefiant gas.

These are a few of the investigations which did not yield all that Tait hoped for; in some instances indeed they were wholly abortive, or did not reach even the stage of being tried.

When any promising idea occurred to his mind, Tait was in the habit of referring it to the judgment of Stokes or Maxwell; and although unfortunately Tait's letters to Maxwell were not preserved, there has survived an amusing rhyming correspondence bearing upon the nature of electricity. In June 1877 Tait thus described the experiment he was thinking of making:

Will mounted ebonite disk
  On smooth unyielding bearing
When turned about with motion brisk
  (Nor excitation sparing)
Affect the primitive repose
  Of + or − in the wire,
So that while either downward flows
  The other upwards shall aspire?
Describe the form and size of coil
  And other things that we may need,
Think not about increase of toil
  Involved in work at double speed.
I can no more, my pen is bad,
  It catches in the roughened page—
But answer us and make us glad,
  THOU ANTI-DISTANCE-ACTION SAGE!
Yet have I still a thousand things to say
  But work of other kinds is pressing,
So your petitioner will ever pray
  That your defence be triple *messing*!

This last Anglo-German pun on the well-known Horatian text is a good example of one of Tait's forms of humour.

The following is Maxwell's reply as preserved in the original letter pasted into Tait's Scrap Book. The annotations are Maxwell's even to the references to Art. 770 at verse 1 and Art. 577 at verse 7. These were pencilled in and refer to Maxwell's *Electricity and Magnetism*, First Edition.

GLENLAIR,
DALBEATTIE,
25 *June* 1877.

Art. 770

The mounted disk of ebonite
  Has whirled before nor whirled in vain,
Rowland of Troy, that doughty knight,
  Convection currents did obtain[1]
In such a disk, of power to wheedle
From her loved north, the pensile needle.

'Twas when Sir Rowland, as a stage
  From Troy to Baltimore, took rest
In Berlin, there old Archimage
  Armed him to follow up this quest
Right glad to find himself possessor
Of the "irrepressible professor[2]."

But would'st thou twirl that disk once more?
  Then follow in Sir Rowland's train
To where in busy Baltimore
  He rears the bantlings of his brain;
As he may do who still prefers
One Rowland to two Olivers[3].

But Rowland,—no, nor Oliver,—
  Could get electromotive force
[Which fact and reason both aver,
  Has change of some kind as its source]
Out of a disk in swift rotation
Without the least acceleration.

But with your splendid roundabout
  Of mighty power, new hung and greasy,
With galvanometer so stout,
  Some new research would be as easy—
Some test which might perchance disclose
Which way the electric current flows.

Take, then, a coil of copper pure
  And fix it on your whirling table,
Place the electrodes firm and sure
  As near the axis as you're able
And strive to learn the way to work it
With galvanometer in circuit.

[1] *Berlin Monatsberichte.*
[2] Sylvester's address to Johns Hopkins.
[3] Heaviside and Lodge.

Art. 577

> Not while the coil in spinning sleeps
>   On her smooth axle swift and steady,
> But when against the stops she sweeps,
>   To watch the light spot then be ready,
> That you may learn from its deflexion
> The electric current's true direction.
>
> It may be that it does not move,
>   Or moves, but for some other reason,
> Then let it be your boast to prove
>   (Though some may think it out of season
> And worthy of a fossil Druid)
> That there is no electric fluid.

There is no evidence that Tait ever began on the line of work here indicated.

Taking a general view of Tait's experimental work we find it characterised by a true physical insight into the essential nature of each problem. Superfine accuracy was never his aim; and perhaps from this point of view some of his investigations lack finish. His methods were in many cases rough and ready, but they were always under complete mathematical control. Having laid down the broad lines of attack on any question he put together his apparatus with little apparent attention to detail; but his intuitions generally led him right. In many cases the first rough arrangement was committed to the care of two of his "veteran" students, in whose hands the final form of apparatus gradually evolved itself as difficulty after difficulty was surmounted. In this way the resourcefulness of the master and the enthusiasm and patient skill of the disciples worked together towards the perfected end. In his published accounts Tait never failed to give full credit to those who helped him in carrying his ideas to fruition.

The most laborious experiments undertaken by Tait were those on the conduction of heat in bars, on the errors of the "Challenger" thermometers, on the compression of liquids and on the laws of impact. In all of these Thomas Lindsay was his righthand man; and the successive bands of students who helped in the work consisted, in a sense, of picked men, for in those days only real enthusiasts ever thought of continuing their laboratory work so as to rank as "veterans." Already I have incidentally named a few of the students of whom Tait made use. Others, however, are equally worthy of mention. Thus the compression work with the "Big Gun" owed much of its success to the labours of R. T. Omond, afterwards superintendent of the Ben Nevis Observatory, and H. N. Dickson, at present Lecturer on Geography in University College, Reading. Again it was largely through the exertions of A. J. Herbertson, now Reader in Geography, Oxford University, and R. Turnbull, now Inspector, Department of Agriculture, Dublin, that the impact apparatus evolved itself from the first rude form to the final perfected arrangement; and the later set of experiments and their reductions were practically carried out by Alexander Shand.

It is interesting to note that many of Tait's students who helped him in research work did not become professional physicists. Under the old system, which present day pigeon-hole organisers rather despise, men had time to put in valuable work which lay outside their official course of study. Tait and his assistants soon saw who were the more resourceful among the laboratory students, and these were quickly enrolled in the unofficial squadron of workers. Under the present system of detailed courses of obligatory work, carefully scheduled for the benefit of the average student, such a method as Tait commanded in his day could hardly be applied. Many more students are trained now than formerly to make physical measurements; and the training is more systematic and thorough; but, with the exception of those who expect to follow out physics in their life career, very few ever come in touch with the stimulus which real research work gives. The day apparently is past for fruitful physical work to be effected in their student days by men who afterwards become clergymen, physicians, geographers, botanists, zoologists, or even engineers.

This account of Tait as an experimental philosopher would not be complete without some reference to the encouragement he gave to any of his students following out researches of their own, which had not been directly suggested by him. In particular, to those of us who desired to prepare an experimental thesis for the Doctorate in science, he gave every facility in the way of accommodation and apparatus. Among the more extended investigations of this independent nature I might mention Ewing and MacGregor's measurement of the electric conductivity of saline solutions, Macfarlane's experiments on the electric discharge through air and other dielectrics, my own work on contact electricity, Crichton Mitchell's study of the rate of cooling of bodies in steady currents of air, and C. M. Smith's experiments on conduction of heat in insulating material.

These and other similar pieces of research by his successive assistants and demonstrators could never have been carried out, had Tait not generously given us unrestricted access to his laboratory. Once we gained his confidence, we could roam at will through the whole department, and appropriate for our own purposes any apparatus which for the moment was not being used. There could be no truer way of encouraging research.

CHAPTER III

MATHEMATICAL WORK

In the preceding chapter Tait's experimental work has been dealt with apart from the other scientific activities of his mind. At no time however did he limit his attention to one problem exclusively ; and while with the aid of his company of voluntary workers he was for the last thirty years of his life busy with experiments in the laboratory, at home in his study he was using his mathematical powers with great effect in all kinds of enquiries. This mathematical and theoretical work may be conveniently classified under three headings : namely, quaternions, mathematics and mathematical physics outside the quaternion method, and the labours incidental to the writing of his more mathematical treatises. The quaternion work will be considered in an appropriate chapter ; another chapter will be devoted to the preparation of Thomson and Tait's *Natural Philosophy* ; and Tait's other literary contributions in book form will have a similar separate treatment. Here, in a somewhat discontinuous manner, I propose to give a general account of the more mathematical of his scientific papers and notes, tracing as far as possible their genesis and their connection with other lines of research.

Passing over his early quaternionic papers in the *Quarterly Journal of Mathematics*, the *Messenger of Mathematics*, and the *Proceedings* of the Royal Society of Edinburgh, we come in 1865 to a purely mathematical paper on the Law of Frequency of Error (*Trans. R. S. E.* Vol. xxiv ; *Sci. Pap.* Vol. i, p. 47). He was led to enquire into the foundations of the theory of errors when he was writing the article Probabilities for the first edition of *Chambers' Encyclopaedia*, his aim being to establish the ordinary law of errors by a " natural process " free from the mathematical complications which characterise the work of authorities like Laplace and Poisson. Starting from a simple case of drawing white and black balls from a bag, he deduced the well-known exponential expression, and then generalised the demonstration. If we except his much later papers on the kinetic theory of gases Tait does not seem to have returned to questions involving the theory of probabilities.

The preparation of the great *Treatise on Natural Philosophy* led his mind into various dynamical problems, such as central forces, the hodograph, and the theory of Action. On these he communicated short notes to the *Messenger of Mathematics* and to the *Proc. R. S. E.* His greatest effort at this time was, however, his paper "On the Application of Hamilton's Characteristic Function to special Cases of Constraint." In it he showed how brachistochrones or paths of shortest time were to be discussed by the same general method which Hamilton had applied to the theory of Action. Most of the investigation was embodied in the second edition of Tait and Steele's *Dynamics of a Particle.* Before publishing the paper Tait took the precaution of asking Cayley if he had been forestalled. Cayley replied:

"I have only attended to the direct problem of Dynamics, to find the motion of a system under given circumstances,—whereas the question of brachistochrones belongs of course to the inverse one...—and I really hardly know anything about it. My impression is that the subject is new."

In his address to Section A of the British Association in 1870 Clerk Maxwell, when referring to the rival theories of light, said

"To understand the true relation of these theories in that part of the field where they seem equally applicable we must look at them in the light which Hamilton has thrown upon them by his discovery that to every brachistochrone problem there corresponds a problem of free motion, involving different velocities and times, but resulting in the same geometrical path. Professor Tait has written a very interesting paper on this subject[1]."

Now this discovery which Maxwell ascribes to Hamilton was really made by Tait in the paper under discussion. Maxwell was usually very accurate in his history, and we can imagine the glee with which Tait found his friend tripping. He would by some merry joke make fun of Maxwell's momentary deviation from the lines of historic truth. Accordingly on July 14, 1871, Maxwell apologised in quaint fashion on an unsigned post card as follows:

"O T' Total ignorance of H and imperfect remembrance of T' in *Trans. R. S. E.* caused $\frac{dp}{dt}$ to suppose that H in his optical studies had made the statement in the form of a germ which T' hatched. I now perceive that T' sat on his own egg, but as his cackle about it was very subdued compared with some other incubators, I was not aware of its origin when I spoke to B. A. When I examined hastily H on Rays I expected to find far more than was there. But the good of H is not in what he has done but in the work (not nearly half done) which he makes other people do. But to understand him you should look him up, and go through all

[1] See Maxwell's *Scientific Papers*, Vol. II, p. 228.

kinds of sciences, then you go back to him, and he tells you a wrinkle. I have done lines of force and = potls. of double tangent galvanometer in a diagram, showing the large uniform field. Is T still in London?"

It was in this paper also that Tait proved his neat theorem "that a planet moving about a centre of force in the focus of its elliptic orbit is describing a brachistochrone (for the same law of speed as regards position) about the other focus," or in other words, "while time in an elliptic orbit is measured by the area described about one focus, action is measured by that described about the other." These statements are intimately bound up with the general theorem connecting brachistochrones and free paths already referred to.

In December 1871 Tait communicated to the Royal Society of Edinburgh a mathematical note on the theory of spherical harmonics (see *Proc. R. S. E.* Vol. VII, pp. 589–596). The interest of the note lies entirely in the simple manner in which certain fundamental relations are deduced. The article seems to have taken form to some extent under the influence of Maxwell, who, on a post card of date Sept. 5, 1871, wrote

"Spherical Harmonics first written in 1867 but worked up from T and T' when that work appeared. Have you a short and good way to find $\iint (\vartheta_i^{(\theta)})^2 dS$? If so make it known at 1ce that I may bag it lawfully as T' 4nion path to harmonic analysis."

Tait seems to have replied by sending a sketch of a new method, for Maxwell on October 23 wrote (again on a post card)

"O. T'! R. U. AT 'OME? $\iint$ Spharc$^2 dS$ was done in the most general form in 1867. I have now bagged ξ and η from T and T' and done the numerical value of $\iint (Y_i^{(\theta)})^2 dS$ in 4 lines, thus verifying T + T"s value of $\iint (\vartheta_i^{(\theta)})^2 dS$. Your plan seems indept. of T and T' or of me. PUBLISH!"

This was followed up ten days later by a fairly long letter bearing upon Tait's notes, the one quaternionic and the other in ordinary analysis. Tait must now have sent his analytical note very much in the form in which it was finally published. Regarding it Maxwell wrote:

11 SCROOPE TERRACE,
CAMBRIDGE,
2 *Nov.* 1871.

O T'

Your notes have ravished me. An interest in $\Sigma \phi a \rho \xi$ being revived this is exactly what is wanted for a quantitative or computative discussion of the symmetrical system considered as depending only on certain symbols i and s.

It seems to have little or nothing to do with your 4nionic reduction which is of course indept. of a selected axis[1].

My method is also indept. of a selected axis, but does not seem to be equivalent to your 4nion reduction which goes by steps.

Murphy is not at all bad in his way and affords a very good specimen of a Caius man working a calculation.

How is it that $\Sigma\phi\alpha\rho\xi$ can be worked only at Caius? See Murphy, Green, O'Brien, Pratt. When I examined here the only men who could do figure of the earth were mild Caius men. All the rest were Prattists if anything.

I think a very little mortar would make a desirable edifice out of your article.

In selecting the absolute value of the constant coefficient of a harmonic we may go on one of several principles.

There then followed a comparison of his own expressions with the corresponding expressions used by T and T' and by Tait. He continued:

The great thing is to avoid confusion. I rather think your value is the best to impress on the mind. It lies between it and $\vartheta_i^{(s)}$ which has a certain claim.

The diggings in $\Sigma\phi\alpha\rho\xi$ are very rich and a judicious man might get up a capital book for Cambridge, in which the wranglers would lade themselves with thick clay till they became blind to the concrete.

But try and do the 4nions. The unbelievers are rampant. They say "show me something done by 4nions which has not been done by old plans. At the best it must rank with abbreviated notations."

You should reply to this, no doubt you will.

But the virtue of the 4nions lies not so much as yet in solving hard questions, as in enabling us to see the meaning of the question and of its solution, instead of setting up the question in $x\ y\ z$, sending it to the analytical engine, and when the solution is sent home translating it back from $x\ y\ z$ so that it may appear as A, B, C to the vulgar.

There appears to be a desire for thermodynamics in these regions more than I expected, but there are some very good men to be found.

You will observe a tendency to bosch in this letter which pray xqs as I have been reading an ill assorted lot of books till I cannot correct prooves.

yours truly

$$\frac{dp}{dt}[2].$$

---

[1] Nevertheless Tait says in his paper that he was led to the method while engaged in some quaternionic researches.

[2] $\frac{dp}{dt} = JCM$, (Maxwell's initials), one expression for the Second Law of Thermodynamics, as used by Thomson in his early papers, and by Tait in his *Historic Sketch*, J being Joule's equivalent, C Carnot's function, and M the rate at which heat must be supplied per unit increase of volume, the temperature being constant.

In a post card to Thomson of date Nov. 5, 1871, Maxwell, after referring to some proof sheets of his book which he had sent to Thomson to revise, remarked :

"Laplace has a clear view of the Biaxal harmonic. T′ has an excellent discussion of Q_i and $\mathfrak{S}_i^{(s)}$ and their relations deduced from their definitions and not from their expansions as Murphy does. Murphy is very clever, but not easily appreciated by the beginner."

This post card found its way finally to Tait and was duly filed along with the other correspondence. The whole correspondence shows the free interchange of thought which went on between Maxwell and Tait and the subtle manner in which each helped the other. We can in many cases infer the nature of Tait's letters which Maxwell was obviously replying to ; but the characteristic language in which these must have been expressed is unfortunately irrecoverable.

For anything of Hamilton's Tait had a profound respect ; and in the "beautiful invention of the Hodograph" he found on more than one occasion a source of inspiration. His hodograph note communicated to the Royal Society of Edinburgh in 1867 contains an elegant geometrical construction in which the equiangular spiral is used with effect to represent motion in a resisting medium. Maxwell practically introduces the whole investigation into the second volume of his *Electricity and Magnetism*, when he is discussing the theory of damped vibrations of a swinging magnetic needle.

The powerful quaternion papers on the rotation of a rigid body and on Green's theorem were communicated to the Royal Society in 1868 and 1870 respectively. They will be most suitably discussed in the following chapter on quaternions. To this period also belongs a quaternion investigation into the motion of a pendulum when the rotation of the earth is taken into account. This is reproduced in the second edition of his *Quaternions*. The paper is called an "Abstract" in the *Proceedings* ; and the closing sentences epitomising other developments imply that Tait had every intention of publishing a complete and elaborate discussion as a *Transactions* paper. For this however he never found leisure. This habit of printing an abstract, indicating the lines of development in a projected large memoir which never saw the light, was one which grew with the progress of the years.

During the early seventies, when the experiments in thermo-electricity were in full swing, nothing very serious was taken up on the mathematical side ; but the game of golf suggested this curious and by no means easy problem ; "When a golf-player is x holes 'up' and y 'to play,' in how many ways may the game finish ?" The paper in which Tait considered the problem is called a question of arrangement and probabilities. He first solved the simpler question as to the number of ways the player who is x up and y to play may win. Let this number of ways of winning be represented by $P(x, y)$. Then starting with $P(x+1, y+1)$, we see that at the first stage the player may win, halve, or lose the next hole, and the number of possible ways of winning will then be represented by $P(x+2, y)$, $P(x+1, y)$, and $P(x, y)$ respectively ; hence follows Tait's fundamental equation

$$P(x+1, y+1) = P(x+2, y) + P(x+1, y) + P(x, y).$$

If then we construct a coordinate scheme with x measured horizontally and y vertically downwards, and place in the position xy the number $P(x, y)$, we can at once pass by simple addition of three consecutive values of x for any one value of y to the values for the next higher value of y. The following is the scheme as far as $y = 5$.

0	0	0	0	0	0	1	1	0	0	0	0	...x
0	0	0	0	0	1	2	1	1	0	0	0	
0	0	0	0	1	3	4	4	1	1	0	0	
0	0	0	1	4	8	11	9	6	1	1	0	
0	0	1	5	13	23	28	26	16	8	1	1	
0	1	6	19	41	64	77	70	50	25	10	1	
	etc.						etc.					

y

The zero positions are enclosed in the double lines ; and the meaning of the entries to the left of the vertical lines is the number of ways in which the player may lose. The unit values on the right and left flanks are determined by the limiting conditions, which show that when x is greater than y, the game is won, so that $P(x, y) = 1$. Similarly, when x is not less than y, the player cannot lose. Hence $P(-x, y) = 0$. These considerations also explain why the fundamental equation given above does not apply to the second last unit on the right of each row. As an example, let a player be 2 up and 4 to play ; he may win in 26 different ways. His opponent who is 2 down and 4 to play may of course lose in the same number of ways. But the number of ways in which the player who is 2 up may lose is only 5. These numbers 26 and 5 give an estimate of the respective probabilities of either player winning. The number of possible draws is obtained from the same fundamental equation, the limiting conditions being $P(x, y) = 1$ when $x = y$, $P(x, y) = 0$ when $x > y$, whether x is positive or negative. The values are represented by the following scheme.

					0	1	0					...x
				0	1	1	1	0				
			0	1	2	3	2	1	0			
		0	1	3	6	7	6	3	1	0		
	0	1	4	10	16	19	16	10	4	1	0	
			etc.			⋮		etc.				

y

Thus when the one player is 2 up and 4 to play, the game may be drawn in 10 different ways, and hence the number of distinct ways in which such a game may end is $26 + 5 + 10 = 41$. These schemes were expressed by Tait in a formula based upon the expansion of the expression $(a + 1 + 1/a)$ raised to the power y.

In a brief paper on a Fundamental Principle in Statics, communicated to the Royal Society on Dec. 21, 1874, Tait compared in a remarkably simple manner the gravitational attraction between the two hemispheres of the earth and the tendency to split across the diametral plane separating these in consequence of the earth's rotation. He thus proved that it was gravitation and not cohesion which kept the material of the earth together. A planet of the earth's mean density and of tensile strength equal to that of steel would be held together as much by cohesion as by gravitation if its radius were 409 miles. I believe this must be the result referred to by Kelvin in a short letter to Tait, which was written from White's workshop in Glasgow, but of which the date unfortunately had been torn off. It runs

Dear T′

I thought as much. It is not the thing I object to but your PFian way of doing it. However enough of that.

I still think your planet the greatest step in dynamics made in the second half of the 19th century.......

I am up to see new electrometers but find them too unfinished.

Yours

T.

Not able to understand the reference to the planet I sent the note to Kelvin himself, who, writing on Oct. 3, 1907, said

"I return my old pencilled letter to Tait, which has come to me enclosed with yours of yesterday. I have no recollection of the wonderful planet.

"PFian meant Pecksniffian. Pecksniff was a great hero of Tait's in respect to his almost superhuman selfishness, cunning, and hypocrisy, splendidly depicted by Dickens."

The only other planetary theorem with which Tait's name is associated is the one already referred to in connection with Action and Brachistochrones; but this comparison between the effects of cohesion and gravitation when first made was just the kind of thing to appeal to Thomson.

Tait's excursions into the field of pure mathematics were not frequent; and his paper on the Linear Differential Equation of the Second Order (Jan. 3, 1876) practically stands alone. It contains some curious results and suggests several lines of further research. The general idea of the paper is to compare the results of various processes employed to reduce the general linear differential equation of the second order to a non-linear equation of the first order. The properties of the operators of the form $\left(\frac{\partial}{\partial x} x \frac{\partial}{\partial x}\right)^n$ are incidentally considered, and the question is asked as to the evaluation at *one step* of the integral

$$\iint \left(\frac{\partial y}{\partial x}\right)^2 dx^2.$$

At the British Association Meeting of 1876, Tait communicated a note on some elementary properties of closed plane curves, especially with regard to the double points, crossings, or intersections. He pointed out the connection of the subject with the theory of knots, on which he was now about to begin a long and fruitful discussion. He was attracted to a study of knots by the problem of the stability of knotted vortex rings such as one might imagine to constitute different types of vortex atoms. Some of these were figured in Kelvin's great paper, which itself was the outcome of Tait's own experimental illustrations of Helmholtz's theorems of vortex motion. The conception of the vortex atom gave an extraordinary impulse to the study of vortex motion, and the following early letter of Maxwell indicates some of the lines of research ultimately prosecuted by Thomson and Tait.

GLENLAIR,
DALBEATTIE,
Nov. 13, 1867.

Dear Tait

   If you have any spare copies of your translation of Helmholtz on "Water Twists" I should be obliged if you could send me one.

  I set the Helmholtz dogma to the Senate House in '66, and got it very nearly done by some men, completely as to the calculation, nearly as to the interpretation.

  Thomson has set himself to spin the chains of destiny out of a fluid plenum as M. Scott set an eminent person to spin ropes from the sea sand, and I saw you had put your calculus in it too. May you both prosper and disentangle your formulae in proportion as you entangle your worbles. But I fear that the simplest *indivisible* whirl is either two embracing worbles or a worble embracing itself.

  For a simple closed worble may be easily split and the parts separated

but two embracing worbles preserve each other's solidarity thus

though each may split into many, every one of the one set must embrace every one of the other. So does a knotted one.

yours truly

J. CLERK MAXWELL.

  Here Maxwell expressed very clearly one of the ideas which Tait finally made the starting point of his discussion of knots. The trefoil knot, the simplest of all knots, was chosen by Balfour Stewart and Tait as a symbolic monogram on the title page of the *Unseen Universe*; and some of the speculations put forward in that work must have been closely connected with the line of thought which found a scientific development in Tait's later papers. It may have been while thinking out the attributes of vortex atoms in an almost frictionless fluid that Tait came to see there was a mathematical problem to attack in regard to the forms of knotted vortex rings.

  If we take a cord or, better still, a long piece of rubber tubing, twist it round itself in and out in any kind of arbitrary fashion, then join its ends so as to make a closed loop with a number of interlacings on it, we get a vortex knot. We may suppose it drawn out and flattened until the crossings have been well separated and reduced to the lowest possible number. Projected on the plane this will appear as a closed curve with a certain number of double points. Hence the fundamental mathematical problem may be thus stated : Given the number of its double points, find all the essentially different forms which a closed continuous curve can assume. Beginning at any point of the curve and going round it continuously we pass in succession through all the double points in a certain order. Every point of intersection must be gone through twice, the one crossing (in the case of the knot) being along the branch which passes above, the other along the branch which passes below. If we lay down a haphazard set of points and try to pass through them continuously in the way described, we shall soon find that only certain modes are possible. The problem is to find those modes for any given number of crossings. Let us begin to pass the point A by the over-crossing branch. We shall evidently pass the second point by an under-crossing branch, the third by an over-crossing again, and so on. Calling the first, third, fifth, etc., by the letters A, B, C, etc., we find that after we have exhausted all the intersections the even number crossings will be represented by the same letters interpolated in a certain order. To fulfil the conditions of a real knot, it is clear that neither A nor B can occupy the second place, neither B nor C the fourth, and so on. This at once suggests the purely mathematical problem :—How many arrangements are there of n letters when a particular one cannot be in the first or second place, nor another particular one in the third or fourth, nor a third particular one in the fifth or sixth, and so on. Cayley and Thomas Muir both supplied Tait with a purely mathematical solution of this problem ; but even when that is done, there still remain many arrangements which will not form knots, and others which while forming knots are repetitions of forms already obtained. These remarks will give an idea of the difficulties attending the taking of a census of the knots, say, of nine or ten intersections—what Tait called knots of nine-fold and ten-fold knottiness. If we take a piece of rubber tubing plaited and then closed in the way suggested above, we shall be surprised at the many apparently different forms a given knot may take by simple deformations. Conversely, what appear to the eye to be different arrangements, become on closer inspection Proteus-like forms of the same. While engaged in this research, Tait came into touch with the Rev. T. P. Kirkman, a mathematician of marked originality, and one of the pioneers in the theory of Groups. Kirkman's intimate knowledge of the properties of polyhedra suggested to him a mode of attack on knots quite distinct from that developed by Tait. Taking advantage of Kirkman's extension of the census to knots of eight-fold and nine-fold knottiness, Tait was able to give in his second paper (1884) all the forms of knots of the first seven orders of knottiness, the numbers being as follows:

Order of knottiness	3	4	5	6	7	8	9
Number of forms	1	1	2	4	8	21	47

A year later in his third paper Tait, basing his enumeration on Kirkman's polyhedral method of taking the census, figured the 123 different forms of ten-fold knottiness. Higher orders have been treated by Kirkman and Little (*Trans. R. S. E.* Vols. XXXII, XXXV, XXXVI, XXXIX).

In his second paper Tait pointed out that with the first seven orders of knottiness we have forms enough to supply all the elements with appropriate vortex atoms.

A curious problem in arrangements suggested by the investigations in the properties of knots was thus enunciated by Tait:

"A Schoolmaster went mad, and amused himself by arranging the boys. He turned the dux boy down one place, the new dux two places, the next three, and so on until every boy's place had been altered at least once. Then he began again, and so on; till, after 306 turnings down all the boys got back to their original places. This disgusted him, and he kicked one boy out. Then he was amazed to find that he had to operate 1120 times before all got back to their original places. How many boys were in the class?"

The answer is 18 (see *Proc. R. S. E.* Jan. 5, 1880; *Sci. Pap.* Vol. 1, p. 402).

In his discussion of knots Tait established a new vocabulary and gave precise meanings to such terms as knottiness, beknottedness, plait, link, lock, etc. He introduced with effect the old Scottish word "flype" which has no equivalent in southern English speech, the nearest being "turn-out-side-in." Clerk Maxwell has described some of Tait's processes in the following rhymes:

(CATS) CRADLE SONG.

By a Babe in Knots.

Peter the Repeater
  Platted round a platter
Slips of silvered paper
  Basting them with batter.

Flype 'em, slit 'em, twist 'em,
  Lop-looped laps of paper;
Setting out the system
  By the bones of Neper.

Clear your coil of kinkings
  Into perfect plaiting,
Locking loops and linkings
  Interpenetrating.

Why should a man benighted,
  Beduped, befooled, besotted,
Call knotful knittings plighted,
  Not knotty but beknotted?

It's monstrous, horrid, shocking,
  Beyond the power of thinking,
Not to know, interlocking
  Is no mere form of linking.

But little Jacky Horner
  Will teach you what is proper,
So pitch him, in his corner,
  Your silver and your copper.

One of Tait's most beautiful self-contained papers is his paper on Mirage (1881), published in the *Transactions* of the R. S. E. (*Sci. Pap.* Vol. 1, No. LVIII). It is worked out as an example of Hamilton's general method in optics. Not only is it an elegant piece of mathematics, but it shows to advantage the clearness of Tait's physical intuition in his assumption of a practically possible vertical distribution of temperature and density capable of explaining all the observed phenomena. A less technical account of the paper on Mirage was published in *Nature* (Vol. XXVIII, May 24, 1883) under the title " State of the Atmosphere which produces the forms of Mirage observed by Vince and by Scoresby." This article is printed below.

In 1886 Tait's attention was strongly drawn to the foundations of the Kinetic Theory of Gases, on which subject he communicated four memoirs to the *Transactions* of the Royal Society of Edinburgh and a fifth (in abstract) to the *Proceedings* within the six succeeding years. His first aim, as indicated in the title, was to establish sure and strong the fundamental statistical propositions in the distribution of speeds and energy among a great many small smooth spheres subject only to their mutual collisions; and the one initial point aimed at was a rigorous proof of Maxwell's theorem of the equal partition of energy. An interesting question carefully considered by

Tait was how to define the Mean Free Path, in regard to which he differed from Maxwell. He also laid stress on the principle that throughout the investigation each step of the process of averaging should not be performed before the expressions were ripe for it. Some of his views are put very succinctly in a letter to Thomson in 1888, just about the time he was printing the third paper of the series. We may regard it as containing Tait's last statement on the question.

38 GEORGE SQUARE,
EDINBURGH, 27/2/88.

O. T.

Ponder every word of this and report.

Since there is absolute social equality in the community called a simple gas, the average behaviour of any one particle during 3.10^{20} seconds is the same as that of 3.10^{20} particles (the content of a cubic inch) for one second.

Hence if n_v be the chance that the speed is from v to $v + dv$, and if p_v be *then* the mean free path; and if C be the number of collisions in 3.10^{20} seconds, we have

$$n_v C$$

as the number of collisions in which the speed is v to $v + dv$, and the path p_v. Thus the whole space travelled over in 3.10^{20} seconds (10^{13} years nearly) is

$$C\Sigma(n_v\, p_v).$$

This consists of C separate pieces. The average of these, i.e. the Mean Free Path, is therefore

$$\Sigma(n_v\, p_v) \quad \dots\dots\dots\dots\dots\dots\dots\dots\dots\dots\dots(1).$$

Also the *interval* between two collisions, when the speed is v, is p_v/v. Hence the whole time spent on C collisions is $C\Sigma\left(n_v\dfrac{p_v}{v}\right)$. This is 3.10^{20} seconds. Thus the average number of collisions per particle per second is

$$\frac{C}{3.10^{20}} = \frac{1}{\Sigma\left(n_v\dfrac{p_v}{v}\right)} \quad \dots\dots\dots\dots\dots\dots\dots\dots\dots(2).$$

Both of these results differ from those now universally accepted. Instead of (1) they, Maxwell, Meyer, Boltzmann etc., give

$$\frac{\Sigma(n_v\, v)}{\Sigma\left(n_v\,\dfrac{v}{p_v}\right)} \quad \dots\dots\dots\dots\dots\dots\dots\dots\dots(\bar{1}),$$

and instead of (2)

$$\Sigma\left(n_v\,\frac{v}{p_v}\right) \quad \dots\dots\dots\dots\dots\dots\dots\dots\dots(\bar{2}).$$

Both are, I think, *obviously* wrong.

Yrs.

There is no record what reply Thomson made to this very clear statement.

Having established the fundamental propositions in the first paper Tait proceeded in his later papers to develop the subject in its application to viscosity, thermal conduction, diffusion, the virial, and the isothermal equations. Certain strictures which Tait in his fourth paper applied to Van der Waals' method of evolving his well-known isothermal equation led to a discussion with Lord Rayleigh and Professor Korteweg (see *Nature*, Vols. XLIV, XLV, 1891–92). While accepting their explanations of Van der Waals' process he was not convinced that the process was valid in the sense of being a logical following out of the virial equation.

On November 23, 1893, Tait reviewed in *Nature* (Vol. XLIX) the second edition of Dr Watson's Treatise on the Kinetic Theory of Gases: and the following paragraphs give very clearly his own view of the significance and aim of his papers on the subject:

"I believe that I gave, in 1886 (*Trans. R. S. E.* Vol. XXXIII), the first (and possibly even now the sole) thoroughly legitimate, and at least approximately complete, demonstration of what is known as Clerk-Maxwell's Theorem, relating to the ultimate partition of energy between or among two or more sets of hard, smooth, and perfectly elastic spherical particles. And I then pointed out, in considerable detail, the logical deficiencies or contradictions which vitiated Maxwell's own proof of 1859, as well as those involved in the mode of demonstration which he subsequently adopted from Boltzmann. Dr Boltzmann entered, at the time, on an elaborate defence of his position; but he did not, in my opinion, satisfactorily dispose of the objections I had raised. Of course I am fully aware how very much easier it is for one to discover flaws in another man's logic than in his own, and how unprepared he usually is to acknowledge his own defects of logic even when they are pointed out to him. But the only attacks which, so far as I know, have been made on my investigation, were easily shown to be due to misconception of some of the terms or processes employed.......

"From the experimental point of view, the first great objection to Boltzmann's Theorem is furnished by the measured specific heats of gases; and Dr Watson's concluding paragraphs are devoted to an attempt to explain away the formidable apparent inconsistency between theory and experiment. In particular he refers to a little calculation, which I made in 1886 to show the grounds for our confidence in the elementary principles of the theory. This was subsequently verified by Natanson (*Wied. Ann.* 1888) and Burbury (*Phil. Trans.* 1892). Its main feature is its pointing out the absolutely astounding rapidity with which the average amounts of energy per particle in each of two sets of spheres in a uniform mixture approach to equality in consequence of mutual impacts. Thus it placed in a very clear light the difficulty of accepting Boltzmann's Theorem, if the degrees of freedom of a complex molecule at all resemble those of an ordinary dynamical system."

The calculation referred to here was given in the first paper as Part v, the earlier parts being concerned with the mean free path, the number of collisions, and the general proof of Maxwell's theorem. Part vi is devoted to the discussion of some definite integrals, and the remaining three parts of the first paper take up the question of the mean free path in a mixture of two systems, the pressure in a system of colliding spheres, and the effect of external potential. In the second paper Tait proceeded to apply the results of the first paper "to the question of the transference of momentum, of energy, and of matter, in a gas or gaseous mixture; still, however, on the hypothesis of hard spherical particles, exerting no mutual forces except those of impact." Before entering on this line of investigation, Tait took occasion to answer certain criticisms which had been made of his methods in the first paper, especially in regard to the number of assumptions necessary for the proof of Maxwell's theorem concerning the distribution of energy in a mixture of a gas. Tait contended however that all he demanded was "that there is free access for collision between each pair of particles, whether of the same kind or of different systems; and that the number of particles of one kind is not overwhelmingly greater than that of the other." In the third paper, a special case of molecular attraction is dealt with. The particles which are under molecular force are assumed to have a greater average kinetic energy than the rest. In terms of this assumption the expression for the virial is developed in the fourth paper, leading finally to Tait's form of the isothermal equation

$$pv = E + \frac{C}{v+\gamma} - \frac{A-eE}{v+a},$$

where C, A, e, γ, a are constants, and E is a quantity which in the case of vapour or gas of small density has the value $\frac{1}{2}\Sigma mu^2$, where u is the speed of the particle of mass m. This average kinetic energy is generally assumed to be proportional to the absolute temperature; but Tait had grave reasons for not accepting this view. He said:

"It appears to me that *only* if E above (with a *constant* added when required, as will presently be shown) is regarded as proportional to the absolute temperature, can the above equation be in any sense adequately considered as that of an *Isothermal*. If the whole kinetic energy of the particles is treated as proportional to the absolute temperature, the various stages of the gas as its volume changes with E constant correspond to changes of temperature without direct loss or gain of heat, and belong rather to a species of *Adiabatic* than to an *Isothermal*. Neither Van der Waals nor Clausius, so far as I can see, calls attention to the fact that when there are molecular forces the mean-square speed of the particles necessarily increases with diminution of volume, even when the mean-square speed of a free particle is maintained unaltered; and this simply because the time during which each particle is free is a smaller fraction of the whole time. But when the whole kinetic energy is treated as constant (as it must be in an Isothermal, when that energy is taken as measuring the absolute temperature), it is clear that isothermal compression must reduce the value of E....

"For the isothermal formation of liquid, heat must in all cases be taken from the group M. This must have the effect of diminishing the value of E. Hence, in a liquid, the temperature is no longer measured by E, but by $E + c$, where c is a quantity whose value steadily increases, as the temperature is lowered, from the value zero at the critical point...."

Putting then $E = Rt$, where t is the absolute temperature, Tait introduced the pressure temperature and volume at the critical point, and threw his equation into the form

$$p = \bar{p}\left(1 - \frac{(v - \bar{v})^3}{v(v + a)(v + \gamma)}\right) + R\left(1 + \frac{e}{v + a}\right)\frac{t - \bar{t}}{v},$$

where the barred letters refer to the critical values. He compared this with the corresponding equations of Van der Waals and Clausius and pointed out that, although they all three agreed in form for the critical isothermal, they could not do so for any other. He then found, by direct calculations from Amagat's results for Carbon Dioxide, that the pressures obtained by his formula for given volumes at the critical temperature agree almost perfectly with the measured pressures, between a range of volume from 1 to 0·0035.

This practically finishes the series of papers on the Foundations of the Kinetic Theory of Gases; for the fifth instalment was printed only in abstract and indicates lines of investigation which were never completed.

For five full years Tait occupied his mind with these researches; and if we except his quaternion work there is no other line of investigation which made such serious demands upon both his mathematical powers and his physical intuitions. Throughout the whole series he is essentially the natural philosopher, using mathematics for the elucidation of what might be called the metaphysics of molecular actions. No writer on the subject has put more clearly the assumptions on which the statistical investigation is based; and apparently he was the first to calculate the rate at which under given conditions the "special state" is restored when disturbed. His abhorrence of long and intricate mathematical operations is strongly expressed more than once. He was convinced of the general accuracy of Maxwell's conclusions; but he could not admit the validity of all his demonstrations. If we may judge from a letter written to him by Maxwell as early as August 1873, Tait had been seeking enlightenment years before he himself thought of tackling the problem. Maxwell's letter consists of a set of numbered paragraphs, 1, 3, 7, 5, evidently in answer to a set of corresponding questions put by Tait. Paragraph (5) runs thus:

"By the study of Boltzmann I have been unable to understand him. He could not understand me on account of my shortness, and his length was and is an equal stumbling-block to me. Hence I am very much inclined to join the glorious company of supplanters and to put the whole business in about six lines."

Maxwell then gave the conclusion of his paper on the Final State of a System of Molecules in motion subject to forces of any kind (*Nature*, Vol. VIII, 1873: *Scientific Papers*, Vol. II, pp. 351–4) and continued:

"In thermal language—Temperature uniform in spite of crowding to one side by forces. Molecular volume of all gases equal. Equilibrium of mixed gases follows Dalton's Law of each gas acting as vacuum to the rest (in fact it acts as vacuum to itself also). In my former treatise I got these results only by way of conclusions. Now they come out before any assumption is made as to the law of action between molecules."

A few months later (Dec. 1, 1873) Maxwell returned to the subject evidently in reply again to Tait. This letter of Maxwell's touches upon a great variety of points, all in reference to Tait's varied activities at the time; and it seems better to give the letter here as a whole with footnote elucidations than to break it up into bits distributed throughout the volume.

Natural Sciences Tripos. 1 *Dec.* 1873.

O T'. For the flow of a liquid in a tube[1], axis z

$$\mu \left(\frac{\partial^2 w}{\partial x^2} + \frac{\partial^2 w}{\partial y^2} \right) = \frac{\partial p}{\partial z} \quad \dots\dots\dots\dots\dots\dots\dots\dots\dots (1).$$

Surface condition
$$\mu \frac{\partial w}{\partial \nu} = \lambda w \quad \dots\dots\dots\dots\dots\dots\dots\dots\dots (2),$$

where ν is the normal drawn towards the liquid. When the curvature is small, (2) is equivalent to supposing the walls to be removed back by μ/λ and then λ made ∞ or $w = 0$. For glass and water by Helmholtz and Pietrowski $\mu/\lambda = 0$.

If so, and if the value of w is $C(1 - x^2/a^2 - y^2/b^2)$,

$$2\mu C \left(\frac{1}{a^2} + \frac{1}{b^2} \right) + \frac{\partial p}{\partial z} = 0, \text{ which gives } C.$$

[1] See Tait's Laboratory Notes (*Proc. R. S. E.* VIII, p. 208): On the Flow of Water through fine Tubes. The experiments were made by C. Michie Smith and myself with tubes of circular and elliptic bore. Tait had asked Maxwell to give him the theory of the phenomenon as a problem in viscosity.

If not, you may write

$$w = A + Br^2 + C_2 r^2 \cos 2\phi + C_4 r^4 \cos 4\phi + \text{etc.},$$

where $x = ar \cos \theta$ and $y = br \sin \theta$ and then

$$2\mu B \left(\frac{1}{a^2} + \frac{1}{b^2} \right) + \frac{\partial p}{\partial z} = 0$$

and you satisfy (2) the best way you can when $r = 1$.

As to Ampère—of course you may lay on d_1 (anything) where d_1 is with respect to the element of a circuit. Have you studied H² on the potential[1] of two elements? or Bertrand? who, with original bosh of his own rushes against the thicker bosches of H²'s buckler and says that H² believes in a force which does not diminish with the distance, so that the reason why Ampère or H² or Bertrand observe peculiar effects is because some philosopher in α Centauri happens to be completing a circuit. $XQq\,D$ [tails][2] as I am surrounded by Naturals and cannot give references.

In introducing 4nions[3] do so by blast of trumpet and tuck of drum. Why should $V.\,\alpha\beta\gamma$ come in sneaking without having his style and titles proclaimed by a fugleman? Why even . should be treated with due respect and we should be informed whether he is attractive or repulsive.

What do you think of "Space-variation" as the name for Nabla?

It is only lately under the conduct of Professor Willard Gibbs that I have been led to recant an error which I had imbibed from your $\theta\Delta cs$, namely that the entropy of Clausius is *unavailable energy*, while that of T′ is *available energy*[4]. The entropy of Clausius is neither one nor the other. It is only Rankine's Thermodynamic Function....

I have also a great respect for the elder of those celebrated acrobats, Virial and Ergal, the Bounding Brothers of Bonn. Virial came out in my paper on Frames, R. S. E. 1870 in the form $\Sigma Rr = 0$, when there is no motion. When there is motion the time average of $\frac{1}{2}\Sigma Rr = $ time average of $\frac{1}{2}\Sigma Mv^2$, where R is positive for attraction.

But it is rare sport to see those learned Germans contending for the priority of the discovery that the 2nd law of $\theta\Delta cs$ is the Hamiltonsche Princip, when all the time they *assume* that the temperature of a body is but another name for the vis viva of one of its molecules, a thing which was suggested by the labours of Gay

[1] The reference is to H(ermann) H(elmholtz)'s electrodynamic investigation which supplied the true criterion in place of the hasty generalisation of § 385 in the first edition of *Thomson and Tait*.

[2] The [tails] are drawn as arrow-headed wiggles of various lengths and forms.

[3] See the chapter on Quaternions for other remarks by Maxwell on Tait's quaternion work. Maxwell was reading Kelland and Tait's *Introduction to Quaternions* which he reviewed in *Nature* shortly after.

[4] Tait suggested in the first edition of his *Thermodynamics* (contracted into $\theta\Delta cs$ by Maxwell) that the word Entropy should be used in this sense. In the second edition he went back to the original meaning as given by Clausius.

Lussac, Dulong, etc., but first deduced from dynamical statistical considerations by $\frac{dp}{dt}$. The Hamiltonsche Princip, the while, soars along in a region unvexed by statistical considerations, while the German Icari flap their waxen wings in nphelo-coccygia amid those cloudy forms which the ignorance and finitude of human science have invested with the incommunicable attributes of the invisible Queen of Heaven....

General [quaternion] exercise. Interpret every 4nion expression in literary geo-metrical language, e.g., express in neat set terms the result of $\frac{\beta}{\alpha}.\gamma$.

$$\frac{dp}{dt}.$$

There is a close association between these remarks by Maxwell in 1873 and some of Tait's own comments in his Kinetic Theory papers published thirteen years later.

In 1896, in a note on Clerk Maxwell's Law of Distribution of Velocity in a Group of equal colliding Spheres (*Proc. R. S. E.* Vol. XXI), Tait published his last views on the subject. He repelled certain criticisms of Maxwell's solution brought forward by Bertrand in the *Comptes Rendus* of that year. Bertrand's enunciation of what he conceived to be the problem attacked by Maxwell, and the enunciation of the problem really attacked, were set side by side; and Bertrand was condemned out of his own mouth. At the same time Tait strengthened the experimental foundations of the argument that the solution of the problem is unique and cannot be destroyed by collisions, by an application of Doppler's principle to the radiations of a gas.

The results of Tait's investigations into the flight of a golf ball have already been detailed (Chap. I, p. 27). A brief sketch of the mathematical method by which he deduced his results is appropriately given here. Tait published two papers on the Path of a Rotating Spherical Projectile, the first in 1893, the second in 1896 (*Trans. R. S. E.* Vols. XXXVII, XXXIX). The foundation of the theory was the assumption that, in virtue of the combination of a linear speed v and a rotation ω about a given axis, the ball is acted on by a force proportional to the product of the speed and the rotation, and perpendicular both to the line of flight and to the axis of rotation. This transverse force acts in addition to the retarding force due to the resistance of the air; and the first problem solved by Tait was the case in which no other than these two forces act. It is easy to show that under the influence of such forces the sphere will move in a spiral whose curvature will be inversely as the speed of translation and whose tangent will rotate with a constant angular velocity. The projection on the horizontal plane of the path of a pulled or sliced golf ball will be very approximately portions of this spiral. The introduction of gravity acting constantly in one direction greatly complicates the problem, which cannot be solved, even to a first approximation, except on the supposition that the path nowhere deviates greatly from the horizontal. To obtain forms of paths at all like those observed, somewhat lengthy numerical calculations require to be made. The method by which Tait builds up the curve is very instructive and is a good example of his insight into the essence of a physical problem and of his capacity in working out a sufficient solution. The practical details will be found in the article on Long Driving reprinted below.

In addition to the greater efforts of his mathematical powers, Tait contributed to the *Messenger of Mathematics*, to the *Proceedings* of the Royal Society of Edinburgh, and latterly to the *Proceedings* of the Edinburgh Mathematical Society, a variety of small notes, many of which he incorporated in the successive editions of his books. These notes were always interesting in themselves and frequently presented old truths from new points of view. In not a few of them his skill as a geometrician comes strongly into evidence. Tait, in fact, was no juggler with symbols; and when taking up a new subject he invariably tried to make of it a geometrically tangible creation; otherwise he would have none of it. Maxwell expressed this view of Tait's mental habitude in a letter in which, replying evidently to a demand of Tait's to consider a problem in conduction of heat, he wrote:

"O T' If a man will not read Lamé how should he know whether a given thing is ν? Again, if a man throws in several triads of symbols and jumbles them up, pretending all the while that he has never heard of geometry, will not the broth be thick and slab? If the problem is to be solved in this way by mere heckling of equations through ither[1] I doubt if you are the man for it as I observe that you always get on best when you let yourself and the public know what you are about."

Of those casual things which Tait threw off largely as mathematical recreations, about a dozen were communicated to the Edinburgh Mathematical Society. The subjects treated of are nearly as numerous as the papers, including plane strains, summations of series, orthogonal systems of curves,

---

[1] "Through ither," an expressive Scottish phrase, meaning lack of method so that things get tangled up one with the other—higgledy-piggledy comes near it. It is often used with reference to a thriftless housewife who has no method but drives through her work anyhow. "Heckling of equations through ither" means assorting the equations in a random manner in the hope that they will be disentangled and simplified.

circles of curvature, attractions, centrobaric distributions, logarithms, etc. The note on centrobaric distributions he afterwards simplified and extended in his booklet on Newton's Laws of Motion, and gave a remarkably simple geometrical proof that the potential of a uniform spherical shell is constant throughout the interior, and varies for external points inversely as the distance from the centre.

The last published paper not connected with quaternions was on a generalization of Josephus' problem (1898, *Proc. R. S. E.* Vol. xxii). The original problem stated simply is to arrange 41 persons in a circle in such a way that when every third person beginning at a particular position is counted out, a certain named one will be left. What position relatively to the first one counted will he occupy? It is said that by this means Josephus saved his life and that of a companion out of a company who had resolved to kill themselves so as not to fall into the hands of the enemy. Josephus is said to have put himself in the 31st place and his friend in the 16th place. Tait's generalization consists in pointing out that, if we know the position of "safety" for any one number, we can without going through the labour of the obvious sifting-out process at once say where the position of "safety" will be if the number is increased by one. This position is simply pushed forward by as many places as there are in the grouping by which the successive individuals are picked out. By successive application of the process, Tait quickly found that if every third man is picked out of a ring of 1,771,653 men, the one who is left last is the occupier of place 2 in the original arrangement. Hence if there were 2,000,000 in the circle the place to be assigned to the last one left after the knocking out by threes is evidently

$$2 + 3 \times (2{,}000{,}000 - 1{,}771{,}653) = 2 + 3 \times 228{,}347 = 2 + 685{,}041 = 685{,}043.$$

When the number reaches 2,657,479 a new cycle will begin with the place of safety in position 1. The general rule given by Tait is:

"Let n men be arranged in a ring which closes up its ranks as each individual is picked out. Beginning anywhere, go continuously round, picking out each mth man until r only are left. Let one of these be the man who originally occupied the pth place. Then if we had begun with $n + 1$ men one of the r left would have been originally the $(p + m)$th, or (if $p + m > n + 1$) the $(p + n - n - 1)$th."

CHAPTER IV

QUATERNIONS

TAIT's quaternion work was unique; and his influence in the development of the calculus was second only to that of the great originator himself. He alone of all Hamilton's contemporaries seems to have been able to grasp the real significance of the method by direct perusal of Hamilton's *Lectures*. The extraordinary seventh "Lecture" bristled with novelties and difficulties. In grappling with these in his later Cambridge days Tait saw the value of quaternions as an instrument of research. But it was not till he was settled in Belfast that he began to make headway.

On August 11, 1858, Dr Andrews wrote Hamilton a note introducing his young mathematical colleague as one who "had been directing his attention of late to Quaternions, and is anxious to be allowed to correspond with you on that subject."

In a cordial response to this letter Hamilton speaks of having recently turned his attention to "differential equations and definite integrals in connection with old but revived researches of my own (I do not mean, just now, those which Jacobi has enriched by his comments)." He enclosed, no doubt to test the powers of his would-be correspondent, a number of questions, some of which Tait answered in his second letter of August 20.

The first letter, of date August 19, must ever be regarded as of great historic importance. It began a remarkable correspondence, which brought Hamilton himself back to the study and further development of the subject, culminating finally in the production of both Hamilton's *Elements* and Tait's *Elementary Treatise*.

After thanking Hamilton for the very kind manner in which he had responded to Andrews' request, Tait continued:

I attacked your volume on Quaternions immediately on its appearance, and easily mastered the first 6 lectures—but the portions I was most desirous of understanding, viz. the physical applications of the method, have given me very considerable trouble; and, but for your offered assistance, I am afraid I should have had to relinquish all hopes of using Quaternions as an instrument in investigation, on account of the time I should have had to spend in acquiring a sufficient knowledge of them.

I have all along preferred mixed, to pure, mathematics, and since I left Cambridge, where the former are little attended to, have been busy at the Theories of Heat, Electricity, etc. Your remarkable formula for $\frac{\partial^2}{\partial x^2} + \frac{\partial^2}{\partial y^2} + \frac{\partial^2}{\partial z^2}$ as the square of a vector form, and various analogous ones with quaternion operators, appear to me to offer the very instrument I seek, for some general investigations in Potentials, and it is therefore almost entirely on the subject of Differentials of Quaternions that I shall trespass on your kindness....

The correspondence thus begun continued week by week with wonderful continuity until July 1859, when Hamilton began to print the *Elements*. The successive letters were numbered (Hamilton's in Roman, and Tait's in Indian, numerals) and copies kept by the writers themselves, so that there might be no difficulty in referring to questions raised by either at all stages of the correspondence.

In his letter of August 20, 1858, Tait mentioned particularly certain difficulties :

Perhaps it is only due to the novelty of the subject, but I have felt at several points that the otherwise known result was (perhaps not necessary but at all events) very desirable, in suggesting the transformation suitable for its proof. As instances I may mention $-b^2$ found in Art. 474 of your Lectures for the value of

$$\rho^2 + 4\,(\iota - \kappa)^2\,S\iota\rho\,S\kappa\rho,$$

and the transformation of the Tractor function for the 2nd integration of the equation of motion of a planet....

Again in Art. 591 I cannot see how you infer that v is a normal vector—when the equation to a surface is put in the form $Sv d\rho = 0$, $Td\rho$ not being indefinitely small, because it seems to me that in such a case v is a vector perpendicular to the chord $d\rho$.

It was in reply to Tait's difficulties regarding the notion of finite differentials that Hamilton wrote the long letter v, which might have been a chapter in a treatise on the fundamental conception of the fluxion or differential method. Hamilton subsequently gave the argument clearly in his second treatise, the *Elements of Quaternions*, developing the whole discussion from the definition :

Simultaneous Differentials (or Corresponding Fluxions) are limits of equimultiples of simultaneous and decreasing Differences.

In this remarkable letter (dated October 11 to October 16, 1858) which occupies 45 closely written pages of large-sized note paper, and is subdivided into 32 paragraphs, Hamilton began by comparing himself to the fox in Chaucer's story, The Nonne Prest, his Tale, and quoted:

> " But, Sire, I did it in no wick(ed) entent:
> Com doun, and I schal telle you what I ment."

"But," continued Hamilton, "it is time to make a prodigious, a mortal leap, and to pass from Chaucer to Moigno. By the way did you ever meet the Abbé?—'a little, round, fat, oily man of God'—who has however been sometimes called, in Paris, 'le diable de M. Cauchy.'

"(2) Your *name* was familiar to me, before Dr Andrews was so good as to propose that we should have some personal acquaintance with each other. But I regret (and perhaps ought to be ashamed) to say, that as yet I have not had an opportunity of reading any of your *works*. However from the specimen sheet which you sent me, along with your first letter, of a book of yours on analytical mechanics, & in which you did me the honour to introduce the subject of the Hodograph, I *collect* that you consider it *judicious, at least* (if not absolutely necessary) *in instruction*, to use *differential coefficients only* & to *exclude differentials* themselves. And perhaps you may have *adopted*, even publicly—as Airy has done, using the (to me) uncouth notation $\int_\theta (\quad)$ for $\int (\quad) d\theta$—the *system* which *rejects differentials*. If so, I can only plead that I am not intentionally, nor knowingly, *controverting* anything which you have published. And if I now quote Moigno, it is merely to show that *I am not wishing to be singular*."

Moigno's book from which Hamilton quoted with criticisms and comments was published in 1840; but before the letter was finished Hamilton's copy of Cauchy's *Leçons sur le Calcul différentiel* (1829) was discovered " buried under masses of papers " in a corner of his library. There (as he expected) he found the inspiration of Moigno's views without Moigno's mistakes. Cauchy is then quoted and shown to treat throughout of differentials, and only in a secondary sense of differential coefficients; and not only so, but Cauchy's differentials may have any arbitrary values and are not essentially infinitesimal. Then followed what must have delighted the heart of Tait.

"(29) Although it was, perhaps, allowed to suppose that you might not have access to Cauchy's *Leçons sur le Calcul différentiel* (1829), which may be out of print, and even that Moigno (1840) might not be in your hands, I must not presume to imagine that a Cambridge man can possibly be unacquainted with the *Principia*. It may, however, be just permitted to *remind* you, that in the Lemmas VII, VIII IX of the 1st Book, Newton's 'intelligantur (or intelligatur) semper ad puncta longinqua produci,' as also his 'recta semper finita' in Lemma VII, and his 'triangula tria semper finita' of Lemma VIII, are conceptions, to which the process of construction proposed in paragraph (16) of the present Letter appears to have much analogy. And in that famous Second Lemma of the Second Book, which is stated by himself, in his appended Scholium, to contain the foundation of his Method of

Fluxions ('methodi hujus generalis fundamentum continetur in lemmate praecedenti')
Newton expressly says...'Neque enim spectatur in hoc lemmate magnitudo momentorum, sed prima nascentium proportio. Eodem recidit si loco momentorum
usurpentur vel velocitates incrementorum ac decrementorum (quas etiam motus,
mutationes et fluxiones quantitatum nominare licet) vel finitae quaevis quantitates
velocitatibus hisce proportionales.' The *finite differentials* of Cauchy & myself, &
doubtless of other moderns, are therefore really the *fluxions* of Newton in disguise;
and I ought to talk, or at least might talk, of *fluxions of quaternions*, and of their
functions.

"(30) Before I was 17 years old, I had diligently studied at least the three first
sections of the 1st Book of the *Principia*....But I think it was about that age, that
I was carried away by the attractions of the French School, & specially by that of
Lagrange. The *Calcul des Fonctions* charmed me, & for several years I supposed
it to be, not merely an elegant and original production of a genius, whose mathematics
almost sublimed themselves into poetry, but a sound and sufficient basis for the
superstructure of the Differential Calculus....But you may possibly be aware that
it is now a long time since I pointed out a fatal defect in the *foundation* of
Lagrange's theory, as set forth in the *Calcul des Fonctions*.......I suppose that no
one *now* contests the necessity of *founding* the differential *calculus* on the *notion* of
limits; at least, if it be desired that the structure should be a weather-proof and
habitable house:—or, in short, good for anything. In *that* respect, at least, though
certainly not in the *notation* of fluxions,—we are all glad to go back to Newton.

"(31) To connect my definition more closely still with Newton's views, we have
only to conceive that, if $r = dq = \Delta q$, the quaternion function, fq, of the quaternion
variable q, GROWS,...and passes, GRADUALLY, by such GROWTH, through the $n - 1$
intermediate stages (of *state*, rather than of *quantity*)

$$f\left(q + \frac{r}{n}\right), \quad f\left(q + \frac{2r}{n}\right), \quad f\left(q + \frac{3r}{n}\right), \text{ &c.,}$$

where n is a large positive whole number, until it ATTAINS at last the state

$$f\left(q + \frac{nr}{n}\right) = f(q + r) = f(q + \Delta q) = fq + \Delta fq."$$

Tait's reply to this long letter was as follows:

Q. C. BELFAST,
October 19th/58.

My dear Sir William Hamilton

Plunged as I now am in the middle of the entrance & Scholarship
examinations for this session, I shall not have for some days the amount of time
requisite for a careful reading of your excellent No. v....

I am tolerably familiar with the works of Moigno—and I quite agree with you
in your estimate of him. Did you ever see his 'Repertoire d'Optique Moderne'?
It is the strangest mixture of valuable matter and utter trash I ever came across.
I should like very much to know your opinion of Cauchy's investigations in the
Undulatory Theory—for I have found it possible by apparently legitimate uses of

his methods to prove almost *anything*. But I have given up these speculations for the present till I see whether I cannot get the requisite command of Quaternions, as I feel that they must inevitably much simplify the investigations....

All that I have to say on the subject of my *School* (though I fancy myself rather a cosmopolitan) as regards *Differentials*, &c., I must beg you to let me reserve for some days till I have comparative leisure again—

Yours very truly

PETER G. TAIT.

While Letter v was in process of construction, Hamilton sent two shorter letters, Nos. vi and vii, relating to other quaternion questions. In the former he discussed the surface of revolution in the form

$$\rho = a^t \phi_u a^{-t},$$

where ϕ_u is any vector function of the scalar u, a is the vector parallel to the axis of revolution and t is a second scalar variable.

Letter vii contained an interesting historic note with reference to the quantity $\dfrac{d\rho}{dt}$:

" I have lately observed that Mr Warren, of Cambridge, as long ago as 1828, in his Treatise on the Geom. Representation of the Square Roots of Negative Quantities,... gives, in his page 119, that *very symbol* $\dfrac{d\rho}{dt}$ to represent a *line* which in *length* and *direction measures* the *velocity* of a moving point....*My* ρ has *no necessary dependence* on *any* sq. root of -1, so long as we are merely *using* it to form such expressions as ρ' or $\dfrac{d\rho}{dt}$ for the vector of *velocity*,...or ρ'' or $\dfrac{d^2\rho}{dt^2}$ for the vector of *acceleration*; where ρ', ρ'' are fairly entitled to be called '*derived functions*' of t, of the 1st and 2nd orders, the *primitive* function being ρ."

In letter 7 of date October 25, 1858, Tait wrote:

I do not intend even today to enter upon the subject of differentials—though I may state that I have re-read with great care your letter No. v, and have quite understood, and agreed with, it—while at the same time I must confess that a good deal of it besides that referring more particularly to Quaternions was new to me.

Towards the end of this letter Tait propounded the problem to find the envelope of the surface $S^2 a\rho + 2 S a\beta\rho = b^2$ when $Ta = 1$. He had given it incorrectly in a postscript to a previous letter, and Hamilton at once saw there must be some mistake. Tait, after making the correction, continued thus:

The first equation represents I suppose a paraboloid and the second was intended (though I presume it is not explicit enough) to mean that a might be *any* unit vector. The question had reference to the finding the locus of ultimate intersections of the series of paraboloids, a problem which arose out of an investigation I was lately making—and which I felt was too much for me at the time—but if you will permit me to withdraw it again for a little, I think I *may* perhaps manage it now.

The problem will be found solved in Tait's *Quaternions*, § 321 (3rd edition), very much as Tait solved it in his letter 8. Hamilton was greatly taken with the question and discussed the geometry of the envelope at great length in his letters XI and XIII. The envelope is a surface of revolution of the fourth degree having the quaternion equation[1]

$$-(V\beta\rho)^2 = \rho^2(\rho^2 \pm b^2),$$

and this Hamilton proposed to call Tait's Surface. It is curious to note that the first solution sent by Hamilton to Tait did not agree with Tait's. By his first method of elimination, in fact, Hamilton introduced a "foreign factor" in the form of a sphere. In the very short letter XII he writes:

"*Your* investigation would look much better in print than my own; for you see that I take no pains, in this correspondence, to put any check on a natural tendency to diffuseness—& scarcely ever copy from a draught, although the style of the composition would thereby be greatly improved, especially in the way of *condensation.*

"It takes, you know, *more pains to* write a *short* than a *long* letter, or essay, on any subject:—not that I pretend to have taken any pains with this short note! but I must tell you, some time or other, of its once costing me half a quire of paper to write a note of *one page* to a lady who wanted my *opinion* on an astronomical manuscript of her own."

Meanwhile along with the prolonged discussion of Tait's Surface in letter XIII Hamilton was continuing his elucidation of the theory of differentials in letter x. After acknowledging receipt of parts of these letters on November 13, 1858, Tait continued in letter 10 in these words:

For a week I have been hard at work trying to deduce the equation to Fresnel's wave-surface by a process *purely* quaternionic—starting from the data employed by Archibald Smith in the Cam. Math. Journal. As yet I have only deduced the directions of the planes of polarization for any wave-front, and the law connecting the velocities of the two rays, and these come out with admirable simplicity. In attempting to find the equation to the surface I have come upon a terrible array of *Versors*. Of the latter I have still a sort of horror arising principally I suppose from my having avoided the use of them on any occasion on which it was possible.

[1] The Cartesian equation is $a^2(x^2 + y^2) = (x^2 + y^2 + z^2)(x^2 + y^2 + z^2 \pm b^2)$.

Hamilton acknowledged the receipt of this letter by sending the first instalment of letter XIV.

OBSERVATORY, *Nov. 17th.* 1858.

My dear Mr Tait

Although X and XIII are still unfinished,—not to mention IX, which is little more than begun,—I am in a mood to commence now a new letter, of a perfectly miscellaneous nature, and free from the tyranny of any fixed idea.

You tell me that you have been making progress with treatment of Fresnel's wave by Quaternions, but that you have not (or had not at the time of writing) completed the investigation. Whenever you have quite satisfied yourself with a result, or set of results, upon that subject, I should prefer you *not immediately communicating* such result, or results, *to me*; because I should like to try, either to re-investigate the equation of the wave, or perhaps to hunt out an old investigation of it, in one of my manuscript books. The fairest, or at least the pleasantest course for *both* of us may therefore be, that we should *agree upon some day* and *each of us on that day post a letter* containing some of our *separate* results.

This suggestion was warmly welcomed by Tait; and in his letter 12 of date Nov. 29, 1858, the following reference was made to the agreement:

You mentioned no day in particular for our exchanging results on the Wave Surface. I have (in a sense) completed my investigations—but they are far from simple—and I suspect strongly that there is some very elementary theorem of Transformation with which I am not acquainted which would immensely simplify them at once. I would therefore, to avoid knocking my head longer against eliminations which at present I find impracticable though I know they must be possible, request you to name as early a day as may be consistent with your perfect convenience, as you then may be able to tell me in a moment the reason of my imperfect success.

At the close of letter x, Hamilton, writing on December 1, fixed December 4 as the day for exchanging confidences on the Wave Surface. On that date accordingly Tait sent Hamilton his investigation along with the following letter:

Q. C. BELFAST,
4th Dec. 1858.

My dear Sir William Hamilton

I have to acknowledge the receipt of the rest of X with PS on two separate occasions, also of pp. 17—28 of XIV....I shall take an early opportunity of expressing my ideas with respect to V and X on the subject of finite differentials[1] —meanwhile, as it is now late, I must explain as I best can the enclosed, which

[1] This expression of ideas seems never to have been given. Other and more important Quaternion developments had to be considered.

with all its deficiencies is the best I could make out of the subject before today when a new idea suggested itself—that of avoiding the fearful eliminations which my method would seem to require in obtaining the equation of Fresnel's Wave Surface. The idea, which I have easily satisfied myself is correct, is to show that surfaces derived from reciprocal ellipsoids are themselves reciprocal.

Meanwhile on December 3 Hamilton began his letter xv on the Wave Surface and dispatched the early sheets of it along with some pages of letter xiv, in which he acknowledged receipt of Tait's

"...note No. 13 together with its very valuable enclosure of two sheets entitled 'Quaternion Proofs of some Theorems connected with the Wave Surface in Biaxal Crystals.'...I have read the first sheet of your Quaternion Proofs, and must say that they appear to me to be *wonderfully elegant* and to exhibit a very remarkable degree of *mastery* (so far) over the *calculus* of Quaternions, used as an *instrument* of *expression* and of *investigation*.

"It would interest me much to know, whether (previous to our present correspondence) you had received ANY assistance from *any other* student of that calculus. Or did you learn *all* that you had acquired from the BOOK itself, combined (no doubt) with *your own* private exercises of various sorts? If the 'Lectures on Quaternions' have been your ONLY teacher, I must consider the result of such a state of things to be not merely creditable to *your* own talents and diligence, but also complimentary to, and evidence of, some (scarcely hoped for) *didactic capabilities* of my volume; which ought to tend to *console* me, under my *artistic consciousness* (as an author) of so many *faults of execution*, that if I could afford the expense of bringing out a *New Edition* I should be more likely to make it a *New Work*....My old friend John T. Graves called my attention about a year ago to a highly favourable, and very eloquent, article in the North American Review for July, 1857, on the subject of the Quaternions, and of my Book. But a conscientious Author wishes rather to be *read*, than to be praised, and therefore I should like to be informed, *what* drew your attention to my Book, and *whether* you had any personal assistance in studying it."

To this request Tait replied in his letter 14 of date December 7, 1858:

With regard to my study of Quaternions I may affirm with some certainty that when I ordered your book, on account of an advertisement in the Athenaeum, I had NO IDEA what it was about. The startling title caught my eye in August '53, and as I was just going off to shooting quarters I took it and some scribbling paper with me to beguile the time....However, as I told you in my first letter I got easily enough through the first six Lectures—and I have still a good many notes I made at that time from which it now seems to me that I had not fully appreciated the simplicity of the method—but had used quaternions generally in the shape

$$(ix + jy + kz)^m = (-1)^{\frac{m}{2}} \left(\frac{ix + jy + kz}{\sqrt{x^2 + y^2 + z^2}} \right)^m$$

and treated i, j, k as imaginaries (like $\sqrt{-1}$) though of course according to their proper laws of combination. For fun I extract *this*

$$U.\,\alpha\beta = -\left(\frac{1}{T(\alpha\beta)}\{i(zy'-yz')+\dots\}\right)^{-\widehat{2\alpha\beta}/\pi} !!!$$

Much of course could not have been made of this, and accordingly on my return to Cambridge I set to read other things, and to write my recently published Treatise on Particle Dynamics. The Theories of Heat, Electricity and Light have since occupied much of my spare time, and it was only in August last that I suddenly bethought me of certain formulae I had admired years ago at p. 610 of your Lectures—and which I thought (and still think) likely to serve my purpose exactly. [The matter which more immediately suggested this to me was a paper of Helmholtz's in *Crelle's Journal* (Vol. LV) which I was reading in July last as soon as we received it, and which put the subject of Potentials before me in a very clear light. The title (in German) I forget—but an MS translation of my own which I have now beside me is headed " Vortex Motion[1]." It refers to the integration of the general equations in Hydrodynamics, when $udx + vdy + wdz$ is *not* a perfect differential.]... So far from having any assistance, save what *you* have so kindly given me, I am not even acquainted with any one who knows aught about quaternions (except Boole of Cork—with whom however I have not exchanged a remark on the subject, and who, I suspect, looks on them in their *analytical* capacity *only*).

So you see that, if there is any credit in my progress, it is *entirely* to your Lectures and Letters that it is due.

Hamilton's letter XIV, which was begun on Nov. 17, and continued at fairly short but irregular intervals till Feb. 5, 1859, when it reached 88 closely written pages, ran on till April 3, in the form of eight postscripts. There seems to be no later reference to Tait's confession of how he began the study of Quaternions; but various sections call for quotation because of the bearing they have on the subsequent history.

In his letter 19, of date Jan. 3, 1859, Tait wrote as follows of Quaternions in general:

About quaternions in general I may remark (as indeed I very frequently feel) that the processes are sometimes *perplexingly easy*—by which I mean that one is often led in a step or two and without (at once) knowing it to the solution of what would be by ordinary methods a work not so much of difficulty as of labour. This however I take it must form one of its great excellencies in the hands of a person very well acquainted with it. A drawback to a beginner, but (as I am gradually being led to perceive) an immense advantage to one well skilled in the analysis, is the enormous variety of transformations of which even the simplest formulae are susceptible; a variety fully justifying a remark of yours (Lectures—Art. 504) which not many months ago used somewhat to puzzle me. If I had gained nothing more

---

[1] The translation was published in *Phil. Mag.* July 1867.

by reading this subject than the facility of making problems and transformations for Examination papers (especially in Trigonometry) and so saving an immense amount of time and trouble, I should have considered myself amply rewarded,—but I hope in time to be able to apply it to perfectly original work (if anything *can* be quite original in these days)....

In the portion of letter XIV which containing his reply to this letter from Tait Hamilton suggested publishing in the *Philosophical Magazine* his own investigations on the Wave Surface, and referred in particular to certain sections of his letter XV which might form the substance of this note.   He said:

"(54) It seems to me that some such *sketch*..., instead of *forestalling your own* communication,—which appears likely to be of *weight* enough to deserve *ampler space* than the pages of a Magazine could afford,—might, on the contrary, serve as a not ungraceful *introduction* to whatever you were disposed to publish afterwards. But let me know...what *your* FEELINGS in the matter are.   I am quite aware that I can implicitly rely on your allowing me *at least as much credit* as you may be of opinion that I deserve; and I think that you have really *made* the subject *your own* by your laborious and (so far as I yet know) successful investigations."

To this Tait replied:

Q. C. BELFAST,
7/1/59.

My dear Sir William Hamilton

Many thanks for your very kind letter containing XIV pp. 57–60 & XV pp. 93, 94, which I received this morning...

I had been casting about as to how I should ask you to do the very thing you have just proposed—as I have, as you will see when you look at the recent sheets of my Quat. Proofs, found one or two things which I believe were given by you for the first time but which I had either not received from you or not read until my own investigations were advanced beyond that point. For instance, I consider that I am not *directly* indebted to you for the quaternion form of the equation[1] to the wave in ι, κ,—though of course you had it years before I knew of such a thing as quaternions at all.   But then, knowing as I do the date of your discovery of that formula, I could not have published my own investigation without specially mentioning that you had communicated it to me, and the latter course it was impossible to follow, as I consider your letters private.

You see then that I was in a difficulty and I should probably have tried at some other matter for a paper to publish, but for your last.   I am delighted at the idea of being introduced to the Phil. Mag. (in which I have never written) in connection with quaternions by you, especially when the *subject* as well as the

---

[1] This is equation 13 in Tait's paper published in the *Quarterly Journal of Mathematics*, May 1859 (*Sci. Pap.* Vol. I, page 7), namely,

$$(\kappa'^2 - \iota'^2)^2 = \{S\,(\iota' - \kappa')\,\rho\}^2 + (TV\iota'\rho \mp TV\kappa'\rho)^2.$$

method owes so much to you. But before venturing to publish under such auspices I must wait for your own opinion on my investigation itself which I think you may find interesting (though cumbrous) as I see on comparing the two it differs so much from yours......

I am delighted that you intend to publish soon, and as I have already said you may make any mention you choose of our correspondence.

The next day, Jan. 8, 1859, Tait continued in a letter which he called PS. to 20:

Having posted 20 this morning, and having a respite of a couple of hours while 3 men are at work preparing our ozone with an electrical machine, I have compared our methods of deducing the equation to the wave.

Your $\phi^{-1}($ $)$ is the *same* as my $(\underline{\quad})$, or, as your $\delta\rho$ is my ϖ, and your u my $\frac{\alpha}{T\varpi}$, all our equations can be at once compared by putting

$$\phi^{-1}\delta\rho = \underline{\underline{\varpi}}$$

(where each member represents the whole elastic force called into play),

$$\phi\delta\rho = \overline{\overline{\varpi}},$$

$$\phi^{-1}\mu = \frac{\alpha}{T\underline{\underline{\varpi}}}, \text{ \&c.}$$

Your symbol has over mine the great advantage of being separable from the subject, so that you can write

$$0 = S\mu^{-1}(\phi^{-1} - \mu^{-2})^{-1}\mu^{-1}.$$

Having thus (as I hope) sufficiently allowed the superiority of *your* notation, I may be permitted to remark that I think *mine* has one advantage as I have applied it, namely, that of introducing *directly* the half of your operator ϕ^{-1}, or what might be written $\phi^{-\frac{1}{2}}($ $)$ which will be what I denote by

$$(\underline{\quad}) \quad \text{or} \quad -aiSi(\quad) - bjSj(\quad) - \text{\&c.}$$

I have not time to examine the point, but I fancy that the introduction of $\phi^{-\frac{1}{2}}$ into your process would make it even simpler than it is.

As to the real question at issue I consider myself not to have used your function ϕ, as though my notation can be interpreted into something of the same kind it wants the peculiar advantage of *concentration* which yours possesses, and which forms *one distinctive feature* of your XV.

Tait developed this new notation in his letters 22 and 23. Hamilton did not immediately reply to this suggestion, other questions which will be referred to in due course having absorbed his attention. On February 5, however, he remarked in [76] of Letter XIV:

" But let me first get off my hands a remark about the *new Form* which you suggest for the equation of the Wave Surface. I read it as

$$T(\rho^2 + \phi^2)^{-\frac{1}{2}}\rho = 1,$$

and on just now *glancing* at your No. 22 received yesterday or the day before, but quite *unexamined* hitherto...I see that the symbol

$$(\phi^2 + \rho^2)^{-\frac{1}{2}}$$

occurs several times. You have therefore probably introduced some *new definition* of the functional symbol and I am not entitled to say that your formula requires any *correction*. Of course we cannot *afford* to part with a certain *liberty* of *notation*. But with *my meaning* of ϕ as developed in my Lectures and Letters, I found, a few minutes ago—the *hint* (as I admit) having been taken from your last letter—that the formula,

$$\{T(\phi^{-1} - \rho^2)^{-\frac{1}{2}} \rho\}^2 = - S\rho (\rho^2 - \phi^{-1})^{-1} \rho,$$

is an *identity*; and therefore that one of my *symbolical forms* of the equation of the wave, namely, the equation

$$1 = S\rho (\rho^2 - \phi^{-1})^{-1} \rho,$$

may be immediately *transformed* to the following

$$1 = T(\phi^{-1} - \rho^2)^{-\frac{1}{2}} \rho,$$

a result which I confess that I had *not expected*, but which (I suppose) agrees substantially with yours....You deserve I think great credit for having *perceived* this transformation...."

Thus we owe to Tait the discovery that the square root of a linear vector function or matrix of the third order enters symbolically into certain expressions exactly like an ordinary algebraic quantity. He was led to this discovery by a comparison of his own special notation with the notation used by Hamilton, who, on his own confession, had never thought of treating the linear vector function in this way. It is not a little curious that, at the time, neither Hamilton nor Tait seemed to have considered the analytical significance of the square root of a linear vector function. This was done in 1870 by Tait whose results, based on kinematic considerations, led to an interesting correspondence with Cayley and a further development of the properties of the matrix (see below, p. 152).

After a good deal of further correspondence on the subject of the Wave Surface, Hamilton communicated his method to the Royal Irish Academy, and Tait published his investigation in the *Quarterly Journal of Mathematics*. Meanwhile, in Hamilton's mind a new project had been forming itself, which was first referred to in paragraph 71 of letter XIV, written on January 21, 1859. Here Hamilton wrote :—

"[71] I must tell you however of a quite different project of mine, which *may* occupy a good part of the present year if a fair share of health is spared me. I want to prepare for 1860—though I do not forget a passage in St James—either a new edition of my Lectures, or what may be better, an entirely *new work*, which might perhaps be called a 'Manual of Quaternions.' In it I *suppress* (decidedly) *more than half* of the existing Book; not that I am ashamed of it, but because I conceive that it has served its purpose: and that what we may call a *working* volume is wanted now.

"I fear that No. XVI of the series of MS will never be completed, or will be brought abruptly to a termination[1]: but I don't think that you require my word,—for you have perhaps already indications enough,—that I possess a number of uncommunicated results, respecting the function ϕ for instance, which will yet throw additional light on the treatment by quaternions of surfaces of the second order....

"[72] January 31, 1859. I see that the enclosed sheet, though not yet sent off, was written ten days ago. I have not even *thought* about the Wave Surface since, much less written a line about it; but I by no means abandon the project of publishing some such *short* paper as I described to you in a former sheet; leaving it to *you* to develope, in whatever form you choose, your *own* independent investigations and results. It really seems to me that there would be some *impertinence* in my having the air of *examining* whether your formulae on that subject are *correct*. You are quite as well able as myself to decide any such point: especially since you have got into the *way* of making *transformations* and of multiplying them. I trust however that it is *not* an impertinence in me to confess that I think (or at all events, hope) that this correspondence has been *useful* to you, in some degree; chiefly by causing you to feel a greater degree of confidence in your *own* powers; as applied to a new subject; and as evincing that whatever obscurity may have been allowed to remain in parts of my printed Lectures, from want of skill of an artistic kind in the author, it has not been fatal to a comprehension of the Book, by such a Reader as yourself; although the particular obscurity (about $d\rho$), which *led* to our correspondence, has not (in my opinion) been at all sufficiently yet removed, by my Letters V and X.

"[73] As to myself I cheerfully confess, that I consider myself to have, in several respects, derived advantage, as well as pleasure, from the Correspondence. It was useful to me, for example, to have had my attention *recalled* to the whole *subject* of the *Quaternions*, which I had been almost trying to *forget*; partly under the impression that nobody cared, or would soon care, about them. The result seems likely to be, that I shall *go on* to write some such 'Manual,' not necessarily a very *short* one,—as that alluded to in a recent paragraph.

"[74] In fact, after pretty nearly filling two books, A. 1858 and T. 1858 with matters relating to the 'Tait Correspondence'—[for 'A' had happened to be *reserved*, although 'B,' 'C,' 'D,' and 'E' (at least) had been stuffed with things connected with De Morgan, and with Definite Integrals &c.—and after a few more letters of the alphabet having been pressed into the service, I used 'Alliteration's artful aid' and made a sudden bound, in honour of *you*, to 'T']—I have lately

---

[1] No. XVI was begun on Dec. 14, 1858, but the greater part was written on Jan. 11, 1859. It was abruptly finished off on Feb. 4, 1859, after a few paragraphs on surfaces of the second order had been put together.

taken possession of a *very large* book, which book I call A. 1859, and which is to relate *entirely* to *quaternions*. As yet, *in it*, I have confined myself to a new discussion of FIRST PRINCIPLES."

Tait's reply to this constituted the greater part of his letter 23. He said :—

Many thanks for your kind and flattering letter....I applaud your purpose of publishing a practical "Manual of Quaternions." I may mention to you that I had been thinking of attempting something of the kind (but of course a very elementary work) if the idea met with your approval—but that was of course *before* I heard that you intended doing anything of the kind yourself. There was one feature of my dawning idea which might suit you—that was to get it printed as one of Macmillan's Cambridge series of which my Treatise on Dynamics forms a portion. It would thus be directly introduced to the largest body of mathematicians in this country....Another feature would have been (and without this no book *takes* in Cambridge) numerous examples of the great simplicity of the new method....I merely mention my own half-developed scheme to show you that I think your present proposal an excellent one, and perhaps to give you a useful hint or two with the object of Quaternionizing my own University.

In letter XVI of date April 10, 1859, Hamilton referred in a remarkably prescient manner to the part which Tait was destined to play in the development of quaternions. He wrote :

"Let me be permitted to *congratulate* YOU (as well as *myself*—most sincerely do I add this last objective case) on your having *taken up* the Quaternions. They will owe MUCH to you; but I think that you will owe *something* to *them*. This may be only the natural vanity of an author; but I believe that an early *appreciation* of genius wins a corresponding appreciation, in its turn, from mankind, for itself; even if not accompanied, as in your case it is, and will be, by independent acts of *discovery*."

These extracts show unmistakably that the mathematical world owes more to Tait than has yet been revealed. It was he who fired Hamilton with the ambition to write his second great Treatise on Quaternions. As we read the correspondence, and especially Hamilton's long chapter-like letters, we see some of the leading features of the *Elements* taking shape. Had Hamilton lived to write the Preface to the unfinished *Elements* he probably would have mentioned explicitly the value of the Tait Correspondence. All we have, however, in published form is a footnote towards the close of the unfinished work, where Tait is spoken of as one "eminently fitted to carry on, happily and usefully, this new branch of mathematical science; and likely to become in it, if the expression may be allowed, one of the chief successors to its inventor."

The following extracts from Tait's letters in March and April of 1859 show how thoroughly he was becoming saturated with the quaternion ideas and methods.

[March 2.] I have added a good many new theorems to the wave investigations, but I fear their importance is nothing particular.

The problem of the wave-front for which there is the greatest angular separation of the rays has only led me to some complicated and almost intractable equations.

I have been led in connection with the wave surface to the study of the curve

$$\rho = \phi^x . \alpha,$$

where ρ (the vector of any point) is a function of the scalar x—α being a given vector and $\phi(\ ) = -aiSi(\ ) - bjSj(\ ) - \&c.$ From this I have got some curious results, but have been stopped short by a difficulty of a kind new to me in Quaternions, while trying to find x from

$$T\phi(\alpha\phi^x\alpha) = T\phi^{x+1}\alpha,$$

ϕ having the same meaning as before....

Here again a new difficulty presented itself—the elimination of m (an arbitrary scalar) between two equations of the form (where $\theta^2 = m^2 + \phi^2$)

$$\begin{cases} T\phi\theta^2\varpi = TV\theta\varpi\phi^2\theta^2\varpi \\ e^2T^2\theta\varpi = 1 + T^2\phi\theta\varpi. \end{cases}$$

You may see that I have my hands pretty full of work—even if the matters in question be of no importance.

[March 18.] I have been working farther at the wave of late and I think am in a fair way to find the equation to the central surface of the second order concentric with the wave which has the closest contact with it at a given point. The difficulty consists in the solution of a functional equation or rather in determining the general value of a certain $\psi^{-1}(o)$, where ψ is a linear and vector function.

I have at last attacked the subject of Potentials which was the cause of my recent (and, this time, successful so far) attempt at the study of Quaternions, and I think I have got the method of applying the calculus to the matter.

I have also been working at some illustrative problems. I met with this in a Cambridge Examination Paper, 'Find the locus of the centre of a sphere which touches two given lines in space.' I modify it into 'Find the locus of the centre of a surface of the second order, whose axes are given in ratio and direction, and which touches two given lines.'

The required locus is given in the form

$$TV\beta\phi(\rho - \alpha) = TV\gamma\phi(\rho + \alpha),$$

where β and γ are the unit vectors along the given lines, 2α is the common perpendicular and ϕ is the function of the surface.

In letter XVIII, dated April 12, 1859, Hamilton returned to the wave surface, and after deducing afresh its equation remarked:

"*Could* anything be simpler or more satisfactory? Do you not *feel*, as well as think, that we are on a *right track*, and shall be *thanked* hereafter? Never mind *when*....

"De Morgan and I have long corresponded *unofficially* and said odd things to each other. He was the very *first* person to *notice* the quaternions *in print*, namely, in a paper on Triple Algebra in the Camb. Phil. Trans. of 1844. It was, I think, about that time, or not long afterwards, that he wrote to me, nearly as follows :— 'I suspect, Hamilton, that you have *caught the right sow by the ear*!' Between us, dear Mr Tait, I think that *we* shall *begin* the SHEARING of it."

Tait replied in letter 31 of date April 13, 1859:

I have just received XVII and XVIII, the latter an hour or two ago.

Your deduction of Fresnel's construction from the symbolic form of the equation to the wave is very elegant. I have given (in a paper which I suppose is now being printed, for it has been sent off ten days or more) a proof of the same, which is a *mere interpretation* of some of the equations which I have written down in deducing that to the wave.

I have recently (as I mentioned in letter 26) come to a seemingly formidable difficulty in Quaternions. It is to find the most general form of linear and vector function ψ from the equation

$$S\rho\psi^2\varpi = S\sigma\varpi - \sigma^2 S\rho\varpi,$$

where $\sigma = (\phi^2 + \rho^2)^{-1}\rho$ and where the scalar and vector constants of the required function ψ involve ρ, σ and the operation ϕ....

In the third PS. to your VIII you mentioned a result of Maccullagh's[1] which I have since found in the Trans. R. I. A. I was lately trying the problem in an extended form. I find for instance the following amongst a host of other results.

(1) If the two lines which move in the planes are not at right angles, let the cosine of their inclination be e, and let the third line be perpendicular to them ; it traces a cone of the 4th order....

(2) If one of the moving lines be a generating line of a cone of the second order, the second lying in a plane which passes through the vertex thereof, and the third perpendicular to the other two, the locus is in general a cone of the 8th order....

While this letter was being penned, Hamilton was beginning his letter XIX, the importance of which demands a full transcription.

---

[1] As given by Hamilton, the problem is, If three rectangular lines so issue from a common origin that two of them move in fixed planes, the third will describe a cone of the 2nd order, whose circular sections are parallel to the two planes.

Obs^v,
April 14th, 1859.

My dear Mr Tait

Although what I am about to write must be very short, and might be marked as PS. to No. XVII, yet, on the whole, I choose to number it as above, partly with a view to encourage myself to write *short* letters.

[1.] There is, as you know, a very important problem of transformation, to which you have alluded, both in early and in recent letters, and of which I by no means *deny* that those letters *may* contain a *sufficient* solution or solutions: for I have hitherto *avoided* to examine them, in connexion with that problem, which I certainly conceived myself to have resolved, about ten years ago, and to which (*as* solved) I *alluded* at the *end* of art. 567, in page 569 of the *Lectures*....

[4.] The problem...haunted me, as it happened, yesterday, while I was walking from the Provost's house to that of the Academy, &c.; and though I wrote nothing down that day I resumed it this morning: and arrived at what you might call, in the language of *your* No. 19, a '*perplexingly easy*' *solution* (in the sense of being very UNLABORIOUS, for I do not pretend that the *reasoning* does not require a close *attention*); not in any way introducing $i\,j\,k$, nor $\alpha\,\beta\,\gamma$ (of an ellipsoid) nor ι, κ, but depending entirely on the properties of the function ϕ. *So simple* does this solution appear, that I hesitate as yet to place entire confidence in it; and therefore, till I have fully written it out—for at present it is partly mental—and have given it a complete and thorough re-examination, I hesitate to communicate it to you. Meantime, however, I must say, that I am *not conscious* of having taken any *hint*, in this investigation, from any of your letters....

[5.] April 15th—I shall just jot down here the enunciation of a few Theorems[1], which I have lately proved (as I think) anew, and which are intimately connected with the question.—

THEOREM I. If $\phi\rho$ be a distributive and vector and real function of a real vector ρ, such that $S\sigma\phi\rho = S\rho\phi\sigma$, (a), then the eq^n $V\rho\phi\rho = 0$, (β), is satisfied by (*at least*) *one real direction* of ρ.

THEOREM II. Whatever be the given and real dir^{ns} of ρ, (*at least*) *two* real and rectangular directions, ρ' and ρ'', can be assigned, for a vector ϖ, which shall satisfy the two eq^{ns} $S\rho\varpi = 0$, (γ), and $S\rho\varpi\phi\varpi = 0$, ($\delta$).

THEOREM III. If ρ and ϖ satisfy the system of the three eq^{ns}, (β) (γ) (δ), then ϖ satisfies (β), or more fully $V\varpi\phi\varpi = 0$, (ϵ).

THEOREM IV. (Extension of Theorem I.) The equation (β) is *always* satisfied by at *least one system* of *three real and rectangular directions*, ρ_1, ρ_2, ρ_3, of ρ.

Proof obvious, from what precedes.

THEOREM V. The functional symbol ϕ satisfies a cubic equation,

$$(\phi + g_1)\,(\phi + g_2)\,(\phi + g_3) = 0, (\zeta),$$

whereof the three roots are always *real*.

---

[1] This is probably what Tait referred to in his paper on the intrinsic nature of the quaternion method (1844; *Sci. Pap.* Vol. II, p. 396), where he states that "one of his many letters to me gave, in a few dazzling lines, the whole substance of what afterwards became a Chapter in the *Elements*."

THEOREM VI. If these roots be also all *unequal*, then the eq^{ns},

$$(\phi + g_1)\rho_1 = 0, \quad (\phi + g_2)\rho_2 = 0, \quad (\phi + g_3)\rho_3 = 0, \quad (\eta),$$

are satisfied by the 3 rectangular directions ρ_1, ρ_2, ρ_3 of Theorem IV, and by *those* directions (or their opposites) *only*.

THEOREM VII. For any *other* vector, $\rho = x_1\rho_1 + x_2\rho_2 + x_3\rho_3$, (θ),

we have
$$\phi\rho = -(g_1 x_1 \rho_1 + g_2 x_2 \rho_2 + g_3 x_3 \rho_3), \; (\iota),$$

and
$$S\rho\phi\rho = -(g_1 x_1^2 \rho_1^2 + g_2 x_2^2 \rho_2^2 + g_3 x_3^2 \rho_3^2), \; (\kappa).$$

THEOREM VIII. Whatever the real scalar, g, and the real vectors, a, a', ... and β, β', ... may be, it is possible to find 3 real scalars, g_1, g_2, g_3, and 3 real and rectangular unit vectors, ρ_1, ρ_2, ρ_3, such that the following shall be an *identical* transformation:

$$g\rho^2 + 2\Sigma Sa\rho S\beta\rho = g_1 (S\rho_1\rho)^2 + g_2 (S\rho_2\rho)^2 + g_3 (S\rho_3\rho)^2, \; (\lambda).$$

THEOREM IX. The data, g, a, β, a', β', ... being still real we have finally this other transformation:

$$g\rho^2 + 2\Sigma Sa\rho S\beta\rho = g`\rho^2 + 2Sa`\rho S\beta`\rho, \; (\mu),$$

without any sign of summation in the 2nd number; and $g`$, $a`$, $\beta`$, can always be made real.

Having written so far, and even had the first sheet of this letter *copied* (into A. 1859), I think that I may now indulge myself in *opening* your letter received this morning....For I have been apprehensive of your anticipating me, or hitting on my old train of thought, before I had (as above) recovered it for myself.

Tait, on April 21, replied:

I was greatly pleased with the transformations in XIX. I can easily prove all your theorems with the exception of the first, i.e. that "$V\rho\phi\rho = 0$ admits of one real solution at least." It is certainly a very elegant mode of attacking the question, and I had never thought of so simple a point of view as the making the normal coincide with the radius vector. But when I try to prove your theorem, I fall back again into the cubic of my letter[1] 30, or at all events a simple case of it,—so that I do not see how you manage to avoid a reference to something or other equivalent to i, j, k.

In a PS. to letter XXII, dated Easter Tuesday, 1859, Hamilton indicated the proof which Tait longed for:

"My Theorem I, of Letter XIX, was proved by showing, on the plan of Lecture VII, Art. 567, that the equation

$$(\phi + g)\rho = 0$$

could be satisfied *without* our having also $\rho = 0$, provided that g was a root of a certain cubic equation. It is not at all necessary, for this purpose, that ϕ should satisfy the functional condition

$$S\rho\phi\sigma = S\sigma\phi\rho,$$

[1] In regard to letter 30 Hamilton had remarked that he liked the look of it. Unfortunately a copy of this particular letter does not seem to have been preserved by Tait.

but as I *assumed* that this condition was satisfied in most, if not in all, of the subsequent theorems, I believe that I thought it convenient to enunciate it at starting. Besides I wrote in some haste."

Hamilton's letter XXIII contains a systematic investigation of the linear vector function, which differs markedly in the details of development from the investigation given in his subsequent book *The Elements of Quaternions*. In its initial stages it resembles Tait's mode of presentation, which Tait himself calls "Hamilton's admirable investigation" (see Tait's *Quaternions*, 3rd edition, §§ 156–159). Writing on May 11, 1859, Tait in letter 33 remarked :

Your No. XXIII (which I received yesterday) was indeed a treat. Nothing could be more beautiful than your method of attacking the equation of the second degree. I have been trying to supply for myself the demonstrations you suppressed and have succeeded completely, though perhaps not elegantly. Thus as

$$\phi^{-1} V\lambda\mu = m V\psi\lambda\psi\mu,$$

assume
$$\psi^{-1} V\lambda\mu = m' V\phi\lambda\phi\mu,$$

and if $m = m'$, your theorem about the interchange of ϕ and ψ is proved. The above equations are evidently equivalent to

$$\phi^{-1} V\psi^{-1}\lambda\mu = m V\lambda\psi\mu$$

and
$$m'\psi V\phi\lambda\mu = V\lambda\phi^{-1}\mu.$$

Multiply together, and equate scalars, and we have at once

$$m' (S\phi\lambda\mu S\mu\psi^{-1}\lambda - \lambda^2\mu^2) = m (S\lambda\psi\mu S\lambda\phi^{-1}\mu - \lambda^2\mu^2)$$

or
$$m' = m,$$

since
$$S\phi\lambda\mu = S\lambda\psi\mu$$

and therefore also
$$S\psi^{-1}\lambda\mu = S\lambda\phi^{-1}\mu.$$

Another curious property of these functions resulting from this last equation is that $\phi^{-1}\psi$ is the *conjugate* of $\phi\psi^{-1}$.

I came upon the following (which seems neat). Generally, whether n be + or − or even = 0,

$$S\psi^n\lambda\psi^n\mu\psi^n\nu = \frac{(S\phi\lambda\phi\mu\phi\nu)^n}{(S\lambda\mu\nu)^{n-1}},$$

which is true (of course) of ϕ also.

What I was most puzzled with was the proof that m (in your notation) is a constant. I saw at once that it could not contain the *tensors* of λ and μ, but I did not feel so sure about the versors. I have satisfied myself on that point by making use of the *distributive* property of ϕ^{-1}.

Six days later in letter 34, Tait made a further reference to the same investigation.

When I came to your equation (31) of XXIII—I tried to prove it for myself—and was so successful that I was just about to send you a note on the subject—when I luckily read on and found that your *luminous thought* had completely anticipated me. Here is my work as it stands in an MSS book.

$$m\psi^{-1}V\lambda\mu = V\phi\lambda\phi\mu,$$
$$\therefore \quad mV\phi^{-1}\lambda\phi^{-1}\mu = \psi V\lambda\mu.$$

Change ϕ into $\phi + g$, &c. and multiply by M,

$$M^2 V(\phi+g)^{-1}\lambda(\phi+g)^{-1}\mu = M(\psi+g)V\lambda\mu$$

or

$$V\Omega\lambda\Omega\mu = M(gV\lambda\mu + mV\phi^{-1}\lambda\phi^{-1}\mu).$$

No letter from Hamilton of date later than July 19, 1859, has been preserved, although there are copies of eight of Tait's own letters to Hamilton ranging from Sept. 7, 1859, to January 14, 1861. From these we gather that Hamilton was absorbed in the preparation of his new book and was keeping Tait steadily supplied with the proof sheets of the earlier chapters. Meanwhile Tait was strengthening himself in the use of the calculus, and in letter 41 of date Sept. 26 gave, very much as it afterwards appeared in his *Treatise*, his quaternion investigation of Ampère's electrodynamic theory. This investigation, especially in the more generalised form in which it was presented in his paper of 1873 on the various possible expressions for mutual forces of elements of linear conductors (*Proc. R. S. E.* VIII; *Sci. Pap.* Vol. I, p. 237), is a good example of the directness with which the quaternion method deals with a general problem[1]. Beginning with a general form of function, involving the relative position and the directions of two current elements, Tait developed the form of this function by a skilful use of Ampère's fundamental experimental laws. In letters 42 and 43 of date Nov. 3, 1859, and March 22, 1860, Tait continued the development of his electrodynamic investigations, pointing out the importance of the vector

$$\beta = \int \frac{V\alpha\alpha'}{T\alpha^3} = \int \frac{dU\alpha}{\alpha}$$

in all investigations connected with the action of a circuit, where α' is the element at the point α of the circuit.

A few months later Tait commenced his Edinburgh career, having been helped thereto by the following testimonial from Hamilton:

Understanding that Professor Peter Guthrie Tait, now of the Queen's College, Belfast, but formerly of St Peter's, Cambridge, is likely to become a candidate for

---

[1] See also Clerk Maxwell's *Electricity and Magnetism*, Vol. II, Chap. II.

the Professorship of Natural Philosophy in the University of Edinburgh, in the event of that office becoming vacant, I consider it to be only just to Mr Tait to attest that, in consequence of a rather copious correspondence between him and me, which has been carried on for somewhat more than a year, on mathematical and physical subjects, including Quaternions, and the Wave-surface of Fresnel, my opinion of the energy and other capabilities of Professor Tait for any such appointment is very favourable indeed.

WILLIAM ROWAN HAMILTON.

OBSERVATORY OF TRINITY COLLEGE,
DUBLIN, *Dec. 10th,* 1859.

Tait's return to Edinburgh and his assumption of new duties meant a considerable break in the line of his mental activities; and it was not till Dec. 4, 1860, that he wrote letter 44 of the quaternion series to Hamilton. A few days earlier he had sent Hamilton a copy of his inaugural address, in which he had referred in glowing terms to the "powers" of Hamilton's "tremendous engine," to the great secret of quaternion applications, which "seems to be the *utter absence of artifice,* and the *perfect simplicity and naturalness* of the original conceptions."

EDINBURGH,
Dec. 4th, 1860.

My dear Sir William Hamilton,

I received your letter this morning and am glad you are pleased with my introductory lecture. Its treatment by others has not been in all cases so lenient, in fact I am now doing battle with at least two opponents, who have vigorously attacked different parts of it. I am sure I am not violating confidence in telling you that one of these attacks is directed against the mention of Quaternions (towards the end of the lecture) as "likely to aid us to a degree yet unsuspected in the interrogation of Nature." The writer, I daresay, is a personal friend of your own— that I do not know—but, at all events while speaking of you with admiration and due courtesy, he protests in the interests of Science against my having published such a sentence as that above quoted...

I was sorry to see from your letter that we must have been completely misunderstanding each other for some time as to my projected publication on Quaternions. In the first place, to prevent all misconception, let me say that when Dr Andrews wrote a note introducing me to you as a correspondent, I had not the slightest idea of ever being the author of a Volume on the subject. So he could know nothing whatever about the matter. And I think you will acknowledge that the whole is a mistake when I tell you that it never entered into my head to write a Book on Quaternions till I was asked by some Cambridge friends to do so, that I at once wrote to you about it, and asked how far it might be consistent with your wishes or plans that I should undertake such a work. In my letter to you, No. 38, I proposed two forms of publication, one a dry practical treatise, very short, *assuming* most of the fundamental laws of Quaternion multiplication, but *stuffed*

18—2

with examples—the other, the examples alone. I went on to say that even the first of these "could not in the least interfere with your (then projected) new work, as it would treat only of the *practice* of the method, and not at all of the *principles*." And I added, "I have not the least intention of publishing a volume on the subject without your approval." When (in XXVII) you wrote in answer to the above "I should prefer the *establishment* of PRINCIPLES being left, at least for some time longer,— say even 2 or 3 years—in my own hands; and I think you may be content to *deduce* the Associative Law from the *rules* of *i, j, k*, etc."—I fancied that you meant me to *give these deductions in print*—beginning from $i^2 = j^2 = k^2 = ijk = -1$ as something established in your Lectures and Manual. When some months or so later, I wrote to you that I had asked Macmillan to advertize for me "An Elementary Treatise on Quaternions, with numerous examples" I had no idea whatever that I was giving you any annoyance......

But (as I have already quoted from 38) I am most desirous to avoid the slightest suspicion of interference with your intentions—and I therefore particularly request you to give me a perfectly distinct idea of your desire in the matter—and my advertisement and form of treatment shall be at once adapted to it. But I regret you did not tell me of this, at once, more than a year ago, when I enclosed a printed copy of Macmillan's advertisement......

Hamilton's reply to this was evidently very satisfactory, for on December 11, 1860, Tait wrote:

I am glad to find that my explanation has been sufficient, for I assure you that I had attributed the slackness of our correspondence of the last year to your having been bored and tired with my continued questions about various old and new points in Quaternions, and had no idea whatever that I had annoyed you in any way by the publication of my unlucky advertisement.

In letter 46, January 14, 1861, Tait acknowledged receipt of proof sheets of the *Elements*, and made further references to his electrodynamic work.

Here the correspondence practically ended. We learn from Tait's preface to his *Treatise* that Hamilton shortly before his death in 1865 urged Tait to push on with his book, as his own was almost ready for publication.

It is pleasing to know that the misconception of the situation which had fretted the mind of the master was entirely removed by the straightforward honest dealing of the disciple.

Broadly speaking the subject-matter of the Hamilton-Tait correspondence may be grouped under five heads.

(1) Quaternion differentials. These are discussed at length in Hamilton's letters v and x, the former of 45 pages having been written between the dates of Oct. 11 and 16, and the latter of 48 pages between

the dates of Oct. 25 and Dec. 2, 1859. The discussion is reproduced in essence in the *Elements*, although much more briefly.

(2) Transformations connected with Fresnel's wave-surface. Tait began the discussion in letter 10 and continued it in many of the subsequent letters down to letter 34. Hamilton took up the theme in letter xiv and elaborated it in letters xv, xv', xv'', which ran on consecutively for 96 pages. Here also the essential parts of the investigations both of Hamilton and Tait will be found in their works. In letter 20 Tait suggested the use of the form $\phi^{-\frac{1}{2}}$ and in letter 23 gave the wave-surface equation in the new form $T\left(\rho^2+\phi^2\right)^{-\frac{1}{2}}\rho=1$; a form whose elegance Hamilton at once recognised and continued thereafter to use.

(3) The theory of the linear vector function. This is chiefly contained in Hamilton's xix, xxiii, xxv, and in Tait's 32 and 33. The essential parts are reproduced in Hamilton's *Elements* and in Tait's *Elementary Treatise*.

(4) The theory of envelopes. This was begun by Tait's problem of the paraboloid cylinder which forms section 321 of his *Treatise* (3rd edition). The problem greatly took Hamilton's fancy. He began the discussion in letter viii, and developed it in elaborate detail by quaternion processes in letters xi and xiii.

(5) The planning of the new treatises on the calculus. Early in 1859 Hamilton began to write his "Manual," which finally appeared in 1866 as the *Elements*, unfortunately incomplete in consequence of the death of the author in 1865. Tait's own treatise was projected during the summer of 1859, but was withheld from publication until Hamilton's work should appear. It was finally published in 1867.

In connection with the preparation of Tait's *Quaternions* the following letter to Sir John Herschel is of considerable interest. Tait had sent Herschel copies of some of his quaternion contributions to the *Quarterly Journal of Mathematics* and, in reply to Herschel's acknowledgement, wrote on Dec. 14, 1864, as follows:

My Dear Sir

I am much obliged by your very kind note just received....
Five years ago, Messrs Macmillan & Co. advertized for speedy publication an "Elementary Treatise on Quaternions" by me; but, as my good friend Sir W. R. Hamilton thought that it might possibly interfere with his forthcoming "Elements of Quaternions" I withdrew it—and have published only the few articles I recently sent you—all of them with *his* approval.

I had no idea that you had been engaged in preparing such a work; and I merely write to say that I shall be most happy if you will persevere in your intention of publishing an elementary volume on the subject. In fact the papers I have sent you contain nearly the *whole* of my researches in the *elementary* part of the theory. I have an immense store of work in MSS relating to its higher applications—but unfit for an elementary treatise.

Since I projected the treatise I have ceased to be a Professor of Mathematics; and with private experiments and the ordinary preparation for the work of my class, I feel that I have barely time enough to contribute my fair share to the "Treatise on Natural Philosophy" which Thomson and I have undertaken. And, as this Treatise is certain to extend to *three* volumes at least, of which (after two years work) not even *one* is yet published, I feel that it may be years before I shall be in a position to write on Quaternions in a carefully considered popular style. I am sure that my old friend Macmillan would be delighted to have the chance of substituting your name for mine in the advertisement, which he has been hopelessly repeating for some years.

But the consent of Sir W. R. Hamilton is absolutely necessary to anyone undertaking the work.

Believe me, my dear Sir,

Yours very truly

P. GUTHRIE TAIT.

Sir J. F. W. Herschel, Bart.

It is certainly remarkable that Herschel at the age of 72 should have thought of such a project.

Only a careful comparison of the pages of Hamilton's and Tait's works could establish to what extent Tait's contributions were essentially original. Their methods were markedly different. Hamilton revelled in geometrical developments of all kinds, the fertility of his mathematical imagination tending at times to make him discursive and almost prolix. Tait's endeavour in all his really original quaternion work was to grapple with physical and dynamical problems. Compare for example the Hamiltonian development of the properties of the linear vector function with the chapter on strains which Tait contributed to Kelland and Tait's *Introduction to Quaternions*—each mode of treatment admirable in its way.

The linear vector function continued to absorb much of Tait's attention up to the very last day of his life. He made important contributions to the theory as well as many interesting applications of its power. See for example papers XV, XXI, XXVI, CXIV, CXX, CXXI, CXXII, CXXIV in the *Scientific Papers*, Vols. I and II—especially the first-named, that on the Rotation of a Rigid Solid.

Unquestionably, however, Tait's great work was his development of the powerful operator ∇. Hamilton introduced this differential operator in its semi-Cartesian trinomial form on page 610 of his *Lectures* and pointed out its effects on both a scalar and a vector quantity. This, it will be remembered, was one of the points especially brought forward by Tait when he began the correspondence with Hamilton. Neither in the *Lectures* nor in the *Elements*, however, is the theory developed. This was done by Tait in the second edition of his book (∇ is little more than mentioned in the first edition) and much more fully in the third and last edition.

From the resemblance of this inverted delta to an Assyrian harp Robertson Smith suggested the name Nabla. The name was used in playful intercourse between Tait and Clerk Maxwell, who in a letter of uncertain date finished a brief sketch of a particular problem in orthogonal surfaces by the remark " It is neater and perhaps wiser to compose a nablody on this theme which is well suited for this species of composition."

In 1870, when engaged in writing his *Treatise on Electricity and Magnetism*, Maxwell sent Tait the following suggestions as to names for the results of ∇ acting on scalar and vector functions :

Glenlair, Dalbeattie,
Nov. 7, 1870.

Dear Tait

$$\nabla = i \frac{d}{dx} + j \frac{d}{dy} + k \frac{d}{dz}.$$

What do you call this? Atled?

I want to get a name or names for the result of it on scalar or vector functions of the vector of a point.

Here are some rough hewn names. Will you like a good Divinity shape their ends properly so as to make them stick?

(1) The result of ∇ applied to a scalar function might be called the slope of the function. Lamé would call it the differential parameter, but the thing itself is a vector, now slope is a vector word, whereas parameter has, to say the least, a scalar sound.

(2) If the original function is a vector then ∇ applied to it may give two parts. The scalar part I would call the Convergence of the vector function, and the vector part I would call the Twist of the vector function. Here the word twist has nothing to do with a screw or helix. If the word *turn* or *version* would do they would be better than twist, for twist suggests a screw. Twirl is free from the screw notion and is sufficiently racy. Perhaps it is too dynamical for pure mathematicians, so for Cayley's sake I might say Curl (after the fashion of Scroll). Hence the effect of ∇ on a scalar function is to give the slope of that scalar, and its effect on a vector function is to give the convergence and the twirl

of that function. The result of ∇^2 applied to any function may be called the concentration of that function because it indicates the mode in which the value of the function at a point exceeds (in the Hamiltonian sense) the average value of the function in a little spherical surface drawn round it.

Now if σ be a vector function of ρ and F a scalar function of ρ

$$\nabla F \text{ is the slope of } F$$

$$V\nabla \,.\, \nabla F \text{ is the twirl of the slope which is necessarily zero}$$

$$S\nabla \,.\, \nabla F = \nabla^2 F \text{ is the convergence of the slope, which is the concentration of } F.$$

Also $\qquad S\nabla\sigma$ is the convergence of σ

$\qquad\qquad\qquad V\nabla\sigma$ is the twirl of σ.

Now, the convergence being a scalar if we operate on it with ∇, we find that it has a slope but no twirl.

The twirl of σ is a vector function which has no convergence but only a twirl.

Hence $\nabla^2\sigma$, the concentration of σ, is the slope of the convergence of σ together with the twirl of the twirl of σ, the sum of two vectors.

What I want is to ascertain from you if there are any better names for these things, or if these names are inconsistent with anything in Quaternions, for I am unlearned in quaternion idioms and may make solecisms. I want phrases of this kind to make statements in electromagnetism and I do not wish to expose either myself to the contempt of the initiated, or Quaternions to the scorn of the profane.

Yours truly

J. CLERK MAXWELL.

A week later (Nov. 14, 1870) Maxwell, when returning Robertson Smith's letter in which the philology of Nabla was discussed in detail, wrote:

"I return you Smith's letter. If Cadmus had required to use ∇ and had consulted the Phoenician Professors about a name for it there can be no question that Nabla would have been chosen on the א ב ג principle. It is plain that Hamilton's ∇ derives itself with all its congeners from Leibnitz' d, which has become consecrated along with $D\,\partial\,\delta$ etc., and a name derived from its shape is hardly the thing.

"With regard to my dabbling in Hamilton I want to leaven my book with Hamiltonian ideas without casting the operations into Hamiltonian form for which neither I nor I think the public are ripe. Now the value of Hamilton's idea of a vector is unspeakable, and so are those of the addition and multiplication of vectors. I consider the form into which he put these ideas, such as the names Tensor, Versor, Quaternion, etc., important and useful, but subject to the approval of the mathematical world....

"The names which I sent you were not for ∇ but for the results of ∇. I shall send you presently what I have written, which though it is in the form of a chapter of my book is not to be put in but to assist in leavening the rest. I shall take the learned Auctor[1] and the grim Tortor[1] into my serious consideration, though Tortor has a helical smack which is distasteful to me but poison to T."

[1] These seem to have been suggestions made by Tait himself, probably more in joke than in serious mood.

It was probably this reluctance on the part of Maxwell to use the term Nabla in serious writings which prevented Tait from introducing the word earlier than he did. The one published use of the word by Maxwell is in the title to his humorous Tyndallic Ode[1], which is dedicated to the "Chief Musician upon Nabla," that is, Tait.

The following letter from Maxwell shows how clearly he had grasped the significance of the quaternion notation.

ARDHALLON,
DUNOON,
Jan. 23, 1871.

Dr T'

Still harping on that Nabla?

You will find in Stokes on the Dynamical Theory of Diffraction something of what you want, this at least which I quote from memory.

1. For all space—your eqn

$$\nabla \sigma = \nabla^2 (\tau + v)$$

where σ is given and τ and v are to vanish at ∞ gives but one solution for τ and one for v, the first derived by integration from $V\nabla\sigma$ and the second from $S\nabla\sigma$ by the potential method, and we then get the result in the form

$$\sigma = V\nabla\tau + \nabla v$$

(because, as Helmholtz has shown (Wirbelbewegung) $S\nabla\tau = 0$). All this is as old as 1850 at least. See Stokes.

Now we leave all space and consider a region Σ within which $\nabla^2 P = 0$ and therefore ∇P has no convergence. Now if a vector function has no convergence it ought to be capable of being represented as the curl of a vector function, or there ought to be a vector σ such that

$$V\nabla\sigma = \nabla P.$$

The simplest case to begin with is of course the potential due to unit of mass at the origin. Find σ and τ for that case! The difficulty arises from the fact that the region in which $\nabla^2 P = 0$ is here periphractic and surrounds completely the origin where this is not true. If we draw a closed surface including the origin then

$$\iint S U\nu\nabla P ds = 4\pi,$$

whereas

$$\iint S U\nu V\nabla\sigma ds = 0, \; necessarily[2].$$

Hence to make it impossible for the region to include the origin we must get rid of periphraxy by drawing a line from the origin to ∞ and defining the region Σ so as not to interfere with this line.

---

[1] Reproduced partly in facsimile at the end of this Chapter.

[2] Because $\iint S d\nu \, V\nabla\sigma = \iiint d v \, S\nabla V\nabla\sigma = 0$ for $\nabla^2\sigma$ is a vector.

We may then write p for $1/r$ and

$$P = -\int_0^\infty S\nabla p d\rho = p,$$

$$\sigma = \int_0^\infty V\nabla p d\rho.$$

If we suppose the line to be in the axis of x, this gives

$$\sigma = i\,(0) + j\,\frac{xz}{r\,(y^2 + z^2)} - k\,\frac{xy}{r\,(y^2 + z^2)}$$

an exceedingly ugly form for a thing derived from so symmetrical a beginning.

But this cannot be avoided if the algebraic sum of the masses is finite.

If it is 0, we may treat it as magnetic matter.

If, in a region Σ' in which there is magnetization, the intensity of magnetization be

$$\Im = iA + jB + kC$$

and if $p = 1/r$, where r is the distance between xyz and $x'y'z'$, then

$$P = \iiint \frac{A\,(x'-x) + B(y'-y) + C(z'-z)}{r^3}\,dx'dy'dz'$$

$$= \iiint S\Im\nabla p dv'$$

or

$$= -\iiint \frac{1}{r}\left(\frac{\partial A}{\partial x} + \frac{\partial B}{\partial y} + \frac{\partial C}{\partial z}\right) dx'dy'dz'$$

$$= -\iiint p S\nabla\Im dv'.$$

Also

$$\sigma = iF + jG + kH$$

where

$$F = \iiint \frac{C\,(y'-y) - B\,(z'-z)}{r^3}\,dx'dy'dz', \text{ \&c.}$$

or

$$\sigma = \iiint V\Im\nabla p dv'.$$

All this occurs in passing from the old theory of magnetism to the electromagnetic.

I have put down a lot of imitations of your jargon mainly that you may check me in any solecism. I think if you are making a new edition of 4nions you should give prominence to the rules defining the extent of the application of symbols such as V, S, T, U, K, &c., which are consecrated letters, not to be used for profane purposes....

What do you make of this?

You say that the constituents of τ are potentials with densities $\frac{1}{4\pi}\frac{\partial P}{\partial x}$ &c. Well, then, take $P = 1/r$ and $\partial P/\partial x = -x/r^3$ &c., then the constituents of τ will be $x/8\pi r$ &c. or

$$8\pi\tau = i\frac{x}{r} + j\frac{y}{r} + k\frac{z}{r}$$

and

$$\nabla\tau = \frac{1}{4\pi}\cdot\frac{1}{r} = P, \text{ a scalar.}$$

In fact of whatever scalar form P be, if $\nabla^2\tau = \nabla P$, $\nabla\tau = P$, a pure scalar. Multiply this by $d\rho$ (a pure vector) and you get a pure vector $d\rho\nabla\tau = d\rho P$. Hence your expression

$$S\int V(d\rho\nabla)\tau = S\int d\rho P = 0$$

because if it is anything at all it is the integral of a vector multiplied by a scalar and that is a pure vector and the scalar part of it is 0. I suppose this is nonsense arising from our being barbarians to one another. Will you therefore be so kind as to give me a code by which I may interpret the symbol $Vd\rho\nabla$, that is to say, tell me what these symbols, thus arranged, ask me to do....

Note—the Vector σ as determined above is such that $S\nabla\sigma = 0$ so that we may truly say $\nabla\sigma = \nabla P$.

In electromagnetism P is the magnetic potential and ∇P is the magnetic force outside the magnet or inside it in a hollow tube whose sides are parallel to the magnetization.

$$\nabla\sigma = \nabla P \text{ outside}$$

$$\text{but inside } \nabla\sigma = \nabla P + 4\pi\mathfrak{J}$$

where $\mathfrak{J}$ is the magnetization. $\nabla\sigma$ is the magnetic force in a crevasse $\perp \mathfrak{J}$.

I have not been able to make much of your τ. I coloured some diagrams of lines of force Blue and red but I must study the astronomer to define the magnetic tints and softness. Sir W. Hamilton (Edin[h]) was partial to redintegration, an operation you should get a symbol for. Among other scientific expressions I would direct your attention to the salutary influence of Demon-stration and Deter-mination, and to two acids recently studied, Periodic and Gallery Thronic acids. The 1st you will find use for. The 2nd is for the L[d] High Commissioner.

Yours J C M

In another letter, of which the opening paragraph has already been given (page 117 above), Maxwell refers to Tait's quaternionic investigations in the stress function. The letter is on a half sheet of note paper and is undated, but was probably written towards the end of 1872. The continuation is as follows:

"I return your speculations on the $\phi(Uv)ds$. Observe that in a magnet placed in a magnetic field the stress function is not in general self-conjugate, for the elements are acted upon by couples. But the $=^n$ of $=^m$ is very properly got as you get it[1].

"Search for a physical basis for

$$S.\nabla^2\sigma\nabla\sigma$$

as a term of the energy developed in a medium by a variable displacement σ. When found, make a note of, and apply to oil of turpentine, eau sucrée, &c., for it brings out the right sort of action on light of all colours. But the mischief is $V\nabla\sigma$, which it is manifest, can be produced without making any physical change inside a body. The very rotation of $\oplus$ produces it. Now $\nabla^2\sigma$ is a vector. Turn it alternately in the direction of $V\nabla\sigma$ and oppositely and you have increase & diminu-

---

[1] See Tait's Note on the Strain Function, *Proc. R. S. E.* 1872; *Sci. Pap.* Vol. I, pp. 196–7.

tion of energy, & therefore a tendency to set like a magnet. The comfort is that $\nabla^2\sigma$ cannot subsist of itself.

"Of course the resultant force on an element is of the form $V\nabla^2\sigma$, and if σ is a function of z only, and $Sk\sigma = 0$,

$$X = -\frac{\partial^2 y}{\partial z^3},$$

$$Y = \frac{\partial^3 \xi}{\partial z^3}.$$

"This is the only explanation of terms of this form in an isotropic or fluid medium, and since the rotation of plane of polarization is roughly proportional to the inverse square of the wave length, terms of this form must exist.

$$\frac{dp}{dt}\text{,,}$$

Thus, on the one hand, we have Tait submitting his quaternionic theorems to Maxwell's critical judgment, and Maxwell recognising the power of the quaternion calculus as handled by Tait in getting at the heart of a physical problem.

Unfortunately Tait's letters to Maxwell have not been preserved; and we can only infer as to the general nature of Tait's replies to Maxwell's constant enquiries regarding quaternion terms and principles. There can be no doubt however that, in introducing the operator ∇ and the Hamiltonian notation associated with it, Maxwell was strongly influenced not only by Tait's masterly paper on Green's and other Allied Theorems but also by his intimate correspondence.

The fundamental properties of ∇ as a differential operator may be expressed very simply in dynamical language. When it acts on a scalar function of the position of a point it gives in direction and magnitude the maximum space rate of change of this function. For example, if u is a potential, ∇u is the corresponding force. Its effect upon a vector quantity is, in general, to produce a quaternion, with its scalar and vector parts. Suppose the vector quantity to be the velocity of flow of a fluid, symbolised by σ; then $\nabla\sigma$ consists of two parts, the scalar and vector parts. The former, $S\nabla\sigma$, represents what Maxwell called the Convergence, indicating a change of density in the fluid at the point where σ is the velocity; and for the latter, symbolised by $V\nabla\sigma$, and measuring in the present case the vorticity, Maxwell's name of Curl has been generally accepted.

It is instructive to read Tait's early papers discussing the properties of ∇, and to follow the growth of his power in dealing with it. At first he was content to begin with Hamilton's trinomial definition, as in the paper of 1862 on the Continuous Displacements of the Particles of a Medium (*Scientific Papers*, Vol. 1, p. 37). But ere long he discovered a less artificial definition, free from Cartesian symbolism. This mode of establishing the theory of ∇ is given in the appendix to his great paper already mentioned, that on Green's and other Allied Theorems (1870, *Sci. Pap.* Vol. 1, p. 136). Here we find developed in an original manner the quaternion integrals through volumes, over surfaces, and along edges, which include as special cases the theorems of Green, Gauss, and Stokes.

In 1868 Tait published an elaborate memoir on the Rotation of a Rigid Body about a Fixed Point (*Sci. Pap.* Vol. 1, p. 86), concerning which, while it was in preparation, he had a good deal of correspondence with Cayley. On August 18, 1868, he sent Cayley the concise quaternion equations of §§ 15, 19, 21, and asked if the results are "merely a shortening of yours"; and on October 17, the same year, he drew Cayley's attention to the Cartesian formulae in §§ 28, 29, and to the fact that, "without integrating Euler's equations at all (and I think from your *second* Report that the problem has always been solved by first finding p, q, r), I find the following equations for w, x, y, z [equation 24 in Paper]." On October 21, Cayley replied :

"The rotation formulae are deducible by an easy transformation from formulae in my paper (*Cam. and Dub. Math. Journ.* Vol. I, 1846)....But the actual form you have given to the formulae is, so far as I am aware, new ; *and a very decided improvement* as reducing the denominator to be of the third order."

For these two quaternion papers Tait was awarded the Keith Prize by the Royal Society of Edinburgh. The Secretary asked Clerk Maxwell to draw up a statement to be read when the prize was formally awarded by the President ; and Maxwell responded with a playful humour which considerably mystified Professor J. H. Balfour. A copy was preserved in Tait's Scrap Book.

(Balfour, having asked Maxwell to write something which could be read at a meeting of the R. S. E. when I was to get the Keith medal, was mystified as follows. P. G. T.)

GLENLAIR, DALBEATTIE, 28/11/70.

Dear Professor Balfour,

I do not presume to inform an officer of the Society with respect to its recent awards. I saw that Tait had got the Keith Prize which is or ought to be known to the public. I have not yet got a copy of the reasons for which it was awarded, so if I coincide with them it does not arise from imitation.

The question seems to be, What is Tait good for ? Now I think him good, first, for writing a book on Quaternions, and for being himself a living example of a man who has got the Quaternion mind directly from Hamilton. I am unable to predict the whole consequences of this fact, because, besides knowing Quaternions, Tait has a most vigorous mind, and is well able to express himself especially in writing, and no one can tell whether he may not yet be able to cause the Quaternion ideas to overflow all their mathematical symbols and to become embodied in ordinary language so as to give their form to the thoughts of all mankind.

I look forward to the time when the idea of the relation of two vectors will be as familiar to the popular mind as the rule of three, and when the fact that $ij = -ji$ will be introduced into hustings' speeches as a telling illustration. Why not? We have had arithmetical and geometrical series and lots of odd scraps of mathematics used in speeches.

Nevertheless I do not recommend some of Tait's mathematical papers to be read as an address to the Society, *ore rotundo*. That on Rotation is very powerful, but the last one on Green's and other allied Theorems is really great.

The work of mathematicians is of two kinds, one is counting, the other is thinking. Now these two operations help each other very much, but in a great many investigations the counting is such long and such hard work, that the mathematician girds himself to it as if he had contracted for a heavy job, and thinks no more that day. Now Tait is the man to enable him to do it by thinking, a nobler though more expensive occupation, and in a way by which he will not make so many mistakes as if he had pages of equations to work out.

I have said nothing of his book on Heat, because, although it is the clearest thing of the sort, it is not so thoroughly imbued with his personality as his Quaternion works. In this however I am probably entirely mistaken, so I advise you to ask Tait himself who I have no doubt could hit off the thing much better than any one.

I remain

yours truly

J. Clerk Maxwell.

It seems appropriate here to reproduce from Maxwell's letters to Tait some extracts bearing upon the quaternion calculus, for which it is clear Maxwell had a profound admiration. The letters of Nov. 2, 1871, and Dec. 1, 1873, have been already given *in extenso* (see above, pp. 101, 115); the following are culled from other letters:

"(Dec. 21, 1871.) Impress on T. that $\left(\frac{d}{dx}\right)^2 + \left(\frac{d}{dy}\right)^2 + \left(\frac{d}{dz}\right)^2 = -\nabla^2$ and not $+\nabla^2$ as he vainly asserts is now commonly believed among us. Also how much better and easier he would have done his solenoidal and lamellar business if in addition to what we know is in his head he had had say, 20 years ago, Qns. to hunt for Cartesians instead of vice versa. The one is a flaming sword which turns every way; the other is a ram, pushing westward and northward and (downward?). What we want a Council to determine is the true doctrine of brackets and dots and the limits of the jurisdiction of operators.

"(Oct. 4, 1872.) How about electromagnetic 4nions as in proof slip 106, 107? I suspect I am not sufficiently free with the use of the Tensor in devectorizing such things as r (distance between two points). The great need of the day is a grammar of 4nions in the form of dry rules as to notation and interpretation not only of $S\,T\,U\,V$ but of . ( ) and the proper position of $d\sigma$, etc. Contents, Notation, Syntax, Prosody, Nablody.

"(Oct. 9, 1872.) I think I had better consecrate ρ [in the Treatise] to its prescriptive office of denoting indicating or reaching forth unto the point of attention (xyz)....Has ρ a name? It is no ordinary vector carrying a point. It is rather a tentacle or feeler which reaches from the subject to the object. Is he the scrutator? I am glad to hear of the second edition of 4nions. I am going to try, as I have already tried, to sow 4nion seed at Cambridge. I hope and trust nothing I have yet done may produce tares. But the interaction of many is necessary for the full development of a new notation, for every new absurdity discovered by a beginner is a lesson. Algebra is very far from o. k. after now some centuries, and diff. calc. is in a mess and $\iiint$ is equivocal at Cambridge with respect to sign. We put down everything, payments, debts, receipts, cash credit, in a row or column, and trust to good sense in totting up.

"(March 5, 1873.) O T'. If, in your surface integrals, ds is an element of surface, is not ds a vector? and does not multiplication by $U\nu$ scalarize it? In your next edition tell us if you consider an element of surface otherwise than as $Vd\alpha d\beta$ where α and β are vectors from the origin to a point in the surface defined by parameters a, b. Here the element of surface is a vector whose tensor is the area and whose versor is $U\nu$. These things I have written that our geometrical notions may in Quaternions run perpetual circle, multiform, and mix and nourish all things. Such ideas are slowly percolating through the strata of Cartesianism, trilinearity, and determinism that overlie what we are pleased to call our minds.

"(Sep. 7, 1878.) Here is another question. May one plough with an ox and an ass together? The like of you may write everything and prove everything in pure 4nions, but in the transition period the bilingual method may help to introduce and explain the more perfect.

"But even when that which is perfect is come that which builds over three axes will be useful for purposes of calculation by the Cassios[1] of the future.

"Now in a bilingual treatise it is troublesome, to say the least, to find that the square of AB is always positive in Cartesians and always negative in 4nions, and that when the thing is mentioned incidentally you do not know which language is being spoken.

"Are the Cartesians to be denied the idea of a vector as a sensible thing in real life till they can recognise in a metre scale one of a peculiar system of square roots of -1?

"It is also awkward when you are discussing, say, kinetic energy to find that to

[1] "And what was he?
Forsooth, a great arithmetician" (*Othello*, Act I, Scene I).
A neat example of Maxwell's ingenuity in literary suggestion.

ensure its being $+ve$ you must stick a $-$ sign to it, and that when you are proving a minimum in certain cases the whole appearance of the proof should be trending towards a maximum.

"What do you recommend for El. and Mag. to say in such cases?

"Do you know Grassmann's Ausdehnungslehre? Spottiswoode spoke of it in Dublin as something above and beyond 4nions. I have not seen it, but Sir W. Hamilton of Edinburgh used to say that the greater the extension the smaller the intention."

We have not the record of Tait's reply to the question of the sign, a question which many later users of vector notations have attempted to answer by simply ignoring one of the distinctive features of Hamilton's calculus. So long as it is a question merely of a concise notation no harm is done; and Maxwell without seriously affecting the symbolic presentation of his theory of electromagnetism might have adopted this method. But he had too great a regard for the founder of Quaternions, and too deep an insight into the inwardness of the quaternion calculus, to allow mere expediency to play havoc with far-reaching fundamental principles.

Meanwhile Tait's activity in developing quaternion applications continued throughout the seventies. In a Note on Linear Differential Equations in Quaternions (*Proc. R. S. E.* 1870; *Sci. Pap.* Vol. 1, p. 153) he struck out on new paths. Here he gave an extremely simple solution of the problem of extracting the square root of a strain or linear vector function.

In a letter to Cayley of date Feb. 28, 1872, Tait gave the Cartesian statement of the problem and continued,—

My quaternion investigation, which is very simple, leads to the biquadratic

$$2\theta = m - \frac{(\theta^2 - m_1{}^2)^2}{4m_2}$$

where m, m_1, and m_2 are known functions of [the elements of the strain]; and from θ the values of [the elements of the square root of the strain] can easily be found.

Thomson and I wish to introduce this into the new edition of our first volume on Natural Philosophy—but he objects utterly to Quaternions, and neither of us can profess to more than a very slight acquaintance with modern algebra—so that we are afraid of publishing something which you and Sylvester would smile at as utterly antiquated if we gave our laborious solutions of these nine quadratic equations.

As I said before the question is of interest in another way (for my Report on Quaternions), for if ϕ be the strain function and ϕ' its conjugate, and if we try to resolve the strain into a *pure* strain followed by a rotation, so that $\phi(\ )=q\varpi(\ )q^{-1}$, I find $\varpi^2(\ )=\phi'\phi(\ )$, so that the pure strain is the square root of the given strain followed by its conjugate.

Cayley replied, March 2, 1872:

"I find that your question may be solved very simply by means of a theorem in my memoir on Matrices, *Phil. Trans.* 1858."

He then proceeded to indicate the steps of a somewhat prolonged process by which the solution might be found; but in a second letter written a few hours later he practically reproduced Tait's process by use of the symbolic cubic, the Matrix symbol M being written instead of the vector function ϕ.

On March 5, Tait wrote:

It is a most singular fact that you seem to have been working simultaneously with Hamilton in 1857–8, just as I found you had been in a very much earlier year...... I have had but time for a hurried glance at your paper on Matrices—and I see that it contains (of course in a very different form) many of Hamilton's properties of the linear and vector function....I send you a private copy of my little article, by which you will see how closely the adoption of Hamilton's method has led me to anticipate almost every line of your last note....There is one point of Hamilton's theory to which I do not see anything analogous in your paper. Expressed in his notation it is that

$$\frac{m_g}{g} S\rho\,(\phi+g)^{-1}\rho \text{ and } \frac{m_h}{h} S\rho\,(\phi+h)^{-1}\rho$$

are *identical*, if we have $gh = mS\rho^{-1}\phi^{-1}\rho$.

The Report referred to by Tait in his letter of Feb. 28 was a Report which, urged by Cayley, he had agreed to prepare for the British Association. Shortly afterwards he asked to be relieved of the task, as it would be of too personal a character, and suggested Clifford as eminently qualified to undertake it. Nothing further seems to have been done.

The quaternion discussion of orthogonal isothermal surfaces was published in 1873 (*Sci. Pap.* Vol. I, p. 176). It is an interesting example of the use of Hamilton's rotational operator $q(\ )q^{-1}$. The opening paragraphs of this paper are not quaternionic, and seem to have been introduced by Tait for the double purpose of showing how he originally began to attack the problem and how much more suggestive and concise the quaternion solution is. In a letter of date July 22, 1873, Maxwell referred in a deliciously humorous manner to the character of Tait's investigations in these words:

"I beg leave to report that I consider the first two pages of Professor Tait's Paper on Orthogonal Isothermal Surfaces as deserving and requiring to be printed in the *Transactions* of the R. S. E. as a rare and valuable example of the manner of that Master in his Middle or Transition Period, previous to that remarkable condensation not to say coagulation of his style, which has rendered it impenetrable to all but the piercing intellect of the author in his best moments."

When this paper was passing through the press Tait had a brief correspondence with Cayley on the nature of his solution. After its publication, Cayley made some interesting comments in a letter of date March 25, 1874. He first reproduced one of his own results which shows that, in order that $r =$ const. may represent a family of orthogonal surfaces, then r considered as a function of $x\,y\,z$ must satisfy a somewhat complicated partial differential equation of the third order. Tait's equation $d\sigma = uqd\rho q^{-1}$, he then pointed out, must be the equivalent of this partial differential equation of the third order. He concluded in these words :

"Do you know anything as to the solution when the limitations [imposed by Tait] are rejected, and imaginary solutions taken account of? Considering simply the equation of the third order and the equation $a + b + c = 0$ [that is $\nabla^2 r = 0$] it would seem probable that there must be a solution of greater generality than the confocal quadrics. I do not see my way to the discussion of the question. The condition $a + b + c = 0$ seems to make no appreciable simplification in the equation of the third order. I admire the equation $d\sigma = uqd\rho q^{-1}$ extremely—it is a grand example of the pocket map."

This comparison of a quaternion formula to a pocket map was quite in accord with Cayley's attitude towards the quaternion calculus. He admitted the conciseness of its formulae, but maintained that they were like pocket maps: everything was there, but it had to be unfolded into Cartesian or quantic form before it could be made use of, or even understood. This view Tait combated with all the skill at his command; and every now and again the two mathematicians had a friendly skirmish over the relative merits of quaternions and coordinates.

Even when they exchanged views on quaternionic problems altogether apart from this central controversial question, their different mental attitude came clearly to the front in their correspondence. This is seen, for example, in the following series of letters.

Dear Tait

In the quaternion $q = w + ix + jy + kz$, assuming
$$\tan \tfrac{1}{2} f = \frac{\sqrt{x^2 + y^2 + z^2}}{w}, \quad (r = \sqrt{(x^2 + y^2 + z^2)} \text{ and } \tfrac{x}{r}, \tfrac{y}{r}, \tfrac{z}{r}, = \cos \alpha, \cos \beta, \cos \gamma)$$
then the quaternion is
$$q = w + ix + jy + kz$$
$$= \frac{r}{\sin \tfrac{1}{2} f} \{\cos \tfrac{1}{2} f + \sin \tfrac{1}{2} f (i \cos \alpha + j \cos \beta + k \cos \gamma)\}$$
and we can interpret the quaternion in a twofold manner, viz., in the first form,

disregarding the scalar part w, as the force represented by the lines x, y, z; and in the second form, disregarding the tensor $r/\sin \frac{1}{2}f$, as a *rotation* f about the axis (α, β, γ).

Then *sum* of two quaternions, quà *force*, is the resultant force.

Product of two quaternions, quà *rotation*, is the resultant rotation.

But is there any interpretation for the sum quà *rotation* or for the product quà *force*? It would be very nice if there were.

We enjoyed our American expedition very much. I was glad to hear from Thomson that he also was going to lecture at Johns Hopkins University....

Yours very sincerely
A. CAYLEY.

CAMBRIDGE,
3rd Nov. 1882.

UNIVERSITY OF EDINBURGH,
4/11/82.

My dear Cayley

I was very glad to get your note, and to hear that you had enjoyed your venturous journey. Thomson's proposal was *quite new* to me! I have not seen him for months.

I am absolutely overwhelmed with work just now; as, besides my University work, and R.S.E. do., I have been virtually forced to give a course of lectures to ladies, and I am writing, against time, a very long article for the *Encyc. Brit.*

Maxwell's death left the staff of the *Encyc.* in a state of great perplexity. He had drawn up a scheme for the scientific articles, and had done the greater part of the work himself. Had he lived, the article "Mechanics" would have been written by him, or entrusted to some competent writer, *two years ago*, at least. As it is, the acting editor discovered, only three months ago, how much had been referred forward to it; and I spent the greater part of my summer holiday in writing it. Seeing it through the press is no joke! And the work of trying to boil down the whole of abstract dynamics into 60 pages has been very heavy.

I fear I misunderstand your questions. Of course I know that Vq is a force and that $V(q+r) = Vq + Vr$; whatever quaternions q and r may be. But, as to rotation, I have always written (after Hamilton)

$$q(\ )q^{-1}$$

where (of course) we need not trouble about the tensor. This gives $qr(\ )r^{-1}q^{-1}$ as the result of $r(\ )r^{-1}$ followed by $q(\ )q^{-1}$; and may be written

$$qr(\ )(qr)^{-1}.$$

Now, in asking about the interpretation of a sum quà rotation, do you mean the effect of $(q+r)(\ )(q+r)^{-1}$? Also, as to the product quà force, do you refer to $V.qr$?

I can easily answer these questions, but I fear I have not caught your meaning....

Before Cayley's reply to this was received, Tait wrote a second note.

UNIVERSITY OF EDINBURGH,
6/11/82.

My dear Cayley

Since I wrote you I have fancied that I ought to have sent you the answers, even if I have misunderstood you.

1. When we deal with a sum of two quaternions, from the rotational point of view, the *ratio of their tensor* plays a prominent part. In fact

$$(q + r)(\quad)(q + r)^{-1} = (qr^{-1})^x r(\quad) r^{-1} (qr^{-1})^{-x}$$

where x is a scalar, which is to be found from an equation of the form

$$\frac{a \sin A}{a \cos A + 1} = \tan xA.$$

This *seems* an answer to your question "Is there any interpretation of the sum quà *rotation*?"

It is the rotation $r(\quad) r^{-1}$ followed by $(qr^{-1})^x(\quad)(qr^{-1})^{-x}$.

Of course it may also be put in the form $(q^{-1}r)^y(\quad)(q^{-1}r)^{-y}$ followed by $q(\quad)q^{-1}$ where y is another scalar found from a transcendental equation.

Compounding these it may also be expressed as

$$\{(qr^{-1})^x rq (q^{-1}r)^y\}^{\frac{1}{2}}(\quad)\{(qr^{-1})^x rq (q^{-1}r)^y\}^{-\frac{1}{2}}$$

which is more symmetrical.

But it can also be expressed as

$$q^l r^m q^l (\quad) q^{-l} r^{-m} q^{-l}.$$

When l and m are found from two equations of the form

$$\frac{2(a \cos \alpha + b \cos \beta)}{b \sin \beta} = c \sin 2l\alpha \sin m\beta + \cos 2l\alpha \cos m\beta,$$

$$\frac{2 a \sin \alpha}{b \sin \beta} = 2c + \sin 2l\alpha \cos m\beta,$$

all the quantities a, b, c, α, β, being known scalars.

Of course the number of such expressions is endless; and I wait further light from you.

2. As to the product quà "force" (as you call it), we have

$$V . qr = Sr . Vq + Vr . Sq + V . VqVr$$

so that the "force" of the product appears as the sum of three forces; two of which are multiples of the separate forces; the other is a force perpendicular to both.

In great haste,
yours truly
P. G. TAIT.

Cayley's letter of the same date which crossed this one was as follows:

Dear Tait

It is only a difference of expression: I say that

$$q = \cos \tfrac{1}{2} f + \sin \tfrac{1}{2} f (i \cos \alpha + j \cos \beta + k \cos \gamma),$$

is the symbol of a rotation because operating in a particular manner with q upon

$ix + jy + kz$ we obtain $ix_1 + jy_1 + kz_1$, the x_1, y_1, z_1 being the new values of x, y, z produced by the rotation: viz. the particular operation is

$$ix_1 + jy_1 + kz_1 = q\,(ix + jy + kz)\,q^{-1}$$

and you say that $q(\quad)q^{-1}$ is the rotation. But of course q, r being the two quaternions, qr in my mode of expression or $qr(\quad)(qr)^{-1}$ in yours, belongs to the resultant rotation.

In my mode of expression[1]

$$T\{\cos \tfrac{1}{2} f + \sin \tfrac{1}{2} f\,(i \cos \alpha + j \cos \beta + k \cos \gamma)\}$$

is equally well with

$$\cos \tfrac{1}{2} f + \sin \tfrac{1}{2} f\,(i \cos \alpha + j \cos \beta + k \cos \gamma)$$

the symbol of the rotation; and my question was is there any interpretation, *in connection with rotations*, of the sum

$$T\{\cos \tfrac{1}{2} f + \sin \tfrac{1}{2} f\,(i \cos \alpha + j \cos \beta + k \cos \gamma)\}$$

$$+ T'\{\cos \tfrac{1}{2} f' + \sin \tfrac{1}{2} f'\,(i \cos \alpha' + j \cos \beta' + k \cos \gamma')\},$$

that is of the sum of any two quaternions

$$w + ix + jy + kz,\ w' + ix' + jy' + kz'.$$

I think therefore you have understood me quite rightly—viz. in asking about the interpretation of a sum quà rotation, I do mean the effect of $(q + r)\,(\quad)(q + r)^{-1}$, and as to the product quà force I do refer to Vqr—and shall be much obliged for the answer.

Believe me, dear Tait, yours very sincerely

A. CAYLEY.

Nov. 6th

PS. I believe it was I who first gave in the *Phil. Mag.* the formula $q(ix + jy + kz)q^{-1}$, showing it was identical with that of Rodrigues for the effect of a rotation—but Hamilton was doubtless acquainted with it.

Tait replied to this the next day:

7/11/82.

My dear Cayley

The note I sent you yesterday, and which I hope you got, will now, I see, *more than* answer your question; which (as I understand it) refers to the sum of two *versors*

$$Uq + Ur$$

---

[1] There is a strong resemblance here between Cayley's symbolism of the rotation involved in the quaternion and the discussion by Klein and Sommerfeld in their *Ueber die Theorie des Kreisels* of what they call "die Quaternionentheorie" (Chap. I, § 7). See Tait's paper "On the claim recently made for Gauss to the Invention (not the Discovery) of Quaternions" (*Proc. R. S. E.* Vol. XXIII, 1889); and "Professor Klein's View of Quaternions, a Criticism," by C. G. Knott (*Proc. R. S. E.* Vol. XXIII, 1889).

(You write a T instead of a U; but the form you adopt—viz.

$$\cos \alpha + (il + jm + kn) \sin \alpha$$

is a versor, its tensor being 1).

Of course in this particular case, the formula I gave you yesterday is much simplified. For instance we have $a = 1$ and $x = \frac{1}{2}$.

Thus $(Vq + Vr)(\quad)(Vq + Vr)^{-1} = (qr^{-1})^{\frac{1}{2}} r (\quad) r^{-1} (qr^{-1})^{\frac{1}{2}}$.

This and indeed the general cases of $q + r$, is easily seen by means of a diagram [proof given by use of a spherical triangle]....

I send with this a copy of an old paper of mine bearing on the question raised in your last....The second of these gives the reference which shows that Hamilton anticipated you about the quaternion rotation.

The third passage refers to what I thought was mine (i.e. putting Rodrigues' expressions in a simpler form) but your letter shows that you also use this *versor* form....

Cayley's reply was:

Dear Tait

Best thanks for the last two letters and the memoir. I am rather glad to find that the formula was first given by Hamilton.

The $(q + r)(\quad)(q + r)^{-1}$ formulae are very curious, but I hardly see as yet what to make of them....

CAMBRIDGE, *8th Nov.* 1882.

Cayley seems to have forgotten to some extent the contents of Tait's paper of 1868.

Towards the end of 1884 an interesting correspondence arose between Tait on the one hand and Cayley and Sylvester on the other in regard to the solution of the quaternion equation $aq = qb$. Sylvester had just published his general solution of the linear matrix equation; and taking a more general view of the quaternion q he obtained what seemed at a first glance to be a different solution from that given by Tait in his *Quaternions*. The analytical theory which admits the possibility of Tq vanishing—a possibility never considered by Tait—is given by Cayley in Chapter VI of the 3rd edition of Tait's *Quaternions*; and parts of this contributed chapter are almost identical word for word with portions of Cayley's letters.

On August 28, 1888, Tait in view of the preparation of this 3rd edition, asked Cayley for suggestions in the way of improvements, especially on the analytical side. Cayley responded immediately with some notes which Tait gratefully accepted. Some weeks later Tait wrote:

Since I returned to Edinburgh I have been considering more closely the question of the new edition of my *Quaternions* and looking up specially Sylvester's papers in the *Comptes Rendus* and the *Phil. Mag.* It seems to me from my point of view (which I think is that of Hamilton) that all these things, excellent and valuable as they are, are not *Quaternions* but developments of *Matrices*. As I understand Hamilton's quest, it was for a method which should *supersede* Cartesian methods, wherever it is possible to do so. Hence *i, j, k*, and their properties, though they were the stepping stones by which Hamilton got his method, are to be discarded in favour of α, *q*, φ, etc.: and no problem or subject is a *fit* one for the introduction of Quaternions if it necessitates the introduction of Cartesian Machinery....

The conclusion from this seems to be that I ought, instead of inserting your contributions in the text of my book as it stands, to make a new chapter "On the Analytical view of Quaternions" (or some such title) in which they will form the spinal column. Therein will naturally assemble all the disaffected or lob-sided members, which are not capable of pure quaternionic treatment but which are nevertheless valuable, like the occipital ribs and the anencephalous heads in an anatomical museum.

Ten days later Cayley replied:

"I...have not yet written out two further notes which I should like to send you for the new Chapter—which (I take it kindly) you do *not* compare with the Chamber of Horrors at Madame Tussaud's....I need not say anything as to the difference between our points of view; we are irreconcileable and shall remain so: but is it necessary to express (in the book) all your feelings in regard to coordinates? One remark: I think you do not give your symbol φ a sufficiently formal introduction: it comes in incidentally through a particular case, without the full meaning of it being shown. The two notes will be on the equation $aq + qb = 0$ and on Sylvester's solution of $aq^2 + bq + c = 0$."

On Oct. 22, 1888, Tait wrote:

I am very glad to know that you will give me two more of them [i.e. the notes]; especially as I found Sylvester's papers hard to assimilate. A considerable part of each paper seems to be devoted to correction of hasty generalizations in the preceding one!

I don't know that my point of view of coordinates is very different from yours, though my sight is vastly inferior. But I can see pretty clearly in the real world, with its simple Euclidean space, by means of the quaternion telescope. Witness a paper of Thomson's which I have just seen in type for the next *Phil. Mag.*; where three *pages* of formulae can easily, and with immense increase of comprehensibility, be put into as many *lines* of quaternions.

In his reply to this letter Cayley, after indicating his desire to see proofs of Tait's Preface to his coming new edition of his *Quaternions*, asked:

"Have you considered how far some of the geometrical proofs are independent of anything that is distinctively Quaternions, and depend only on the notion of $ix + jy + kz$, with i, j, k as incommensurable imaginaries not further defined?"

It was not till the summer of 1889 that the third edition began to be printed; and this naturally led to a renewal of the correspondence on quaternionic subjects. Writing on June 15, 1889, Tait drew Cayley's attention to a new problem which had been interesting him.

In looking over the *Trans. R.S.E.* for your notes for the *Fortschritte d. Math.* I suppose you saw Plarr's paper on the form of the spots which a blackened ellipsoid would make if it were made to slide about in the corner of the ceiling.

I have been trying to simplify the analysis, and have reduced the question to one of mere elimination:—but it is still very complex.

With the view of studying what any point of the ellipsoid does, I had a very true ellipse cut out of thick sheet brass in my laboratory, and have traced the curves (of the 12th degree?) made by a pencil passed through various holes in it when it slides between two perpendicular guide-edges.

This was the beginning of Tait's discussion of the glissettes of the ellipse and hyperbola. In reference to the problem Cayley remarked:

"I abstracted Plarr's paper, but it did not seem to me that he had got out much of a result—not that I saw my way to doing it better. It is a very good question, and a very difficult one. The plane question ought to be much easier tho' I fancy even that might be bad enough. I shall be very glad to see your curves."

On November 21, 1889, Tait referring to the glissettes wrote:

Connected with the curves I sent you in summer there is a very curious theorem which may, perhaps, be new to you. *They can be traced by a point in the plane of a hyperbola which slides between rectangular axes.*

A month later, Dec. 21, Tait wrote:

My dear Cayley

Thanks for your second splendid volume, which has come just in time for my brief vacation, and contains in an accessible form the Quantics, which I have long wished to read properly.

The same post brought me a specimen copy of *Quaternions*, with various colours of cloth to choose from. Brick red seems to be the most taking bait, so when you get it you will have something striking to look *at* if not *into*.

You will see, in a few days, in the *Phil. Mag.* another plea for Quaternions as *the* physical calculus, par excellence. Perhaps it may lead to an increased sale of my volume.

Have you ever considered the locus of intersection of two normals to an ellipse which are perpendicular to one another?

I showed the R.S.E., on Monday last, an Ellipse and a Hyperbola separately tracing the *same* glissette. The uninitiated were much puzzled to see it, as the one curve merely oscillates while the other turns complete summersalts, and they could not conceive that the same curve could be traced by a point of each. But it comes merely to this:—that [in the parallelogram linkage $OABA'$ which was sketched in the letter] the ellipse describes B about O virtually by the two links OA, AB; while the hyperbola does it by the other two sides of the parallelogram. The centre, A, of the ellipse has a to and fro motion through a limited angle, while A' (the centre of the hyperbola) goes completely round.

A later letter from Tait gave a further investigation of this problem very much as it appeared in the published paper (*Proc. R. S. E.* Dec. 1889; *Sci. Pap.* Vol. II, p. 309), which Cayley characterised as "very interesting."

On January 24, 1890, after acknowledging the receipt of Tait's *Quaternions* and a copy of the *Phil. Mag.* paper on the Importance of Quaternions in Physics (*Sci. Pap.* Vol. II, p. 297), Cayley renewed the old discussion in these words:

"Of course I receive under protest ALL your utterances in regard to coordinates. Really, I might as well say, in analytical geometry we represent the equation of a surface of the second order by $U=0$; compare this with the cumbrous and highly artificial quaternion notation $Sp\phi p = -1$. But you cannot contend that this last equation by itself contains the specification of the constants which determine the particular quadric surface; and the fair parallel is between your quaternion equation and $(*\oint x, y, z, 1)^2 = 0$: and if you say yours is shortest, I should reply, mere shortness is no object, or again there is nothing easier than to use a single letter to denote $(x, y, z, 1)$. Again, for a determinant

$$\begin{vmatrix} x & y & z \\ x' & y' & z' \\ x'' & y'' & z'' \end{vmatrix}$$

there is here absolutely nothing superfluous, the determinant depends upon nine quantities which have to be specified: and these are not simply a set of nine, but they group themselves in two different ways into 3's as shown by the lines and columns."

Tait replied as follows:

38 GEORGE SQUARE,
EDINBURGH,
25/1/90.

My dear Cayley

I might say with a great rhetorist, "I am not careful to answer thee in this matter":—but I think that most of your remarks seem to be based on ignoration of the *Title* of my little paper. It is the use *in Physics* that I am speaking of. $U=0$ is just as expressive in quaternions as in any other calculus, i.e. it is, in all,

T.   21 a blank form to be filled up. But $Sp\phi p = -1$ is strictly kinematical, and defines an ellipsoid (or other central conicoid) with reference to a strain—in this case a pure strain—the conception of which is vividly realistic.

$$(A, B, C, D, E, F \,\rangle\!\langle\, x, y, z)^2 = 1$$

gives no *physical* suggestion at all.

I should have said, in my paper, that we have to thank Cartesian processes for the *idea* of an Invariant. In pure quaternions you have them always, so that they present no feature for remark. ρ itself is an Invariant just as much as ∇ is. But what do you say to my little three term formula (on p. 95) which is equivalent to 189 Cartesian terms?

Cayley seems to have made no immediate reply to this letter; but on June 6, 1894, in a short note on other matters he threw in the remark:

" I wish you would tell me in what sense you consider Quaternions to be a *method*: I do not see that they are so, in the sense in which coordinates are a method; and I consider them rather as a theory."

Tait replied on June 10, 1894:

As to your question about Quaternions, I fear that I do not quite catch your meaning, so far at least as regards the technical distinction between a " Theory" and a " Method." From my point of view, Quaternions are a mode of representing geometrical or physical facts in such a clear way that one can see their mutual relations and their consequences. They assist me in these in the same *sort* of way as a figure or a model does, in the case of a knot or a complex surface:—or as an experiment of a crucial kind does. In fact they help one to think. I look upon them as contrasted with, rather than as related to, numerical work whether by logarithms or by definite integrals. These in themselves do not help you to *think*, though they are vitally important when you wish to *measure*; and though the working of them out may require very much thought. I fear this, in its turn, will not be very comprehensible to you:—but I have not been in the habit of dealing with such classes of questions; or at all events of trying to express, in language, my notions about them.

The discussion now entered upon a more definite phase, and on June 18, Cayley wrote:

" Considering coordinates and quaternions each as a *method* I should formulate the relation between them as follows: We seek to determine the position of a variable point P in space in regard to a fixed point O. Thro' O draw the rectangular axes Ox, Oy, Oz.
" Then

(coordinates) the position is determined by the coordinates x, y, z.

(quaternions) the position is determined by means of the vector α ($= OP$).

"But then what do you mean by the vector a? I mean

$$a = ix + jy + kz,$$

so that the knowledge of a implies that of the coordinates....

"But your claim for the superiority of quaternions rests, as I understand it, on the non-necessity of any explicit use of the equation in question $a = ix + jy + kz$, or $a = -iSia - jSja - kSka$....

"As to the modus operandi, if in regard to the points (x, y, z) and (x', y', z') one has to consider the combinations $yz' - y'z, zx' - z'x, xy' - x'y$, I consider these directly as the minors of the matrix

$$\begin{vmatrix} x, & y, & z \\ x', & y', & z' \end{vmatrix}$$

—whereas you represent them (in what seems to me an artificial manner) as the components of $Va\beta$...."

Tait's reply I give in full, since it presents in the briefest possible form the fundamental principles of quaternions as Tait regarded them.

38 George Square, Edinburgh,
19/6/94.

My dear Cayley

In the very first paper I published on Qns. (*Mess. Math.* 1862) I said "the method is independent of axes...and takes its reference lines solely from the problem it is applied to." Unless under compulsion, I keep to a, and do not write either $ix + jy + kz$ or $-iSia$—&c.

Hamilton said (*Lectures*, p. 522) "I regard it as an inelegance, or imperfection, in quaternions, or rather in the state to which it has hitherto been unfolded, whenever it becomes or *seems* to become necessary to have recourse to" x, y, z, &c.

Unfortunately like all who have been brought up on Cartesian food I now and then think of a as $-iSia$—&c. (Hamilton himself was a terrible offender in this way: his i, j, k, was almost a fatal blot on his system). But I *know* that I ought not to do so, *because* a better way is before me. Thus:

(P.S. What follows is, I see, Prosy. But it is necessary.)

Position is essentially relative (though *in physics* direction may be regarded as, in a sense, absolute) so we must have an origin. ρ then, or OP, I look on as $P - O$, the displacer which takes a point from O to P. Should it subsequently be displaced to Q we have

$$Q - O = (Q - P) + (P - O).$$

Hence all the COMPOSITION laws of Vectors. And of course the notion of repetition of any one displacer, so that we get the idea of the tensor, and of the unit vector.

To COMPARE vectors, we may seek their quotient, or the factor which will change one into the other. There are two obvious ways of looking at this.

(*a*) The first is mathematical rather than physical. Here we introduce the idea of a factor, such that

$$\sigma/\rho \times \rho = \sigma.$$

And we see at once that it consists (or may be regarded as consisting) of two independent and commutative factors, its Tensor and its Versor.

But then comes the life of the whole; the recognition of the fact that when POQ is a right angle, the versor of σ/ρ may be treated as in all respects lawfully equivalent to the unit vector drawn perpendicular to ρ and to σ. Thus every unit vector is a quadrantal versor, and conversely. And, further, every versor is a power of a unit vector. Thus, if the angle QOP be, in circular measure, A and if T be the unit vector above defined, we have

$$U\frac{\sigma}{\rho} = T^{2A/\pi} = \cos A + T \sin A.$$

Thus the separation of σ/ρ into the sum of its scalar and vector parts.

(*b*) The second is physical rather than mathematical. We think of vectors in a homogeneously strained solid, and if OP be strained into OQ we write $\sigma = \phi\rho$.

We recognise the conjugate strain ϕ' and see the criterion of the pure strain in $\phi = \phi'$, as well as the general relation

$$m\phi^{-1}V\rho\sigma = V\phi'\rho\phi'\sigma$$

where m is the factor by which volume is increased. Thus we have the linear and vector function, or (as you call it) the Matrix, with its fundamental characteristic.

Finally we have Nabla, which is defined by the equation

$$Sd\rho\nabla = -d$$

expressing total differentiation so far as the shift, or displacement, d is concerned.

In all this there is no reference whatever to anything Cartesian:—and no more need there be such in any application or development of these principles. And I have always not merely allowed but proclaimed that, in the eyes of the mathematician, Qns. have the fatal defect of being confined to Euclidian space. But this is one of their great recommendations to the physicist....

I should like to know at your convenience when and how the notion of the Matrix came to you:—and whether Hamilton's simple case of it was an anticipation or an application of the general theory.

In response to this, Cayley sent Tait an article he had written out for the *Messenger of Mathematics* on "Coordinates versus Quaternions," remarking in a covering letter, "I do not know what has made me write it just now, but it puts on record the views which I have held for many years past and which have not been before published."

He also expressed his dissatisfaction with Tait's sarcastic reference to "Trilinear Coordinates" in the Preface to his *Treatise on Quaternions*, and added in a postscript to his letter:

"I certainly did not get the notion of a matrix in any way through quaternions: it was either directly from that of a determinant; or as a convenient mode of expression of the equations

$$x' = ax + by$$
$$y' = cx + dy.\text{"}$$

In his reply Tait suggested that Cayley should communicate his note to the Royal Society of Edinburgh, and then continued :

Of course I do not agree with you, in fact we are as far as the poles asunder in regard to your main point. There we must continue to differ......

I scarcely think you do me justice in giving *without its context* the remark [in the Preface to Tait's *Quaternions*, 1st, 2nd, and 3rd editions] "such elegant trifles as Trilinear Coordinates." I think that you will see that the context very considerably modifies the scope of the remark: so much so, in fact, that while I am still of the opinion I expressed I am *not* prepared to use the phrase "elegant trifles" even about Trilinear Coordinates (of Quadplanar Coords. I said nothing) without some such qualification, or setting, as it has always had in my *Preface*.

Cayley replied :

CAMBRIDGE, 25*th June*.

Dear Tait

Thanks for your letter. I am quite willing that the paper should be read at the R.S.E.—did you mean also that it should be published in the Proceedings?— if you did I am quite willing to let this be done instead of sending it to the *Messenger*. Please make the reference to the preface of the 1st as well as the 2nd and 3rd editions— and make any additions or explanations to show the context of the "elegant trifles." I was bound to refer to quadriplanar coordinates, because the comparison is between Quaternions, which refer to three-dimensional space, and the Cartesian coordinates x, y, z or in place thereof the quadriplanar coordinates, x, y, x, w. Of course you see my point. I regard the trilinear or quadriplanar coordinates as the appropriate forms including as particular cases the rectangular coordinates x, y or x, y, z—and bringing the theory into connexion with that of homogeneous forms of quantics—and as remarked in my last letter, it is only in regard to these that the notion of an invariant has its full significance; so that trilinear coordinates very poor things, Invariants a grand theory, is to me a contradiction.

In a long formula of Gauss which you quote for its length, you give the expanded form of a determinant—the expression is the perfectly simple one

$$\begin{vmatrix} x'-x, & y'-y, & z'-z \\ dx & dy & dz \\ dx' & dy' & dz' \end{vmatrix} \div (\text{dist. } P, Q)^3$$

Do you put any immediate interpretation on $a^2 = $ scalar, or consider it merely as a necessary consequence of the premises?

Believe me, dear Tait, yours very sincerely

A. CAYLEY.

Cayley's paper *On Coordinates versus Quaternions* and Tait's reply *On the Intrinsic Nature of the Quaternion Method* were published side by side in the R. S. E. *Proceedings*, Vol. xx, pp. 271–284. As each of them expressed it in the correspondence, they differed fundamentally. Cayley thought in quantics and coordinates; Tait laid hold of each physical quantity as an entity for which the quaternion notation supplied the complete mental image.　To Cayley the quaternion of Hamilton was an algebraic complex which he and Sylvester regarded as a matrix of the second order.　For Tait the quaternion was a quantity obeying certain laws, and yielding by its transformations endless physical interpretations. These interpretations were of little interest to Cayley; just as the general solution of the linear matrix equation had small attractions for Tait.

It is a misfortune that in this remarkable correspondence on things quaternion Tait's letters to Maxwell have not been preserved.　Towards Hamilton Tait was the loyal disciple, eager to have the master's help at all stages, and always ready to give him the fullest credit as the prime source of every luminous thought.　This deep loyalty no doubt prompted Tait to destroy certain of Hamilton's later letters, which did not show the great man at his best.　To Cayley Tait turned as to the embodiment of mathematical wisdom and knowledge.　In spite of their fundamental difference of outlook on quaternion fields—a difference which gradually emerged as they corresponded on the subject—Tait had the greatest confidence in Cayley's mathematical intuitions.　Once when questioned as to his opinion of Cayley's discoveries in pure mathematics, he remarked :—"Cayley is forging the weapons for future generations of physicists."　But for Maxwell Tait had not only unstinted admiration as a man of science; he had for him a deep strong love which had its roots in common school life and grew, strengthened and ripened with the years.　He understood to the full Maxwell's intellectual oddities, his peculiar playful humour, his nobility of character, and the deeper thoughts which moved his mind but rarely found expression.　Maxwell's letters, which Tait preserved with the greatest care, imply an equivalence of correspondence on Tait's side, the general nature of which may in certain cases be guessed, but the exact terms of which are no longer accessible.　Just as Tait placed implicit confidence in Maxwell's physical intuitions, so Maxwell accepted the leadership of Tait in quaternion symbolism and interpretation.　He once playfully remarked that he envied Tait the authorship of the quaternion paper on Green's Theorem ; and the extracts given above indicate how much he was influenced by Tait when preparing his great work on Electricity and Magnetism. Tait's quaternion work was indeed the necessary precursor of the quaternion symbolism and nomenclature which Maxwell introduced into his book.

Tait brought out the real physical significance of the quantities $S\nabla\sigma$, $V\nabla\sigma$, ∇u. Maxwell's expressive names, Convergence (or Divergence) and Curl, have sunk into the very heart of electromagnetic theory. His suggested word Slope has been replaced by Gradient or Grad, a word of more general etymological intelligibility. But the point is that Maxwell was led to see the far-reaching importance of these conceptions only after they had been presented by Tait in their simple direct quaternion guise. Lamé, Green, Gauss, Stokes, Kelvin, and others had the ideas more or less disconnectedly in their minds and utilised them in analysis; but it is through Hamilton's calculus alone as developed by Tait that the important space relations, Gradient, Divergence, and Curl, appear as parts of a whole. It was Tait who taught Maxwell this deep-lying truth; and it was Maxwell who spread the good news by his epoch-making treatise on electricity. Most later workers have been content to take the names and the separate conceptions, and reject the central idea embodied in the quaternion operator ∇. It should not be forgotten, however, that these conceptions were first concisely symbolised and fully discussed in their physical significance by Tait, and remain as a rich legacy from him through Maxwell to the non-quaternionic world. Maxwell gave them names, "rough hewn" he called them in his letter to Tait, whom he invoked as a "good divinity" to "shape their ends properly so as to make them stick." He was their sponsor, but Tait was their parent. Probably very few now using these terms, or their equivalents, in electromagnetic literature have realised the debt they owe to Tait, who first polished the facets of the ∇ diamond. Rough and uncut it passed to him from Hamilton; and now all the scientific world more or less unconsciously benefit by its radiance. Here then is one outstanding result of Tait's quaternion labours.

The many vector quantities which call for consideration in modern electrical theory demand some form of vector notation. This was first realised by Maxwell, who, guided by Tait, adopted Hamilton's vector symbolism. Later writers have in many cases followed Maxwell in the spirit but not in the letter. There have arisen in consequence some six or seven distinct systems of vector notations, which are also called systems of vector analysis. The common elements in these rival systems are, with one exception, also common to the quaternion system, which is demonstrably a real analysis and not simply a notation. So far as mere symbolism is concerned, there is little to choose among these various systems. But whatever be the principle of notation adopted, whether a modified Hamiltonian or Grassmannian, these systems are used as Maxwell used quaternions. Tait inspired Maxwell to use the quaternion vector symbolism. All vector analysts follow Maxwell in substituting for the diffuse Cartesian symbolism a more compact and graphic vector notation. In imitating Maxwell they become disciples of Tait: and once more we realise the close historic connection between Tait's quaternion labours and the developments of modern vector analyses applied to physical problems.

It was indeed for the sake of physical applications that Tait made himself master of the quaternion calculus. The conditions under which his *Elementary Treatise*[1] was prepared have already been described; and in judging of the merits of the book, especially in its first edition, we must bear in mind that Hamilton's expressed wishes considerably tied Tait's hands. It was necessary for the sake of the student that the foundations of the calculus should be established in one of the several ways which Hamilton himself had already indicated. But Tait's aim, as indicated in the Preface to his *Treatise*, was to bring out the value of Quaternions in physical investigations.

In the earlier chapters (I refer at present only to the first edition of the *Treatise*) there was of course little scope for Tait to show any originality of treatment. The first chapter in which he began, as it were, to beat out his own path, was Chapter V, on the solution of equations. In discussing the properties of the linear vector function he followed a line suggested by Hamilton in one of his letters; but he followed it out in his own way. In Tait's eyes the linear vector function was a strain; and to a reader acquainted with the theory of strains it is abundantly evident that, even when explicitly confining himself to the purely mathematical side of the question, Tait had the strain conception vividly before his mind. In this early chapter he emphasised those properties which became all important in the later chapters on Kinematics and Physics.

The linear vector function continued to occupy Tait's attention throughout the remaining years of his life; and many interesting applications were added in the second and third editions of his book. These usually appeared, in the first instance, as notes to the Royal Society of Edinburgh. He never found time to put his investigations into the form of a complete memoir. All he was able to accomplish was a series of abstracts giving

[1] First edition, 1867; 2nd edition, 1873; 3rd edition, 1890.

the main results, and indicating various lines of investigation and gaps to be filled in by subsequent research. The last connected set of notes on the linear vector function began to appear in May 1896, six years after the publication of the third edition of the *Treatise*, and continued till May 1899 (*Sci. Pap.* Vol. II, pp. 406, 410, 413, 424). During his last illness Tait spoke several times of the importance of the linear vector function, regarding which he felt that some great advance was still to be made. On July 2, 1901, two days before his death, he was able to put some jottings on a sheet of paper, which he handed to his son asking him to place it in a safe place as it contained the germ of an important development. These notes were published in facsimile by the Royal Society together with a commentary in which I indicate their relation to previous papers.

The general aim of these later papers is to classify and analyse strains with special reference to related pure strains, those, namely, which are unaccompanied with rotation. One of the most interesting of the results is the theorem that any strain in which there are three directions unaltered can be decomposed into two pure strains; and, conversely, two pure strains successively applied are equivalent to a strain in which there are three directions unchanged but not in general at right angles to one another.

It is more particularly in reference to the development of the properties of ∇ that the second and third editions of the *Treatise* show marked advance upon the first. As regards the extent and variety of the physical applications the second edition is indeed an entirely different book. This was the edition which was translated into German by Dr G. v. Scherff (Leipzig, Teubner, 1880) and into French by Gustave Plarr (Paris, Gauthier-Villars, 1884).

There was a suggestion as early as 1871 to prepare a German translation to be published by Vieweg. But the project was not carried out. Helmholtz, in a letter to Vieweg of date November 19, 1871, spoke of Tait's *Quaternions* in these words:

"As regards Tait's Quaternions, it is indeed an ingenious and interesting mathematical book. But it uses a method of mathematical research which, so far as I know, has hardly been taken up in Germany. When perhaps some enthusiast (*Liebhaber*) working his way into it, will undertake the translation, the book will, I feel almost certain, find a great sale. It is keen and penetrating, full of new ideas and conceptions, and one can speak of its scientific value only with the highest recognition. But it lies somewhat removed from the usual paths of mathematical study, and the method has not as yet furnished new results of a kind to attract attention."

These words were spoken of the first edition, which was strictly, in accordance with its title, an Elementary Treatise. But it is in the later editions that Tait displays his strength. The two chapters on Kinematics and Physical Applications abound in numerous illustrations of the power and flexibility of the calculus. The last chapter, which extends to 101 pages in the third edition, passes over nearly the whole range of mathematical physics, from statics and kinetics of bodies through optics and electrodynamics to the series of remarkable sections dealing with the operator Nabla, ∇. Here we find treated gravitational and magnetic potential, hydrodynamics, elasticity, varying action, brachistochrones, catenaries, etc. These later sections are not easy reading. They suffer from what Maxwell playfully called " the remarkable condensation, not to say coagulation, of his style," and cannot be fully appreciated by a student who has not already made some acquaintance with the subjects taken up. It should be remembered, however, that this was exactly what Tait had in mind. His aim was not to write a quaternion treatise on mathematical physics, but to show forth the power and conciseness of the quaternion method when applied to important physical problems. With descriptive letter-press interpolated after the manner of scientific treatises and the details of the symbolism worked out, the last chapter of Tait's *Elementary Treatise on Quaternions* would form a most admirable text book for advanced students in applied mathematics. The greater generality of the quaternion attack as compared with the usual methods introduces some striking novelties, which the ardent student would do well to follow up.

APPENDIX TO CHAPTER IV

Maxwell's "Tyndallic Ode" was dedicated to Tait as the chief musician upon Nabla. As with several of Maxwell's clever rhymes, it was no doubt suggested by some of Tait's own utterances. It is at any rate certain that there would have been no Ode if there had been no Tait. Maxwell could not indeed write to his old school mate without indulging in some quaint fancy or hidden joke; and Tait was wont to respond in similar fashion. In this appendix I reproduce the original Ode as it was handed to Tait in the first instance; and then give a later letter, in which a new verse is added to the Ode, and in which other matters are touched on in Maxwell's inimitable way.

The Ode is a humorous imitation of the style of the popular scientific lecturer. Two copies were preserved by Tait. The first rough draft, consisting of four verses, was written in pencil on paper which bears the printed inscription " British Association, Edinburgh." It was evidently dashed off by Maxwell in the B. A. Reception Room during the Edinburgh meeting of 1871. The heading is

> "To the Chief Musician upon Nabla
>
> A Tyndallic Ode
>
> Tune—The Brook."

This was the version which appeared at the time in *Nature* (Vol. IV, p. 261), where it is spoken of as a paper read before the "Red Lions"!

The other copy, also in Maxwell's hand-writing, is in ink, and seems to have been written the same evening. It must be regarded as the true complete original with its seven verses, the first four of which show several well-marked textual improvements upon the earlier pencilled draft.

The peculiar interest of this copy lies in the heading which is elaborately written in Hebrew, in all probability by W. Robertson Smith, to whom the name of Nabla for the inverted Delta was due. This Hebrew title is after the manner of the Hebrew Psalter and is a literal translation of the title given in the first draft. The evidence for the personality of the Hebrew scribe is complete. It is recorded that T. M. Lindsay and W. R. Smith read a paper before Section A *On Democritus and Lucretius, A Question of Priority in the Kinetical Theory of Matter* (*B. A. Reports* 1871, *Transactions of the Sections* p. 30); and Principal Lindsay tells me that Robertson Smith was continually in the company of Tait and Maxwell during the Meeting of the British Association, that ∇ was the source of many jokes, and that there is little doubt the Hebrew inscription is from the "reed" of Robertson Smith.

The order of the verses is that indicated by pencil in the original, and differs from the order given in Lewis Campbell's *Life of Maxwell*. There are also some textual variants.

לַמְנַצֵּחַ עַל־הַגֶּבֶל לְנַחַת

מִזְמוֹר לִבְנֵי שָׁנְדָל שִׁיר:

I come from fields of fractured ice
Whose wounds are cured by squeezing,
They melt — they cool, but in a trice
Grow warm again by freezing.
Here, in the frosty air, the sprays
With fern-like hoarfrost bristle;
There liquid stars their watery rays
Shoot through the solid crystal.

2

I come from empyrean fires,
  From microscopic spaces,
Where molecules with fierce desires
  Shiver in hot embraces.
The atoms clash, the spectra flash,
  Projected on a screen
The double D, magnesian b
  And Thallium's living green.

3

We place our eye where these dark rays
  Unite in this dark focus;
Right on the source of power we gaze
  Without a screen to cloak us.
Next, where the eye was placed at first,
  We place a disk of platinum;
It glows, then puckers like to burst:
  By Jove, I'll have to flatten him!

4

I light this sympathetic flame
　My slightest wish that answers.
I sing, it sweetly sings the same,
　It dances with the dancers.
I shout, I whistle, clap my hands,
　I stamp about the platform,
The flame responds to my commands
　In this form and in that form.

5

This crystal tube, the electric ray
　Shows optically clean,
No dust nor haze within, but stay,
　All has not yet been seen.
What gleam is this of heavenly blue,
　What wondrous form appearing,
What mystic fish, what whale, that through
　The ethereal void is steering!

6

Here let me pause, these passing facts—
　These fugitive impressions
Must be transformed by mental acts
　To permanent possessions.
Then summon up your grasp of mind—
　Your fancy scientific,
That sights and sounds, with thoughts combined,
　May be of truth prolific.

7

Go to! prepare your mental bricks,
　Bring them from every quarter,
Firm on the sand your basement fix
　With best asphaltic mortar.
The pile shall rise to heaven on high
　To such an elevation
That the swift whirl with which we fly
　Shall conquer gravitation.

$$\frac{dp}{dt}.$$

The following letter written to Tait immediately after the Belfast meeting of the B. A. in 1874, when Tyndall delivered the presidential address, gives an additional verse to the Ode as well as other quaint imaginings.

GLENLAIR,
27th Aug. 1874.

O. T'. B. A. Trans. 1874.

In the *expected* presidential address the following has been inserted as an antipenultimate.

On the atmosphere as a vehicle of sound.

What means that thrilling drilling scream!
  Protect me! 'tis the Siren—
Her heart is fire! her breath is steam!
  Her larynx is of iron!
Sun! dart thy beams. In tepid streams
  Rise, viewless exhalations!
And lap me round, that no rude sound
  May mar my meditations.

Phil. Trans. 1874, p. 183.

Notes of the actual address are enclosed[1].

The effect on the British Ass. is described in the following adaptation of H. Heine.

Tune—Loreley.

I know not what this may betoken
  That I feel so wonderful wise,
The dream of existence seems broken
  Since Science has opened mine eyes.
At the British Association
  I heard the President's speech,
And the methods and rules of creation
  Seemed suddenly placed in my reach.

My life's undivided devotion
  To Science I solemnly vowed,
I'd dredge up the bed of the ocean,
  I'd draw down the spark from the cloud;
To follow my thoughts as they go on
  Electrodes I'd plunge in my brain,
Nay, I'd swallow a live entozoon
  New feelings of Life to obtain.

O where are those high feasts of science?
  O where are those words of the wise!
I hear but the roar of Red Lions,
  I eat what their Jackal supplies.
I meant to be so scientific,
  But science seems turned into fun;
And this with their roaring terrific
  These old red lions have done.

[1] This refers to the clever rhyming epitome of Tyndall's address, which appeared at the time in *Blackwood's Magazine*, and will be found in Lewis Campbell's *Life of Maxwell.*

The following instance of domestic evolution was submitted to Mr Herbert Spencer who was present in Section A.

"The ancients made enemies saved from the slaughter
Into hewers of wood and drawers of water;
We moderns, reversing arrangements so rude,
Prefer ewers of water and drawers of wood."

Mr Spencer in the course of his remarks regretted that so many members of the Section were in the habit of employing the word Force in a sense too limited and definite to be of any use in a complete theory of evolution. He had himself always been careful to preserve that largeness of meaning which was too often lost sight of in elementary works. This was best done by using the word sometimes in one sense and sometimes in another and in this way he trusted that he had made the word occupy a sufficiently large field of thought. The operations of differentiation and integration which appeared, from the language of previous speakers, to be already in some degree familiar to members of the Section, were, he observed, essential steps in the normal progress of evolution. It gave him great pleasure to learn that members of Section A were now turning their attention to these processes. He was also glad to see how entirely the Section concurred with his view of nervous action as a wave of accumulation, and he hoped they would also direct their attention to the mode in which the exhausted nerve recuperates its energy by absorption of heat from the neighbouring tissues which form its environment. In Professor Tait's new edition of his work on Thermodynamics he had no doubt this subject would be ably treated.

Mr Spencer, whose speech was throughout one of the most didactive, exhaustive and automatic efforts ever exerted, then left the Section.

In Section B, Prof. W. K. Clifford read a paper on Chemical equations[1]. The equation

$$XX + LL = 2(XL)$$

was the first selected. He observed that both the constituents of the left member were in the liquid state and that though the resultant might not be familiar to some members, he could warrant it $2XL$. From an equation of similar form

$$H_2 + Cl_2 = 2HCl$$

he deduced by an easy transformation

$$H^2 - 2HCl + Cl^2 = 0,$$

whence by extracting the square root

$$H - Cl = 0 \text{ or } H = Cl,$$

a result even more remarkable than that obtained by Sir B. C. Brodie.

$$\frac{dp}{dt}.$$

[1] Clifford's paper *On the General Equations of Chemical Decomposition* was read before Section A; but the Title only is given in the B. A. *Reports* (1874). An abstract appeared in *Nature*, Sept. 24, 1874, and was reprinted in the Preface to the *Mathematical Papers*, p. xxv. Maxwell's amusing parody has a striking superficial resemblance to the original.

CHAPTER V

THOMSON AND TAIT

"T and T'" or Thomson and Tait's Natural Philosophy.

THE publication of Thomson and Tait's *Natural Philosophy* was an event of the first importance in the history of physical science. No more momentous work had been given to the world since the days of the brilliant French mathematicians, Laplace, Lagrange, and Fourier. Thoroughly familiar with the mathematical methods invented and developed by these great writers, Thomson and Tait conceived the project of an all-embracing treatise on Natural Philosophy, in which physical conceptions and mathematical analysis would be rationally blended in an harmonious interpretation of the phenomena of Nature. The intention, it is true, was realised only in part. The first volume appeared in 1867, and a second edition greatly enlarged was issued, Part I in 1879, Part II in 1883. But in the Preface to Part II the authors announced that "the intention of proceeding with the other volumes is now definitely abandoned." No reasons were given; simply the fact was stated.

Fortunately there was no longer the same necessity for a continuation of the work on the extensive scale originally imagined. Since the appearance of the first edition, other important works had been published covering a large part of the domain of Natural Philosophy. Clerk Maxwell's *Electricity and Magnetism* and Lord Rayleigh's *Theory of Sound* were the most conspicuous of these; and Thomson's own *Reprint of Papers on Electrostatics and Magnetism* supplemented in a striking manner the doctrines inculcated in the *Natural Philosophy*. In all treatises published since its appearance the impress of "Thomson and Tait" is clearly seen. Nevertheless, the world of science must ever lament that the two Scottish Professors did not put in type the sections on Properties of Matter, frequently mentioned in the first edition, and usually with reference to a particularly attractive part of the subject.

Occasionally in conversation Tait would refer to the manner in which the great work, familiarly known as "T and T'," took form and grew, and to the amusing difficulties which frequently arose, especially when proof-sheets were mislaid. Some of the earlier reminiscences have been fortunately preserved as contemporaneous history in Tait's letters to Andrews. Those letters were all kept, and through the kindness of the Misses Andrews I am able to give in Tait's own language the genesis and early development of the *Natural Philosophy*.

The first quotation is taken from a letter of date Dec. 18, 1861:

I told Slesser [Tait's successor at Belfast] to tell you that I had agreed to write a joint book on Physics with Thomson. In fact I had nearly arranged the matter with Macmillan, when Thomson, to my great delight, offered to join.

We contemplate avoiding the extreme details of methods which embarrass the otherwise excellent French books (*vide* Jamin, Daguin, etc.) and which, though they may have led their authors to results, are not those that would generally be used in verification. Also we propose a volume, quite unique, on Mathematical Physics. I know of no such work in any language—and in fact have acquired all my knowledge of the subject by hunting up papers (often contradictory, and more often unsatisfactory) in Journals, Transactions, Proceedings, etc. Such a book is one I would willingly have paid almost any price for during the last ten years—but it does not yet exist. And I think that Thomson and I *can* do it.

We shall commence printing as soon as we have made arrangements with the Publisher, for our first two volumes will contain simply the essence of the Glasgow and Edinburgh Experimental Lectures blended into (I hope) an harmonious whole. A little difficulty arises at the outset, Thomson is dead against the existence of atoms; I though not a violent partisan yet find them useful in explanation—but I suppose we can mix these views well enough...

The incidental remark as to Thomson's disbelief in atoms reads strange in these days when we recall how much of Kelvin's later work had to do with the ultimate constitution of matter[1]. For example, before the decade was finished the Vortex Atom, which was suggested to Thomson by Tait's smoke-ring illustrations of Helmholtz's theory of vortex motion (see above, p. 69), had been launched on its chequered voyage in the sea of molecular speculation.

Through the kindness of Lady Kelvin I am able to give the following extract from a letter from Tait to Thomson of date Jan. 6, 1862:

I like your draft index to Vol. I very well. I have made a few insertions in it, and may perhaps make more before I send it back redrafted, which I will soon

---

[1] See however Thomson's paper on "The Nature of Atoms" (*Proc. Manchester Lit. and Phil. Society*, 1862) quoted in Larmor's *Aether and Matter*, p. 319.

do. Meanwhile I think we may tell Macmillan that the illustrations will in the main be diagrams and not wood engravings (i.e. sketches)......

You see [from an enclosed letter] the information he wants as to advertising. I wish you would give me a hint or two of your likings and dislikes on such a delicate point. If we can settle on the nature of, and constitution of, our *Preface*, I think a sort of précis of it would do very well for Advertisement. I think also that we might begin even now to point out as looming in the future our *Great* work, which so far as I know will be *unique*; of course I mean the *Principia Mathematica*, whatever be the title it is to bear. We may gain considerable credit, and perhaps profit, by the present undertaking; but the other will go over Europe like a statical charge. Don't you think it would be prudent to warn the profane off such ground by a timely notice?

Such as this

In preparation, by the same authors,

A MATHEMATICAL TREATISE ON NATURAL PHILOSOPHY,

containing the elements of the mathematical treatment of Elasticity, Capillarity, Electricity, Heat etc., etc., or anything tending to such a purpose....

A fortnight later (Jan. 20, 1862) Tait detailed to his Belfast friend more concerning the coming book. In reply to a demand for information regarding Heats of Combination, Andrews had referred to the discrepancies between his measurements and those made a little later by Favre and Silbermann; and Tait replied in his turn:

My immediate occasion for information on Heat of Combination (for my lectures) is over, but I am sure Thomson and myself will have particular pleasure in putting you right with regard to Favre and Silbermann, etc. But as matters are at present arranged—that will be in our *second* volume. I will give you here a short index of the proposed Vol. I, with which we are busily engaged. I may merely tell you that I don't feel any alarm on the point you mentioned in your first letter—Thomson has thought far longer, and far more deeply, about matter than I have.

The major part of the writing will be done by me as Thomson feels a repugnance to it which is not common. I have already sent him two chapters, and a general abstract of a *section*; and he speaks of them in the very highest terms....

Here we are, Vol. I.

Section I. Chap. I.   Introductory.

II.   Matter, Motion, Mass, etc.

III.   Measures and Instruments of Precision.

IV.   Energy, Vis viva, Work.

V.   Kinematics.

VI.   Experience (Experiment and Observation).

Section II. Abstract Mechanics (*Perfect* solids, fluids, etc.).

      Chap. I. Introductory (I have written this and will let you see it soon).

          II. Statics.

          III. Dynamics (Laws of Motion, NEWTON. Did you ever read his Latin? Do).

          IV. Hydrostatics and Dynamics.

Section III. Properties of Matter, Elasticity, Capillarity, Cohesion, Gravity, Inertia, etc., etc. (This is to be mine.)

Section IV. Sound.

Section V. Light.

This will give you as good an idea as I yet possess as to the contents of our first volume. All the other physical forces will be included in Vol. II, which will finish up with a great section on the *one* law of the Universe, the Conservation of Energy. No mathematics will be admitted (except in notes, and these will be more or less copious throughout the volume, being printed in the text but in smaller type). But we shall give very little in that way as my great object in joining Thomson in this work is to have *him joined to me* in the great work which is to follow, on the Mathematics of Nat. Phil., which I do not believe any living man could attempt alone, not even Helmholtz.

On September 9, of the same year, when Thomson seems to have been holidaying in Ireland, Tait wrote to Andrews:

Pray impress on Thomson that he should get home again as soon as possible and get into harness else we cannot begin printing in October as was arranged. I think the *first* chapter, at least my part of it (for I have not got Thomson's yet), will please you. It has greatly pleased myself. It is all about Motion, Actual and Relative, and such matters as Rotations, Displacements, etc., and I hope to make the large type part of it intelligible even to savages or gorillas.

It thus appears that, during the few months intervening between the dates of these letters, the plan of the book had somewhat altered. Instead of being Chapter V, Kinematics is to form Chapter I. This was the arrangement finally adopted.

Tait at once began to prepare his manuscript. On the fly-leaf of one of the large quarto volumes of blue tinted paper which he used in the early days for lecture notes and all kinds of scientific work we find the inscription "Mainly written in 1861–2 for T and T' (rough beginning)," and then below "Since used (1885) for K. T.[1] of Gases." Pasted to the fly-leaf are the two halves of a sheet of foolscap containing a table of contents similar to but differing in detail from the scheme sent in the

---

[1] That is, "Kinetic Theory."

letter to Andrews quoted above. The manuscript proper begins, however, with Kinematics on page 1 ; and on assigned pages throughout the book introductory sentences on other branches of the subject are given in Tait's clear, strong hand-writing. The paragraphs on kinematics are the fullest and in many cases are the very paragraphs which appear almost verbatim in the published pages of "Thomson and Tait."

Page 21 of this MS book is reproduced on the opposite page slightly reduced in size in the ratio of 23 : 30. It will be seen to correspond very nearly word for word with portions of paragraph 48 in the *Treatise*, and is given in illustration of the remarks just made, and also as an excellent example of the legibility of Tait's manuscript.

In another similar volume marked with the same date 1862 there is a well planned series of paragraphs on the Properties of Matter, which no doubt were intended to be the large type portion of Division III of the second volume referred to so pointedly in the Preface to the First Edition. These sections, although never printed as part of "T and T'," were afterwards utilised by Tait in his book, *Properties of Matter*.

The same volume also contains the original draft of the sections on Experience and on Measures and Instruments. Here again many of the sentences are exactly as they appear in the *Treatise*, "T and T'."

The name "T and T'" was applied by the authors themselves long before there was any hope of the book being published ; and from 1862 onwards till 1892, when "Kelvin" displaced "Thomson," T and T' were the usual forms of address and signature in their letters.

The book progressed slowly. That it progressed at all was due to Tait's never flagging energies and determination. The original plan of preparing a somewhat elementary work on Natural Philosophy to be followed by a treatise on Mathematical Physics was ultimately given up, and the *Treatise* when it appeared in 1867 was a kind of combination of the two types of book at first conceived.

Some notion of how the book took shape may be gathered from the following extracts from letters written by Tait to Thomson. Writing on March 30, 1863, Tait said :

I think you are unwise in your suggested alteration of the Book. Attractions come naturally and nicely in Prop. Matt.—but *not sooner*[1]. The fact is, we have

[1] It is interesting to note that Tait, in his MS book on Properties of Matter, introduces sections on potential after the account of the Cavendish Experiment.

Yours truly
P. G. Tait.

¶ As an excellent instance of relative motion we may consider that of the moving point with reference to the other in ¶ . And since to do so we must impress upon each in addition to its previously velocity, another equal & opposite to that of the second point, we see at once that the problem becomes the same as the following. A boat crossing a stream is impelled by the oars with uniform velocity and always towards a fixed point in the opposite bank — but it is also carried down stream at a uniform rate. Determine the time of crossing, and the path described. Here, as in the former, there are three cases figured below. In the first the boat, moving faster than the current, reaches the desired point — in the second, the velocity of boat & stream being equal the boat gets across {after an infinite time} describing a parabola, but does not land at the desired point. In the third case its proper velocity being less than that of the stream, it never reaches the other bank, and is carried indefinitely down stream. It is of course understood that we omit here all reference to resistance of fluids &c. so that {the problem is a purely imaginary one in its new form.} We leave this as an exercise, merely noting done so much small print (per cent.) in what is already printed, that we must defer further violent manifestations of such a tendency to a later stage—excepting always what *must* come in Hamiltonianism and birds of a like feather....

As to the R. S. E. I shall give notice of two papers—one by you on certain kinematical and dynamical theorems, another by myself on a certain most important quaternion transformation.... The quat. theorem in question is this—and is really of the greatest value (as witness all your, Maxwell's, Neumann's, etc., attempts at potential function expressions for distributions of magnetism, electricity static or kinetic, etc., etc.),

$$- \nabla^2 = \left(\frac{\partial}{\partial x}\right)^2 + \left(\frac{\partial}{\partial y}\right)^2 + \left(\frac{\partial}{\partial z}\right)^2,$$

$$S. \delta \rho \nabla \left(\frac{V a \rho}{T \rho^3}\right) = S. a \nabla \delta \left(\frac{1}{T \rho}\right),$$

which contains all about potential and will go into the Book in splendid style.

P.S. Send the Sph. Harcs. *soon* and let me see Digitalis[1] on Newton as connected with Energy before a fortnight is out as I shall be arranging my lecture on Energy for the R. S. E.

Two days later Tait continued:

6 G. G. E.
1/4/63.

Dear T.,

You have (no doubt) got the R. S. E. billet, and you have seen that I am responsible for some "Kinematical & Dynamical Theorems" supposed to be sent by you.

I shall give them Twist—but you ought to send me, *by Monday morning at latest*, a few sentences upon Greed and Laziness which will make the affair more complete.

Kirchhoff pitches into you & Stokes in the last Pogg. & also in the April Phil. Mag., while in the latter T″ [i.e. Tyndall] lets me off without comment!!!

Yrs. Ever,
T′.

P.S. Answer by *instant return of post*. Is there to be a small volume by October or not? Our Coll. Calendar is printing, and my enunciation is wanted on Monday—so *say at once* & I shall put T & T′ (small) instead of Goodwin.

The kinematical and dynamical theorems referred to here were duly published in very brief form in the *Proceedings R. S. E.*, Vol. v. They constitute the nuclei of several paragraphs in the *Treatise*. The few sentences on Twist become §§ 120–123; and the dynamical theorems, which

---

[1] *Possibly* a complex pun: Digitalis = poison = Poisson. The lecture referred to was an address before the Royal Society of Edinburgh on the Conservation of Energy. An outstanding feature was the discussion of Newton's Second Interpretation of Lex III (see below p. 191); and it is possible that Thomson may have remembered a reference to it in one of Poisson's many memoirs.

are simply enunciated in the R. S. E. paper, become §§ 311 and 312. Tait nicknamed these theorems Greed and Laziness, namely,

(1) *Greed:* Given any material system at rest, and subjected to an impulse of any given magnitude and in any specified direction, it will move off so as to take the *greatest amount* of kinetic energy which the specified impulse can give it.

(2) *Laziness:* Given any material system at rest. Let any parts of it be set in motion suddenly with given velocities, the other parts being influenced only by their connection with those which are set in motion, the whole system will move so as to have *less* kinetic energy than belongs to any other motion fulfilling the given velocity conditions.

The postscript indicates that there was still some hope of publishing the more elementary book before the other.

The following letter touches on a variety of subjects more or less connected with the book:

6 G. G. E.
24/4/63.

Dear T.,

I have been working up my Quat. article for Proc. R. S. E. I find that in your "Mechanical Representation of Electric etc. Forces" you need never have taken the *rotations* of the solid—you can always get *displacements* suiting any of the forms of force, and they are in fact simpler than yours as given in the Math. Journal. I will send you a proof to-morrow so that you may advise on it before it goes to press.

The whole mystery of Electromagnetism lies in the operator $S + \frac{1}{2} V$, as I pointed out to Sir W. R. H. four years ago—but I cannot get a closer insight, at least at present....

Also Stewart and I have simultaneously struck on an idea of which "Conservation of Energy" is a particular case—and we intend to develop from it some tremendous consequences—numbers of which have already been booked and talked over. A *third* volume of the book will be required for it, if it do not in fact destroy the necessity for the first two. It is a gushing, gasping, idea. But, more anon. As it will at once obliterate everybody who pretends to science but is not acquainted with it, we propose shortly to initiate *you*, in order that you may not be lost in the stramash that will follow its publication.

Gray promised me Sheet 7 (with the cuts) to-night—but it has not come yet, and of course won't to-night. Look out for it on Monday.

Also look sharp to Chap. II for I have no doubt you see the propriety of putting Sph. Harcs. & Quatns. in an appendix at the end of the Vol.

Ponder this.

In Statics and Kinetics we should in the large volume keep to *general* theorems and leave Balance and Screws and Atwood's Machines, etc., for the small book—I mean in the *Second* Division of the Book—Abstract Dynamics....

These extracts together with the letters which have been published in Prof. Sylvanus Thompson's *Life of Lord Kelvin* (Vol. 1, pp. 453–464), show that the final plan of the book must have been agreed upon late in 1862 or early in 1863. Tait had already written a first draft of large parts of the book; but Thomson could not be got to apply himself steadily to the task. Like a sleuth-hound Tait was always on his colleague's tracks begging for copy or for proofs. When in May 1864 Thomson visited Creuznach for the sake of his wife's health, he received the following letter, whose immediate purpose was to bring him to his desk and pen.

6 G. G. E.
20/5/64.

Dear T.,

Send *something* for #. I have only written the examples *twice* yet, and the Kinetics of a particle but *once*....

Rather a curious thing to-day. Thunderstorm. Divided-ring (very low charge) going, and discharging every two minutes, when a thunder-cloud was close, by a spark from a fine wire 1 mm. from the electrode. Abscissae G. M. T. (as supplied by Her M. A. Royal for Scotland). Ordinates—potential at class-room window—1 inch = 500 zinc-copper pairs.

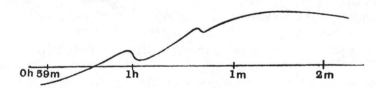

Note the effect of the gun[1]—1st on cloud north of the Coll.; then (after more than 30 s.) on one to the south....

When you ARE at Xnach, I hope you won't spend ALL your time with Dellmann, but will devote at least 4 h. per day to Kinetics of Rigid (about point) and Elastic Bodies (including Fluids). If you *do*, we can easily appear in August, and have the S. B. [= small book] ready for November.

I send you a proof of my Energy[2] signed for press—the publishers want me to make a little volume of it and its predecessor (with mathematical developments). I said I would be guided by you and E. *They* said Engineers had urged them to get it done. Ponder O. T. and C. E. on this, as well as on other subjects[3].

[1] The Edinburgh Time Gun is fired every day, except Sunday, at 1 p.m. Greenwich Mean Time.

[2] One of the articles in the *North British Review*, which finally became the second chapter of Tait's *Sketch of Thermodynamics*.

[3] The initials are difficult of interpretation. They *may* refer to Rankine the well-known Professor of Civil Engineering in Glasgow University. Possibly C.E. means see Engineers.

It is now clear that there is no hope of the small book appearing first.

The references in these letters to Quaternions show that there was in the early days an intention to have an appendix on quaternionic treatment of physical problems. There were obvious practical difficulties in the way of introducing a mathematical method which in spirit and notation was so unlike that in ordinary use. It could not have been done without a preliminary statement of the purely mathematical laws of combination and transformation. Then again Thomson was not merely not interested but emphatically antagonistic. In a letter to Cayley in 1872, quoted above (p. 152), Tait mentioned incidentally that Thomson "objects utterly to Quaternions"; and Kelvin in 1901 in a letter to Chrystal wrote regarding Tait:

"We have had a thirty-eight year war over quaternions. He had been captured by the originality and extraordinary beauty of Hamilton's genius in this respect; and had accepted I believe definitely from Hamilton to take charge of quaternions after his death, which he has most loyally executed. Times without number I offered to let quaternions into Thomson and Tait if he could only show that in any case our work would be helped by their use. You will see that from beginning to end they were never introduced."

The implication here is that Tait was unable to show that the use of quaternions would be of any advantage. It should be remembered, however, that Tait's own most important applications and developments were of later date than the years preceding 1867 when "T and T'" was published; and that, although Tait was himself fully convinced of the value of quaternions as an instrument of research, it was a very different matter to get Thomson to look at the matter from his point of view. Rather than risk the collapse of the whole enterprise Tait relinquished what must have been at one time a keen desire.

Meanwhile the book grew like an organism. With the exception of appendices and parts introduced bodily from Thomson's published papers, the manuscript first supplied to the printer was largely contributed by Tait. Then, in proof, certain paragraphs, whose subject matter seized the mind of Thomson, developed in a wonderful way. New sections and extensions of old sections were added, and were altered, pruned and expanded by both writers, until it was difficult to say how much was due to each.

When the *Treatise* finally appeared in 1867, after six years' preparation, it was at once seen to be a great work; and within two years the edition was sold out. The reviews were all highly complimentary, partly perhaps for the reason that the ordinary reviewer very soon found himself out of his depth. The *Athenaeum* of Oct. 5, 1867, thus chronicled its advent:

"A professor of Glasgow and one of Edinburgh, both well known to the world of science, have produced 750 pages as a first volume only. We defer description until we see the whole. The mathematical part is of a formidable character, and of the most modern type. The authors are thoroughly up to their subject, and have strong physical as well as mathematical tastes. The size of the volume is the fault of the subject and not of the authors, who have, so far as we have looked closely, kept down details. If anything they have not sufficiently diluted the mathematical part with expanded demonstration. But what of that? The higher class students for whom this work is intended are rats who can gnaw through anything: though even *their* teeth will be tried here and there, we can tell them."

The last sentence suggests the touch of De Morgan.

The *Engineer* of Nov. 1, and the *Medical Times and Gazette* of Nov. 16, 1867, both journals of scientific standing, described at some length the contents of the book, the latter ending with the sentences:

"Should the three succeeding volumes at all come up in value to the present one, Thomson and Tait's Natural Philosophy will deserve to take place with Newton's *Principia* and Laplace's *Mécanique Céleste*. This is strong language, but not too strong."

A long and somewhat discursive review appeared in the *Scotsman* of Nov. 6, 1868, fully a year after the publication of the *Treatise*. The reviewer was, however, keenly alive to the real merits of the contribution made by Thomson and Tait to the scientific literature of our time. Referring to the authors he wrote:

"They are to a certain extent a happy complement of each other—the one being deeply speculative but slightly nebulous in the utterance of his original thoughts, as often happens with profound thinkers; the other, though not deficient in originality, being clear, dashing, direct and practical.... What they exactly know they state in a plain intelligible manner.... What they do not know they do not pretend to explain....

"The authors zealously adhere to Newton, and they restore his methods and doctrines where they can—not without reason. They do not much pretend to originality—indeed they do not pretend to it at all. They quite openly and frankly lay hold of every mathematician or philosopher who has anything useful to them, and pillage him outright—always, of course, with the most handsome acknowledgements. Lagrange, Laplace, Fourier, Euler, Gauss, Joule, d'Alembert, Liouville, and the Irish (at least not the Scotch) Sir W. Hamilton are all laid under contribution, and the hard heavy scientific slab which each other had dug for himself out of the big bottomless quarry of Nature, as his personal title to immortality, is looked at carefully, found to be suitable, seized hold of, and at once, after a few chips and modifications (if necessary) laid into this big pyramid, in the climbing of which the youth of this country will hereafter have free scope for trying their strength, and looking down from the sides and the top of it upon this puzzling intricate physical world....

"The world of which they give the natural philosophy is real and not ideal. It is not the abstract world of Cambridge examination papers—in which matter is perfectly homogeneous, pulleys perfectly smooth, strings perfectly elastic, fluids perfectly incompressible....—but it is the concrete world of the senses, which approximates to but always falls short alike of the ideal of the mathematical as of the poetic imagination."

The review finishes with a reference to the doctrine of the dissipation of energy, and a mild expression of wonder as to what the " natural theologian " will make of it.

To a real student of physical science conversant with the text-books of the middle of last century Thomson and Tait's *Treatise on Natural Philosophy* must have come as a revelation. Instead of the usual approach through Statics, with Duchayla's proof of the parallelogram of forces, he found himself introduced to what the authors called Preliminary Notions, arranged in four chapters under the titles Kinematics, Dynamical Laws and Principles, Experience, Measures and Instruments. In opening with the subject of Kinematics or the Geometry of Motion, Thomson and Tait followed to some extent the example of certain recent French writers, such as Delaunay and Duhamel; but, in thus discussing motion and displacement apart from dynamical relations, they carried out the idea much more thoroughly than had ever been done by their predecessors. In other English books of the day there was absolutely nothing like it. In the first few paragraphs the geometrical conceptions of curvature and tortuosity are treated in a novel and elegant manner as illustrative of motion along a curve or line of changing direction. Even the familiar quantities, velocity and acceleration, are discussed in a fresh way ; but it is in the later sections of the first chapter that originality of treatment rivets the attention of the careful reader. Simple Harmonic Motion, which is the foundation of all kinds of wave-motion, is the avenue by which Fourier's Theorem is approached. Composition of angular velocities, the rolling of curves and surfaces on one another, curvature of surfaces, etc., etc., fall into line as the subject develops. An altogether unique part is to be found in §§ 135 to 138 inclusive (somewhat extended in the second edition).

24—2

Once, in conversation with Kelvin, I learned incidentally that these sections had been inspired by Bolzani, professor of Mathematics in Kasan. Writing subsequently for fuller information, I received from him the following letter of date February 11, 1904:

"I learned a good deal from Bolzani in the course of a few days that he stayed with us in Arran, some time about 1860[1]. He gave me a very clear statement regarding Gauss' *curvatura integra* in relation to a normal through the boundary, which I had never before seen in print or learned otherwise. He did not give the name Horograph, but with the aid of my colleague Lushington I devised it and put it into T and T'. I have never seen it elsewhere since that time."

The discussion of the curvature of surfaces leads naturally to the question of flexibility and developability of surfaces; and then follows an important series of paragraphs on strain or change of configuration. A comparison of the sections in "T and T'" with the corresponding sections in Tait's *Quaternions*, which was published in the same year, is very instructive in the light of what has just been said regarding Thomson's attitude to Quaternions. In a few lines the quaternion author gives what in the larger book requires paragraphs and even pages. Tait had a great liking for the theory of strains, and usually discussed it in considerable detail in his Advanced Class lectures.

The last sections of the first chapter deal with degrees of freedom, and conditions of restraint—a subject for which Kelvin had a particular fondness. Probably no instrument of any importance was ever constructed by him which did not contain some neat example of geometrical constraint. The illustrations given in the *Treatise* are fundamental and far-reaching.

True to their general plan the authors finish the chapter on Kinematics with a brief discussion of generalised components of position, velocity, and acceleration, preparing the way for the great dynamical developments of the next chapter.

In an important appendix, Thomson and Tait give (*a*) an Extension of Green's Theorem, (*b*) Spherical Harmonic Analysis. In the latter of these, which covers 20 pages in the first edition (extended to 48 pages in

---

[1] The date was probably 1862; for among Tait's correspondence is a letter from Bolzani of date November 2 of that year, written from Hull on board the "Volga" as he was leaving for home. In this letter he thanked Tait for various kindnesses and for a copy of a quaternion paper. He also asked Tait to get for him as early as possible a copy of Hamilton's new work on quaternions.

the second edition) new ground is broken in the treatment of what had till then been called "Laplace's Coefficients."

In a letter to Cayley of date February 4, 1886, Tait makes a curious remark which touches on the genesis of this remarkable appendix. He had been poking fun at Sylvester and Thomson for their raptures over new ideas, and concluded:

> Thomson, about 1860, used to lay down the law that the smallest *experimental* novelty was of more value than the whole of *mathematical* physics. Then he met with an accident which prevented his experimenting for a whole winter. During this period he became acquainted with Spherical Harmonics, and then his fundamental dictum was wellnigh reversed!

Tait of course is having his joke; but the facts implied are probable enough. Tait's intimacy with Thomson began in 1860 before his appointment to the Edinburgh Chair. Thomson was at the time very busy with electrical work largely the outcome of his labours in furthering the development of ocean telegraphy; and Tait must have been impressed with the eagerness of his new friend in all kinds of experimental work. Then came the accident which lamed Thomson for life; and necessity forced him for a time to devote his energies to mathematical investigations in the famous green-backed note books which were ever afterwards his inseparable companions. It is not improbable that the new treatment of spherical harmonics was begun and to some extent developed during these months of 1860–61.

In Maxwell's letters to Tait there are several passing references to this section of the "Archiepiscopal Treatise" as Maxwell playfully called it, in humorous reference to the fact that the Archbishops of York and Canterbury were at the time also Thomson and Tait. Thus on Dec. 11, 1867, he wrote:

> "I am glad people are buying T and T'. May it sink into their bones! I shall not see it till I go to London. I believe you call Laplace's Coeffts. Spherical Harmonics. Good. Do you know that every Sp. Harm. of degree n has n axes? I did not till recently. When you know the directions of the axes (or their poles on the sphere) you have got your harmonic all but its strength. For one of the second degree they are the poles of the two circular equipotential lines of the sphere. I have a picture of them."

Again, on 18 July, 1868, he wrote:

> "How do T and T' divide the Harmonics between them? I had before getting hold of T and T' done mine for electricity, but I should be delighted to get rid of

the subject out of the book except in the way of reference to T and T'. My method is to treat them as the neighbourhood of singular points in potential systems, those of positive degree being points of equilibrium, and those of negative degree being infinite points."

And then followed a brief outline of the method subsequently developed in *Electricity and Magnetism*, in which the harmonic of the *n*th degree is determined by its *n* axes and a constant fixing its strength.

Thomson and Tait's treatment of Spherical Harmonics is essentially physical, their object being—to quote their own phrase—"the expression of an arbitrary periodic function of two independent variables in the proper form for a large class of physical problems involving arbitrary data over a spherical surface, and the deduction of solutions for every point of space." It is this object which guides them in their analytical work; and through it all it is abundantly clear that the theory of the potential is ever present to their minds. In no true sense can the appendix be regarded as a sustained piece of mathematical reasoning. The convergency of the series is practically assumed, or, let us say, left to be proved by the reader. But the combined mathematical power and physical intuition are shown at every stage; and the use of the imaginary linear transformation, a distinct novelty in 1867, leads to an elegant and simple deduction of useful forms. Further on in the book in the sections on statics the authors give other useful developments; and it is then that they introduce their names, Zonal, Sectorial, and Tesseral Harmonics, according to the character of the nodal circles on the spherical surface.

Passing on to Chapter II of Division 1 we find it devoted to the discussion of Dynamical Laws and Principles. One feature of the early sections of this chapter is specially emphasised by the authors: it is a return to Newton. This means in the main two things. In the first place Newton's Laws of Motion are given in Newton's own words and the whole fabric of dynamics is raised on them as the sufficient foundation. In the second place, by adopting the Newtonian definition of force as being measured by the change of motion produced, Thomson and Tait get rid of the wearisome proof of the parallelogram of forces which was one of the marked features of the text-books of the middle of last century. In fact, as Newton showed, the composition and resolution of concurrent forces follow immediately from the second law of motion. Tait frequently remarked that Thomson and he "rediscovered Newton for the world." They also seem to have been the first to point out clearly the significance of Newton's second interpretation of his Third Law. I have heard Tait tell the story of the search after this interpretation. "The Conservation of Energy," he said to Thomson one day, "must be in Newton somewhere if we can only find it." They set themselves to re-read carefully the *Principia* in the original Latin, and ere long discovered the treasure in the finishing sentences of the Scholium to Lex III.

Considerable portions of the earlier sections of the dynamical chapter are, as they themselves point out, simply paraphrased from Newton. But a marked feature of the discussion is the introduction of a new terminology at once precise and suggestive. Moreover, old words are used with clearly defined meanings, and never used except with these meanings. The Conservation of Energy occupies the first place, and the terms Kinetic Energy and Potential Energy give a new unity to the whole treatment. The Moment of Inertia is defined, in the first instance, in terms of kinetic energy. The conditions of equilibrium are established on their true kinetic basis. The principle of virtual velocities and d'Alembert's principle, which the older writers regarded with such reverence, are shown to be special enunciations of the great laws of energy. Gradually, by almost imperceptible advances along several lines of comparatively simple dynamical reasoning, the way is paved for the entrance into the shrine of Lagrange's generalised coordinates; and thence into the spacious temple of Hamiltonian Dynamics. At the time the book was being planned Thomson seems not to have studied the more recent developments of Lagrange's dynamical method; and his own mathematical methods were based largely on those of Fourier. As Tait once epigrammatically put it, "Fourier made Thomson." Tait used to tell how, when the chapter on Dynamical Principles was being sketched, he remarked, "Of course we must bring in Hamilton's dynamics." Thomson having expressed unfamiliarity with Hamilton's theory, Tait rapidly sketched it on a sheet of foolscap. Thomson was enraptured, took the sheet off with him to Glasgow, and in a short time had the sections written out very much as they appeared in the first edition of the *Treatise*.

The typical examples chosen to illustrate the power of Lagrange's generalised equations are of great variety and interest. They include the gyroscopic pendulum and the hydrokinetic problems of solids moving through perfect fluids. This latter class of problem had been imagined by Thomson as early as 1858; but it was in the pages of "T and T'" that the demonstrations were first given. From § 331 to § 336 of the first edition[1] these hydrokinetic problems are brought into touch with physical questions of far-reaching import. The discussion concludes with the remark that "it must be remembered that the real circumstances differ greatly, because of fluid friction, from those of the abstract problem, of which we take leave for the present." At a much later date Kelvin published a few papers on closely connected problems. In these as in other papers, as well as in the *Treatise*, constant use is made of Thomson's general theorem of "Laziness" as Tait called it (see above p. 182).

Chapters III and IV of Division I, on Experience and on Measures and Instruments respectively, are brief, touching only on the more fundamental aspects of the subject.

Division II (Abstract Dynamics) opens with a short introductory chapter of a few pages, followed by Chapter VI on Statics of a Particle. The academic problem familiar to all students of statics is quite ignored. Four short paragraphs give the solution of the general problem of a particle in equilibrium, under the action of given forces, some of which may be forces of constraint; and the remaining part of the chapter is occupied with the theory of attractions according to the Newtonian Law of the inverse square. By the easy steps of Kelvin's geometrical extensions of Newton's original demonstrations of the attraction of a spherical shell, the reader is led by simple physical reasoning, with the minimum of mathematics, to the enunciation and proof of Green's problem of the unique surface distribution of matter satisfying the assigned potential values at the surface. Attractions of ellipsoids and centrobaric bodies generally are discussed in considerable detail; and this leads to the simpler applications of spherical harmonics to problems of attraction. It may be mentioned in passing that some of the demonstrations are simplified in the second edition.

Chapter VII is devoted to the equilibrium of solids and fluids, flexible cords being included as constituting a kind of intermediate case between the equilibrium of perfectly rigid solids and the deformation of perfectly elastic solids. The transition from the flexible cord to the elastic wire may seem at first sight abrupt, but it is not really so when regard is had to the fundamental kinematical considerations involved. The whole mode

---

[1] These correspond to §§ 320–325 of the second edition, the sections on Action which originally preceded the hydrokinetic applications having been placed after them in the new edition.

of treatment is novel and original in the highest degree. It is instructive to see how by means of suitable and sufficiently obvious assumptions the elastic wire and elastic thin plate can be discussed without the preliminary laying down of the fundamental elastic principles. These are introduced, in due course, as the authors pass on to the general theory of the equilibrium and deformation of elastic solids. At the same time, a student reading these sections for the first time will find them very hard to understand, unless he has already some knowledge of the general theory of elasticity.

In the sections on Elasticity, the important practical problems of twist and bending of bars are completely solved, and then the solutions are indicated of more general cases, such as the elastic deformation of spherical shells and solid spheres subject to given surface tractions and body forces. These problems afford interesting examples of the use of the spherical harmonic analysis developed in the Appendix to the First Chapter. The discussion ends with a promise of further illustrations under "Properties of Matter," a promise which curiously enough still appears in the second edition, in spite of the prefatory remark that the book was not to be carried to completion.

An important application of the laws of deformation of an elastic sphere is made to our earth under the deforming stresses due to sun and moon; but, before this problem can be attacked, it is necessary to investigate the deformation in a spheroid of incompressible fluid. This consideration forms the transition from elasticity to hydrodynamics, and, after a brief discussion of the more simple questions of equilibrium and flotation, Thomson and Tait enter upon the final great series of problems connected with the equilibrium of spheroidal rotating masses of fluid, the theory of the tides and tidal stresses, and the closely related question of the rigidity of the earth. "To promote an intelligent comprehension of the subject," all the polar harmonic elements of the 6th and 7th orders are worked out, tabulated, and represented graphically. The various kinds of tidal influence are then investigated; and it is shown by combined use of hydrostatic and elastic principles that if the earth were only as rigid as steel it would as a whole yield to the tidal action of sun and moon by an amount equal to about one-third of the yield in the case of no rigidity. This concluding, highly original, part is greatly extended by Sir George Darwin in the second edition of the book.

The distinctive feature of the *Treatise* when compared with the earlier great works on the same branches of Natural Philosophy, such as Lagrange's

Mécanique Analytique, is the prominence given to motion rather than to equilibrium. Dynamical Laws and Principles are regarded as Preliminary; and then, as the first part of Abstract Dynamics, we are introduced to Statics of particles, of solids, and of fluids. In a letter of 1863 (quoted above, page 181) Tait protested against Attractions being brought in before Properties of Matter, of which they formed a natural part. It had become customary—and the custom still persists—to treat Attractions and the theory of the Potential as part of Statics of Particles. And yet, physically, attraction means motion; and equilibrium is maintained only by the introduction of balancing forces of another type, cohesion, adhesion, surface constraint, etc. Tait's suggestion to treat Gravitation as a property of matter was certainly more in line with a truly logical arrangement than was the plan ultimately adopted. But Thomson had already published a series of beautiful geometrical demonstrations on the subject; and no doubt he saw a splendid chance of utilising this material and of working out far-reaching problems of terrestrial dynamics. In other words, the problem of the earth's rigidity dominated to a marked degree the composition of the whole section of Abstract Dynamics which constituted Division II. Only in this way can we explain the unusual scope of Chapter VII. No doubt Lagrange, in his classical work, had included under Statics general discussions of the equilibrium of flexible elastic wires and thin plates; but Thomson and Tait were the first to bring within the limits of one chapter the laws of equilibrium of perfect fluids at rest or in steady rotation, and of solids ideally rigid or deformable. Much of this chapter indeed belongs as truly to the prospective section on the Properties of Matter as to the more abstract branch of dynamics. Profound though the influence of "Thomson and Tait" has been on the teaching and coordination of the principles of Natural Philosophy, no later writers have followed their example in devoting a single chapter to ordinary Statics of extended bodies, strings, and flexible wires, to Hydrostatics, and to Elasticity.

As regards the division of labour in the production of the book, I think there are strong indications that there is more of Tait's initial work in the earlier than in the later portions of the volume, and that Thomson's hand is particularly in evidence in the last chapter. Each was the other's severe critic, and many a sentence must have undergone great internal changes under the chiselling pen of each in turn.

It would be easy to find, especially in the more elementary parts, some faults of logical presentation.   For example, very soon after the publication of the work Maxwell in a letter to Tait made the following criticism of some of the statements in paragraphs 207 and 208.

"207. Matter is *never* perceived by the senses.   According to Torricelli quoted by Berkeley 'Matter is nothing but an enchanted vase of Circe, fitted to receive Impulse and Energy, essences so subtle that nothing but the inmost nature of material bodies is able to contain them.'...

"208. Newton's statement is meant to distinguish matter from space or volume, not to explain either matter or density.

"Def.   The mass of a body is that factor by which we must multiply the velocity to get the momentum, and by which we must multiply the half square of the velocity to get its energy.

"Hence if we take the exchequer pound as unit of mass (which is made of platinum) and if we find a piece of copper such that when it and the exchequer pound move with equal velocity they have the same momentum (describe experiment) then the copper has a mass of one pound.

"You may place the two masses in a common balance (which proves their *weights* equal), you may then cause the whole machine to move up or down.   If the arm of the balance moves parallel to itself the *masses* must also be equal.

"Some illustration of this sort (what you please) is good against heresy in the doctrine of the mass.   Next show examples of things which are not matter, though they may be moved and acted on by forces, (1) The path of a body, (2) Its axis of rotation, (3) The form of a steady motion, (4) An undulation (sound or light), (5) Boscovich's centres of force.   Next things which are matter such as the luminiferous aether, and if there be anything capable of momentum and kinetic energy."

But faults of the kind indicated were like spots in the sun.   The greatness of the book became more evident the closer it was studied.   Since the days of Newton's *Principia*, no work on Natural Philosophy of anything like the same originality had been produced in England.   Thomson and Tait's *Treatise* must ever rank with the classical works of Lagrange and Laplace.

It was not long before Helmholtz took steps for the preparation of a German translation; and in 1871, after some delay on account of the Franco-German War, this translation was finally published under the combined authority of Helmholtz and Wertheim.   A few sentences from Helmholtz's Preface will indicate his own view of the value of the *Treatise*.

"The present volume will introduce to the physical and mathematical German public the beginning of a work of high scientific significance, which will, in the most excellent fashion, fill in a very perceptible gap in the literature of the subject...

"One of the authors, Sir William Thomson, has long been known in Germany as one of the most penetrating and ingenious of thinkers who have applied themselves to our Science.   When such a one undertakes to lead us, as it were, into the workshop of his thoughts and to reveal the way in which he looks at things, to disentangle the guiding threads which have helped him in his bold intellectual combinations to master and coordinate the intractable and tangled material, we can but feel towards him the liveliest gratitude. For this work which would indeed overstrain the powers of a single much occupied man, he has found in P. G. Tait, Professor of Natural Philosophy in Edinburgh, a highly fit and gifted collaborator. Only perhaps by such a happy union could the task as a whole have been completed."

Helmholtz also speaks of the difficulty of finding German equivalents for some of the new scientific terms invented by Thomson and Tait. That verbal difficulties other than scientific troubled the minds of the German translators is clear from the following letter from Tait to Helmholtz:

17, DRUMMOND PLACE, EDIN.
20/5/71.

Dear Professor Helmholtz,

I have postponed my answer to your letter till I could catch Thomson, so that the answer to your queries should come from him as well as from me. Now that he has got a yacht he goes off for three weeks at a time; and is now on his way to Lisbon, and perhaps Gibraltar. However, he quite agrees with the following answers.

1. To "scull a boat" has two meanings[1]. In a *sea*-boat it means to work a single oar at the stern like the tail of a fish. In a light racing boat on a river or lake it means to work two oars in the rowlocks, one with either hand. *This* is what we mean in the text. Thomson won the "sculls" in this sense when he was a student in Cambridge.

2. To "run up on the wind" means to turn the ship's head to the wind.

3. To "carry a weather helm" means to put the tiller (Helmholtz?) to windward so that the rudder goes to the other side, tending to turn the ship's head *from* the wind, providing she is moving fast enough to make the rudder act.

As to your proposed changes on the new matter, I have no doubt whatever that they will be improvements, and we have reason to rejoice that you are kind enough to make them.

Thomson desires me to say that, while we all regret you cannot come to the B. A. Meeting, he will not forego your company on a cruise if it is possible to get hold of you, and will make arrangements to start at *any* time that may suit you, from August to November *inclusive*. But the best plan would be for you to try to get to Edinburgh in time for the last day or two of the Meeting, which lasts a *week* usually. This year we hold in Edinburgh the Centenary of Sir Walter Scott just at the *end* of the Association Meeting; and it is probable that, if you care at all about his writings, you would be interested in seeing the collection of things connected with him which will be exhibited then. If Sir William be not ready to start, you might come with me to St Andrews, where my wife will be delighted

---

[1] The reference is to § 336 in the first edition, considerably extended in § 325 in the second edition.

to see you, and where you may learn (at its head-quarters) the mysteries of GOLF! I have secured a house there, and so has my brother-in-law, Crum Brown, for the months of August and September. It appears that Huxley also has done the same, so that we may take to scientific discussions in the intervals of exercise.

I forgot to say that, with us, there are but rarely masts in canal-boats, and therefore the point of attachment is *not* usually in the axis of the boat. Perhaps you might put into *the German translation* the qualifying clause "provided the rope be not attached to a point in the axis of the boat," which is not necessary for an English reader.

All your other surmises are correct. I have told Vieweg to publish by instalments if he likes. I hope *you* will write a Preface to the first instalment, so that the weight of your authority may be brought to bear upon the reception of the new terms introduced[1].

While the first volume of the *Treatise* was shaping itself to a finish, the authors had also in mind the more elementary book for the use of the ordinary student attending the Natural Philosophy classes in Edinburgh and Glasgow. The intention was to make large use of the same material for the two kinds of book, the less mathematical portions being common to both.

There is fairly strong evidence that Division I, including Kinematics, Dynamical Laws and Principles, Experience, and Measures and Instruments was to be in the first instance Tait's domain; while Thomson was to be mainly responsible for Division II, namely, Abstract Dynamics, including Statics, Attractions and Elasticity. Tait very soon put together sufficient material to form a small pamphlet which he printed for the sake of his students in 1863. In the Edinburgh Calendar for the session 1862–3, he had already inserted the following optimistic reference,—"as a text-book on the general subject of the lectures, in case the forthcoming volume by Professors Thomson and Tait did not appear before Christmas, one of the following may be named, etc." Next year's Calendar announced that "In October 1863 there will be published the first volume of a Treatise on Natural Philosophy by

---

[1] As indicated in this letter, Helmholtz did not reach Scotland in time to be present at the British Association Meeting. He spent some days with Tait in St Andrews, but, unlike Huxley, did not yield to the fascinations of golf. In a letter to his wife of date August 20, 1871, Helmholtz gave his impressions of St Andrews life: "St Andrews hat eine prächtige Bai, feine Sandfläche, die dann mit einer scharfen Kante in grüne Grasflächen übergeht....Es ist grosses Leben von Badegästen, eleganten Damen und Kindern, Gentlemen in sporting Costümen, welche golfing spielen....Mr Tait kennt hier nichts anderes als golfing. Ich musste gleich mit, die ersten Schläge gelangen mir, nachher traf ich entweder nur die Erde oder die Luft....Tait ist eine eigenthümliche Art von Wildem Mann, lebt hier, wie er sagt, nur für seine Muskeln, und erst heute am Sonntag, wo er nicht spielen durfte, aber auch nicht in die Kirche ging, war er zu vernünftigen Gegenständen zu bringen."

Comparison of the paragraphs of the *Sketch of Elementary Dynamics* (1863), the *Treatise on Natural Philosophy* (1867), and the *Elements of Natural Philosophy* (1873).

Sketch	Treatise	Elements
Dynamics 1–3 A Kinematics	Chap. I Kinematics 1, 2	Chap. I Kinematics 1–3
4–47	3–5, 14, 15, 17, 19–29, 31, 32, 35, 36, (235), 41–43, 45, 53–58, 79–81, 83, 86, 90, 92, 93, 95, 96, 99, 102, 105	7–9, 16, 18, 20–25, 28–36, 38, 39, 43–48, 54–58, 63, 70–75, 91–93, 95, 97, 98, 100, 103, 104, 106–108, 110, 113, 116
B Dynamical Laws and Principles	Chap. II Dynamical Laws and Principles	Chap. II Dynamical Laws and Principles
48–95	208, 210–214, 216, 217, 220, 223–230, 281, 234–241, 244–247, 251, 253–256, 258–267, 268–273, 278, 281	174, 176–180, 182, 183, 185, 188–195, 198, 201–207, 210–213, 217, 219–222, 224–233, 240–245, 250, 235
96	289 and 292	254 and 257
97	451 (in Chap. v)	404 (in Chap. v)
C Statics	Chap. VII Statics of Solids and Fluids	Chap. VII Statics of Solids and Fluids
98	551	
99	572, Example II	592, Example II
100	572, Example III	592, Example III
101	586	592, Final Sentences
D Kinetics 102–108 109 110	unrepresented	Appendix Kinetics (*a*) to (*g*) (52 in Kinematics) (*h*)
E Hydrodynamics	Chap. VII Statics of Solids and Liquids	Chap. VII Statics of Solids and Liquids
111–118	742–745, 748, 750–752, 762	684–687, 690, 692–694, 703
119	unrepresented	unrepresented
120	753	694
121, 122		(*j, k*, in Appendix)

[etc., etc.]; and also an elementary book on the same subject for less advanced students will soon appear (a portion having been specially printed with the title *Elementary Dynamics*)." Year after year the same announcement was printed in the Calendar, the date of appearing of the forthcoming volume being simply pushed on one more year! At last in the 1868–9 Calendar, "will be published" was changed to "was published"; but the book on *The Elements of Natural Philosophy* was still far from its final form.

Now there is not the least doubt that this small pamphlet of 44 pages, which was published in 1863 by Maclachlan and Stewart, an old Edinburgh firm whose shop faced the College, was almost entirely the work of Tait. With a few exceptions the paragraphs still exist in their earliest draft in his MS note book; and are reproduced practically verbatim in the large type of the *Treatise*. Of the few which are not represented in the *Treatise*, the majority treat of parts of the subject which lay outside the scope of Volume I. The pamphlet was called *Sketch of Elementary Dynamics*, and was issued under the joint names of W. Thomson and P. G. Tait. Its contents are indicated in the first column of the Table on the opposite page.

Simultaneously with the publication of this pamphlet in Edinburgh, Thomson brought out a pamphlet in Glasgow, under the title "*Elements of Dynamics*, Part I, edited, with permission, by John Ferguson[1], M.A., from notes of lectures delivered by William Thomson." With the exception of three short introductory paragraphs and four later paragraphs upon Gauss' absolute unit of force and Kater's measure of gravity at Leith Fort, this pamphlet differs *in toto* from the one published in Edinburgh. It is concerned almost wholly with Statics; and the treatment is that of the old days before the "return to Newton." Paragraphs 22 to 70, covering Chapters I and II and half of Chapter III are unrepresented in "T and T'," and are indeed quite out of sympathy with the whole mode of treatment of the *Treatise*, giving in familiar fashion the now superseded Duchayla's proof of the parallelogram of forces. In the Edinburgh *Sketch* the whole thing is disposed of in one paragraph in the true Thomson-and-Tait style. Paragraphs 71 to 124 and 128 and 166 (covering Chapters III to VIII) in the Glasgow pamphlet are reproduced verbatim in Chapters VI and VII of the *Elements of Natural Philosophy* as published in 1873, but have no place whatever in the large *Treatise*. Serious students of the *Elements* were never greatly attracted by these sections. They did not seem to fit in well with the rest of the

[1] Now Professor of Chemistry in Glasgow University.

treatise. They constituted a careful logical setting forth of what might be called the analytical geometry of forces and couples, necessary perhaps for solving problems, but not demanding much thought on the part of the man familiar with elementary Cartesian methods. This Glasgow pamphlet then may be considered to be Thomson's earliest attempt to contribute his share to the elementary book as originally planned. Its first, second and third paragraphs are identical in language with the corresponding parts of the Edinburgh *Sketch*; paragraph 16 on the formula for gravity is the same as paragraph 222 in the *Treatise*; and paragraphs 17 to 21 (the last being on Kater's pendulum measurements) correspond with 57 and 58 in the Edinburgh *Sketch*, and 223 and 226 in the *Treatise*. With the exception of these few paragraphs the Glasgow pamphlet forms no essential part of the *Treatise on Natural Philosophy*. The fact that the Edinburgh *Sketch*, though bearing the names of both authors, was quite unknown among Glasgow students, and is not mentioned in Professor Sylvanus Thompson's bibliography at the end of Lord Kelvin's *Life*, proves that it was practically the work of Tait.

It is this Edinburgh *Sketch*, accordingly, which must be considered to be the earliest published form of the real "Thomson and Tait." I can best exhibit this by comparing in tabular form (see page 198) the corresponding parts of the three distinct publications, the *Sketch*, the *Treatise*, and the *Elements of Natural Philosophy*. In many cases the original paragraphs of the *Sketch* are simply reproduced; in other cases they are expanded, but in the expanded form the original sentences exist practically unchanged. Of course in the *Treatise* the expansion takes in addition an analytical form.

Thus we see that with the exception of the part on Kinetics and the last two paragraphs on Hydrokinetics, which lay quite outside the scope of Volume I, the Edinburgh *Sketch* passed bodily into the *Treatise*, forming indeed the nucleus about which the first and second chapters of the work crystallised. The great work having been completed, the large type parts of the first four chapters, along with the Kinetic and Hydrodynamic parts of the Edinburgh *Sketch* of 1863, were pieced together so as to form a pamphlet of 120 pages. This was issued in 1867 by the Clarendon Press for the use of the students both in Edinburgh and Glasgow, and was followed in 1868 by a second edition ("not published") of 138 pages. The Glasgow pamphlet was then partly incorporated in the way already indicated; and with the addition of large type parts from Division II of the *Treatise*, the

Elements of Natural Philosophy took its final form and was issued in 1873. In the Appendix a few typical kinetic problems appear exactly as they were originally written down in Tait's Edinburgh *Sketch* of 1863. With the exception of 43 pages devoted to the composition and resolution of forces and couples, and essentially reproduced from the Glasgow pamphlet, the *Elements of Natural Philosophy* of 279 pages is simply an abridgement of the *Treatise on Natural Philosophy*.

To the earnest capable student it was and still is a mine of wealth; but what hours of misery it caused to many a hapless youth! I remember a student from abroad coming in the summer session and asking Tait what he should read so as to prepare himself for the Natural Philosophy class in the succeeding winter. Tait with a smile took up " Little T and T'," said that this was the text-book in use, and recommended the youth to read the first ten pages and come back to report progress. In a few weeks he returned in the direst distress. He had read and re-read the introductory sections, and his mind was an absolute blank!

Indeed, except for the senior students and for men looking forward to Honours, the *Elements* was to a large extent a sealed book and was indeed in certain parts more difficult than the larger work. It was too concise for the ordinary average student, who never got deep enough into the subject to appreciate its aim and scope. The authors wrote with their eye on what was to come; the average student was content with what little knowledge he could gain now. J. M. Barrie, in his brightly written *An Edinburgh Eleven*, calls the book "The Student's first glimpse of Hades." The sentence in the Preface of the *Elements* which speaks of it as being "designed more especially for use in schools and junior classes in Universities" must have been penned with a chuckle on the part of Tait at any rate.

Clerk Maxwell reviewed the *Elements* in *Nature*, March 27, 1873 (Vol. VII), and expressed his

"sympathy with the efforts of men, thoroughly conversant with all that mathematicians have achieved, to divest scientific truths of that symbolic language in which mathematicians have left them, and to clothe them in words, developed by legitimate methods from our mother tongue but rendered precise by clear definitions, and familiar by well-rounded statements."

He did not however appraise the work from the point of view of the tiro.

The first edition of the large *Treatise* was very soon sold out; and the authors had to consider the preparation of a new edition of Volume I, before any progress had been made in getting ready the other promised volumes. It was not, however, till 1875 that by mutual agreement the original contract with the Clarendon Press was cancelled, and a new arrangement was made with the Cambridge University Press for the publication of a second edition. In this edition many sections were considerably expanded; but on the whole the large type portions remained unaltered. The addition of new matter compelled the issue of the book in two volumes known as Part I and Part II. The First Part was published in 1879, and the Second in 1883.

In the foregoing analysis of the contents of the first edition, incidental references have been made to the corresponding parts of the second edition. Without in any way exhausting the list of important changes, I might mention the following subjects as having received new or improved treatment; Spherical Harmonics, Lagrange's generalised coordinates, the ignoration of coordinates, Hamilton's general dynamic theory, gyrostatic action, attraction of ellipsoids and the tides, tidal stresses and strains, due to the influence of the sun and moon.

As the preparation of the new edition proceeded, it gradually became clear to both authors that, judging from past experience, they could hardly hope to accomplish the task on which they had entered with so much enthusiasm in 1861.

Kelvin thus explained the situation in his obituary notice of Tait communicated to the Royal Society of Edinburgh:

"The making of the first part of 'T and T'' was treated as a perpetual joke, in respect of the irksome details, of the interchange of drafts for 'copy,' amendments in type, and final corrections in proof. It was lightened by interchange of visits between Greenhill Gardens, or Drummond Place, or George Square, and Largs, or Arran, or the old or new College of Glasgow; but of necessity it was largely carried on by post.....About 1878 we got to the end of our Division II on Abstract Dynamics; and according to our initial programme should then have gone on to Properties of Matter, Heat, Light, Electricity, Magnetism. Instead of this we agreed that for the future we could each work more conveniently and on more varied subjects, without the constraint of joint effort to produce as much as we could of an all-comprehensive text-book of Natural Philosophy. Thus our book came to an end with only a foundation laid for our originally intended structure."

That they even completed the first volume of the projected treatise was largely due to the indefatigable zeal of Tait in keeping Thomson to his share of the task.   In the revision of the proof sheets for the second edition, Thomson's whole method of working seriously exercised the printers.

Any new aspect which opened up to his mind as he read the pages of the first edition led at once to expansion and interpolation, sometimes of the most alarming dimensions.   The great sheet on which the original page was pasted became covered with the new matter; bits were pasted on like wings bearing precious symbols; while, not unfrequently, the discussion overflowed into extra sheets, subsection after subsection being piled on regardless of space and proof correction charges.   Difficult indeed was the proof reading in such circumstances; and both Kelvin and Tait felt always deeply grateful to Professors Burnside and Chrystal and Sir George Darwin for the aid they gave in the final correction of the sheets.   The last-named, in fact, added several sections to the second volume on the problems of tidal action.

Richer and fuller and more complete in many respects though the second edition was, it could not excel in beauty of printing the original first volume, which though finally published in Oxford was printed by Constable of Edinburgh.   The authors refer to this very pointedly in the preface, as well as to the great care with which the diagrams were made.

No finer tribute to the remarkable influence of "T and T'" could be penned than that with which Clerk Maxwell enriched the pages of *Nature* (Vol. xx), when he reviewed the new edition, Vol. I, Part I, shortly before his death.   A few quotations form a fitting conclusion to this sketch of the genesis and growth of one of the most important scientific publications of the Nineteenth Century.

"What," asked Maxwell, "is the most general specification of a material system consistent with the condition that the motions of those parts of the system which we can observe are what we find them to be?   It is to Lagrange in the first place that we owe the method which enables us to answer this question without asserting either more or less than all that can be legitimately deduced from the observed facts. But though the method has been in the hands of mathematicians since 1788, when the Mécanique Analytique was published, and though a few great mathematicians, such as Sir W. R. Hamilton, Jacobi, etc., have made important contributions to the general theory of dynamics, it is remarkable how slow natural philosophers at large have been to make use of these methods.

"Now however we have only to open any memoir on a physical subject in order to see that these dynamical theorems have been dragged out of the sanctuary of profound mathematics in which they lay so long enshrined, and have been set to do all kinds of work, easy as well as difficult, throughout the whole range of physical science.

"The credit of breaking up the monopoly of the great masters of the spell, and making all their charms familiar to our ears as household words, belongs in great measure to Thomson and Tait. The two northern wizards were the first who, without compunction or dread, uttered in their mother tongue the true and proper names of those dynamical concepts which the magicians of old were wont to invoke only by the aid of muttered symbols and inarticulate equations. And now the feeblest among us can repeat the words of power and take part in dynamical discussions which but a few years ago we should have left for our betters.

"In the present edition we have for the first time an exposition of the general theory of a very potent form of incantation, called by our authors the Ignoration of Coordinates. We must remember that the coordinates of Thomson and Tait are not the mere scaffolding erected over space by Descartes, but the variables which determine the whole motion...."

There then followed a remarkably clear statement of the conditions under which certain kinds of coordinates not only may be ignored but ought to be ignored, and the final illustration was in these words :

"There are other cases, however, in which the conditions for the ignoration of coordinates strictly apply. For instance, if an opaque and apparently rigid body contains in a cavity within it an accurately balanced body, mounted on frictionless pivots, and previously set in rapid rotation, the coordinate which expresses the angular position of this body is one which we are compelled to ignore, because we have no means of ascertaining it. An unscientific person on receiving this body into his hands would immediately conclude that it was bewitched. A disciple of the northern wizards would prefer to say that the body was subject to gyrostatic domination."

CHAPTER VI

OTHER BOOKS

As a writer of scientific books Tait was eminently successful. His books were the outcome of his daily work, and are stamped throughout with the vigour and clear-mindedness of their author. The special characteristics of each will appear as we proceed.

His first venture as an author was in conjunction with W. J. Steele, his college friend and competitor in the great Tripos contest. Tait and Steele's *Dynamics of a Particle* was begun soon after their graduation. It was published in 1856 and at once caught on as a useful Cambridge text-book. Steele's early death compelled Tait to write by far the greater part of the book, Steele's contribution being only about one-tenth of the whole. In a truly chivalrous spirit Tait continued to the end to bring out the successive editions under the joint names, leaving intact the portions contributed by his lamented friend.

The second edition appeared in 1865. Meanwhile a remarkable revolution in the whole dynamical outlook had been effected by Thomson and Tait working together in view of the publication of their *Natural Philosophy*; and to bring "Tait and Steele" into harmony with the new conceptions the second chapter had to be completely recast. This second edition was, therefore, the first book published in which the "return to Newton" was fully effected. The contrast between the two editions is well brought out in the remark made by Chrystal (*Nature*, July 25, 1901) that

"the first edition of Tait and Steele's 'Dynamics'...does not...contain either of the words *work* or *energy*. In its original form it was founded on Pratt's 'Philosophy,' and written on old-fashioned Cambridge lines, which knew not of Lagrange or Hamilton."

Already in 1856 Tait knew enough of Hamilton's work to introduce the Hodograph, but as he confesses in the preface to the second edition he had not himself in 1855 "read Newton's admirable introduction to the *Principia*." When exactly Tait made full acquaintance with Newton's dynamical foundation we cannot tell; but, if we may judge from the following sentences in his inaugural lecture at Edinburgh, it must have been during 1860 at the latest:

That godlike mortal, as Halley does not scruple to call him, who, finding the very laws of motion imperfectly understood, in a few years not only gave them fully and accurately, and devised a mathematical method of almost unlimited power for their application, but explained most of the phenomena of the Solar System including Tides, Precession and Perturbations (though this is but one part of his contributions to Natural Philosophy)—and who was only, after repeated solicitations, persuaded that he had anything worthy to offer to the world, will remain to all time the beau-ideal of magnificent genius and devoted application, alike unstained by vanity and unwarped by prejudice.

In this inaugural lecture we find also an absolutely clear account of the meaning of the Conservation of Energy. A few quotations will make this clear.

When we talk of the *Conservation of Force* as a principle in Nature, it is to be carefully noted that we do not mean force in the ordinary acceptation of the word—and, indeed, the principle is now better known as the *Conservation of Energy*. As this is a matter of very considerable moment I shall treat of it with a little detail. *Energy* may be *Actual*[1] or *Potential*. Actual Energy belongs to moving bodies...... Potential Energy......belongs to a mass or a particle in virtue of its *position*, and is in general *work which can be got out of it on account of that position*.......Supposing that you have now an idea as to the meaning of these two terms, I give the principle of the *Conservation of Energy* as it has been put by Professor Rankine to whom these terms are due—

"In any system of bodies, the sum of the potential and actual energies of the bodies is never altered by their mutual action."

It is abundantly evident that before Tait entered on his Edinburgh career he had already thought deeply on the new doctrine of energy. When the time came for the second edition of *Dynamics of a Particle*, the foundations of the subject were at once brought into line with these modern views, although in most other respects the book did not materially change its character. It is amusing to read Tait's own pencilled annotation on the first page of the preface to the second edition, "Very poor book—others poorer."

As edition after edition was called for, Tait often felt a strong inclination to recast the whole work. Lack of the necessary leisure combined with

[1] The term *Kinetic* was not invented till two years later.

the desire of preserving more or less intact Steele's original contribution to the treatise no doubt stood in the way. It is certain, however, that Tait would never in his later days have put together a book of the type of "Tait and Steele." Written by him when he had just emerged triumphantly from the Tripos examination it was meant for students with like ambitions. Important examples were fully worked out and numerous exercises appended to the chapters for the eager student to sharpen his weapons on. The ground covered was considerable; and when compared with other similar books *Dynamics of a Particle* in its later editions maintained its position both for accurate treatment of important classes of problems and for the great amount of original matter it contained. The treatment was analytical throughout; but here and there, as edition succeeded edition, Tait inserted some of his neat geometrical demonstrations. The contents are fairly well indicated by the titles of the chapters: Kinematics, Laws of Motion, Rectilinear Motion, Parabolic Motion, Central Orbits, Elliptic Motion, Resisted Motion, Constrained Motion, General Theorems (Action, Brachistochrones, etc.), Impact. The chapter on Central Orbits, especially in the numerous examples of extraordinary laws of attraction adjusted to give integrable solutions, reflects the Cambridge School of Wrangler trainers of half a century ago. For these and other artificial problems invented for examination purposes Tait had during the last forty years of his life a genuine horror.

It is curious to note that, as *Dynamics of a Particle* was the first book Tait gave to the world, so the last piece of composition he penned was the preface to the seventh edition (Nov. 7, 1900). It finishes with the sentence,

> Meanwhile I once more despatch the Veteran on a campaign, with a few necessary patches on his battered harness.

In little more than a month ill-health compelled him to lay aside his own harness and weapons, and before many months had passed his battle of life had closed.

The years immediately preceding 1867 were remarkably productive from the literary point of view. Thomson and Tait's *Natural Philosophy* was launched on a wondering world; and as if this did not give scope enough for his busy pen Tait found time to instruct the mathematicians and natural philosophers in the use of Quaternions and the history of

Thermodynamics. The *Natural Philosophy* and Tait's *Quaternions* are discussed in appropriate chapters. Here we shall consider more carefully his contributions to the history of the doctrine of Heat and Energy. For work of this kind Tait was admirably fitted. Having no claim to be regarded as one of the founders of the modern theory of energy, he was early in closest touch with four of the great pioneers, Joule, Helmholtz, Thomson, and Rankine. He witnessed the striking development of the fruitful ideas of last century, helped in no small way to forge and fix appropriate nomenclature, and probably did more than any other single man to spread a knowledge of the true meaning of the first and second laws of Thermodynamics. Year after year he led his two hundred students round the Carnot cycle, and impressed upon them his weighty reflections on energy and matter; but to a still larger audience he appealed through his writings, and one of the most characteristic of his elementary books is his *Sketch of Thermodynamics*. This book was the direct outcome of a controversy with Tyndall as to the historic development of the theory of heat.

To write complete history when it is in the making is probably a human impossibility. No man, however talented and well informed, can see at one and the same time all the influences at work. Nor can he trace accurately the manner of the working. The personal equation necessarily enters in. The contemporary historian is apt to be biassed, though it may be unconsciously. The mental picture will depend upon the observer, just as witnesses of the same scene do not always tell the same story. Such general considerations must be borne in mind when we consider Tait's contributions to the history of modern science.

At the same time it must be remembered that Tait diligently read the literature of any subject in which he was specially interested; and that his knowledge and appreciation of the real significance of the far-reaching work of the early half of last century were probably unsurpassed. His great intimacy with Thomson and Maxwell, two of the geniuses of our time, brought him into immediate contact with the springs of physical thought.

Bearing this in mind, no later historian can pass by Tait's attitude on the history of the development of the doctrine of energy without a careful consideration of the reasons for this attitude. These reasons Tait himself gave frankly and fully.

Of the many facts in the history of energy three only became matter of serious controversy, namely, the true place to be assigned to Mayer as one of the founders of the theory of heat, the sufficiency of Clausius' axiom as a basis for the Second Law of Thermodynamics, and the claim of Clausius to the Entropy integral.

The occasion which led to Tait's incursion into historic fields was a lecture on "Force" delivered by Tyndall in 1862 in the Royal Institution. The subject was what we now call "Transformations of Energy," and in several particulars corresponded curiously with Tait's own inaugural lecture of 1860. In developing the subject Tyndall, however, proclaimed that "the striking generalisations" laid before his audience were "taken from the labours of a German physician named Mayer." This no doubt was true; but the manner in which the work of Mayer was lauded seemed to imply that the examples given of energy transformations were peculiarly Mayer's and had been imagined by no other natural philosopher, and that Mayer's priority claims had been hitherto altogether overlooked. This implication was certainly not true, and controversy was naturally roused. Joule, who was himself one of the first to bring Mayer's work to the notice of scientific men, sent a dignified protest to the *Philosophical Magazine*; and, as a corrective to what they regarded as erroneous history, Thomson and Tait communicated an article on Energy to *Good Words* for October 1862. This article is historically important as the occasion on which the term Kinetic Energy saw the light. Tyndall replied in the February and June numbers of the *Philosophical Magazine* for 1863, and Tait's rejoinders appeared in April and August.

Following up this controversy there appeared two unsigned articles in the *North British Review* for 1864—one in February on the Dynamical Theory of Heat, the other in May on Energy. In the preface to his *Thermodynamics* (1867) Tait refers to these reviews as from his pen—indeed they constitute with necessary changes a large part of the book. The earlier article traces the modern theory of heat from the independent experiments of Rumford and Davy in 1798 through the remarkable reasoning of Carnot (1824) and the epoch-making experiments of Joule (1840–48) to the final establishment of thermodynamics at the hands of Clausius, Rankine, and Thomson; and then, being avowedly a review of Verdet's *Exposé de la Théorie Mécanique de la Chaleur* (1863) and Tyndall's *Heat considered as a Mode of Motion* (1863), closes with appreciations and criticisms of these books, the criticisms being mainly in regard to the history of the subject. The second article on Energy is a review of some of the writings of Joule, Mayer, Helmholtz, and Verdet.

Shortly after the appearance of these reviews, Verdet wrote Tait a letter of date August 6, 1866, in which he referred to some of Tait's criticisms :

"En demandant à l'éditeur de le North British Review de me faire connaître l'auteur des articles sur la Théorie mécanique de la Chaleur je n'avais d'autre désir que de savoir qui je devais remercier des éloges donnés à mon exposé de cette théorie. Je vous prie d'être persuadé que j'y ai été fort sensible et que je le suis davantage depuis que j'en connais l'auteur. Je ne fais aucune difficulté de reconnaître plusieurs inexactitudes historiques que j'avais commises en attribuant à M. Favre une expérience de Joule et en oubliant que Joule avait dès 1843 indiqué le principe des relations entre la physiologie animale et la théorie mécanique de la chaleur. Je m'explique difficilement cette dernière omission, car dès 1852, lorsque j'avais inséré dans les Annales de Chimie et de Physique un extrait du premier travail de Joule, j'avais terminé cet extrait par une traduction du passage dont il s'agit. Au reste une lettre que M. Joule voulut bien m'écrire lorsque je lui envoyai il y a deux ans un tirage à part de mes Leçons me fait croire qu'en somme il a été satisfait de la manière dont j'ai parlé de ses travaux. Personne ne les admire plus que moi ; et je n'hésiterais pas à signer tout ce que vous en dites de ses articles.

"Permettez moi après cette déclaration de conserver mon opinion sur Mayer et de croire qu'une aussi grande découverte que celle de la Conservation de l'Energie peut suffir à la gloire de deux et même de trois inventeurs (car il ne me paraît pas possible de passer Colding sous silence)....

"J'espère, Monsieur, que j'aurai quelque jour le plaisir de vous remercier de vive voix de tout ce que vous avez dit de flatteur pour moi, etc."

Tait had a high opinion of Verdet as an eminent physicist, who, he was wont to remark, had just fallen short of the level of genius.

With fuller knowledge afterward gained of the early work of Mohr, Séguin and Colding, Tait in his second edition of *Thermodynamics*, as well as in *Recent Advances* and in *Heat*, somewhat modified his original sketch ; but his contention to the end was that, however ingenious the views advanced by these other pioneers, their work was not to be put on the same plane with that of Joule as regarded either the soundness of the theory or the accuracy of the experimenting.

Mayer's earliest pamphlet of 1842 was a discourse based on the mediaeval dogma "Causa aequat effectum," whence he deduced the prime property of all Causes, Indestructibility. He defined *Kraft* to mean very much what we now call potential energy, illustrated its transformability into motion by a discussion of falling bodies, defined heat as *Kraft* and not as motion, drew an analogy between compression of air and fall of bodies, and deduced a value for the mechanical equivalent of heat on the assumption that the heat generated in suddenly condensing a gas was equal to the work done in condensation. Tait maintained that such a mingling of truth and error could not be accepted as a sound basis for the true doctrine of thermodynamics; that it was doubtful if in any particular which could be accepted as sound physics Mayer had anticipated others; that his argument in regard to the experiment of heating air by compression involved a gratuitous assumption which might or might not be true; and that already Joule, in his remarkable experiments on the production of heat by electricity and friction (1840 to 1843), had in an irreproachable scientific manner elucidated the true nature of heat.

Mayer's later pamphlets of 1845 and 1848 contained, as Tait pointed out in the second chapter of his *Thermodynamics*, many beautiful examples of the law of transformations. So also did Mohr's papers of 1837, Grove's *Correlation of the Physical Forces* (1842), Joule's papers and lectures between 1840 and 1847, Colding's publications (1840) and Helmholtz's great memoir of 1847. Indeed, as early as 1834, Mrs Somerville in her *Connection of the Physical Sciences*, had called attention to the generality of such transformations. In fact the notions of transformability and of the equivalence of heat and work were in the air. The time at last was ripe for the full comprehension, appreciation, and development of the much earlier experiments of Davy and of Rumford. It is not surprising that several minds of the first order were pondering over the significance of these and related phenomena, each investigator approaching the subject in his own way and to a large extent independent of the others. Yet by the great majority of their contemporaries the early work of these true philosophers was not fully appreciated. Mohr's first paper, "Ueber die Natur der Wärme," appeared in 1837 in Liebig's *Annalen* (Vol. XXIV); but Poggendorf declined an expansion of this paper, which was however accepted for publication in Baumgartner's and v. Holger's *Zeitschrift für Physik*, a publication of comparatively limited circulation. Mayer's 1842 paper, "Bemerkungen über die Kräfte der unbelebten Natur," made its *début* in Liebig's *Annalen*, just five years later than Mohr's similar paper. Neither of them seems to have had any traceable influence in moulding contemporary scientific thought. Helmholtz apparently knew nothing of them when he wrote his tract in 1847; and

Joule does not seem to have become acquainted with Mayer's essay till about 1850.

When in 1876 Tait had his attention drawn to Mohr's paper of 1837, he sent a translation of it to the *Philosophical Magazine* (Vol. II, 5th Series, p. 110) and added the following statement:

About the time when Colding and Joule took up the experimental investigation of Energy at the point where it had been left by Rumford and Davy, there were published a great many speculations as to the nature of Heat and its relation to work. Several of these speculations, especially those of Mayer and Séguin, have been discussed, and at least in part reprinted, in the Philosophical Magazine. It is right, therefore, that the same journal should recall attention to the above paper, which was recently pointed out to me by Professor Crum Brown, and contains what are in some respects the most remarkable of all these speculations.

Singularly enough, it is not even referred to by Mayer, though his much belauded earliest paper appeared only five years later and in the very same journal. It contains, in a considerably superior form, almost all that is correct in Mayer's paper; and, though it contains many mistakes, it avoids some of the worst of those made by Mayer, especially his false analogy and his *a priori* reasoning.

Polarisation of Heat is ascribed to Melloni instead of Forbes; the calculation from the compressibility and expansibility of water is meaningless; and the confusion between the two perfectly distinct meanings of the word *Kraft* is nearly as great as that which some modern British authors are attempting to introduce into their own language by ascribing a second and quite indefensible meaning to the word *Force*.

On the other hand, several of the necessary consequences of the establishment of the Undulatory nature of Radiant Heat are well stated; and the very process (for determining the mechanical equivalent of heat by the two specific heats of air) for which Mayer has received in some quarters such extraordinary praise—though it is in principle, albeit not in practice, utterly erroneous—is here stated[1] much more clearly than it was stated five years later by Mayer.

As regards the experimental determination of the dynamical equivalent of heat, Tait's position is practically upheld in the calm judgment given by

---

[1] Mohr's argument is: "If any species of gas is heated more strongly it strives not only to increase the number of its vibrations, but also to enlarge their amplitudes. If one prevents this expansion, it appears as increased tension. One would require therefore a smaller quantity of heat to warm a gas shut in by firm walls than a gas contained in yielding walls, since, if heat be the cause of the expansion, just as much heat must become latent as there would be cold developed if the gas were allowed to expand as much as before but without supply of heat." In an earlier section he has already stated that "Heat is an oscillatory motion of the smallest parts of bodies....Heat appears as 'Kraft'...the expansion of bodies by heat is a force-phenomenon of the highest kind."

Sir George Stokes when Mayer was awarded the Copley Medal by the Royal Society of London. Commenting on this determination, Stokes wrote:

"This was undoubtedly a bold idea, and the numerical value of the mechanical equivalent of heat obtained by Mayer's method is, as we know, very nearly correct. Nevertheless it must be observed that one essential condition in a trustworthy determination is wanting in Mayer's method; *the portion of matter operated on does not go through a cycle of changes.* Mayer reasons as if the production of heat were the sole effect of the work done in compressing air. But the volume of the air is changed at the same time, and it is quite impossible to say *a priori* whether this change may not involve what is analogous to the statical compression of a spring, in which a portion or even a large portion of the work done in compression may have been expended. In that case the numerical result given by Mayer's method would have been erroneous, and *might* have been even widely erroneous. Hence the practical correctness of the equivalent got by Mayer's method must not lead us to shut our eyes to the merit of our countryman Joule in being the first to determine the mechanical equivalent of heat by methods which are unexceptionable, as fulfilling the essential condition that no ultimate change of state is produced in the matter operated upon."

Whatever view may be taken of the question, one thing is clear. It was Tyndall's eulogy of Mayer which led to the writing of Tait's *Sketch of Thermodynamics.* The first and second chapters, to a large extent reproductions of the articles in the *North British Review,* were printed privately for class use in 1867 under the title *Historical Sketch of the Dynamical Theory of Heat*; and the book complete in three chapters appeared in 1868.

When Tait was preparing his *Thermodynamics* for the press he asked Maxwell for some hints. Maxwell's reply of date Dec. 11, 1867, was very characteristic and of great interest as being probably the first occasion on which he put in writing his conception of those fine intelligences—Maxwell's Demons as Kelvin nicknamed them—who operating on the individual molecules of a gas could render nugatory the second law of thermodynamics. He wrote:

"I do not know in a controversial manner the history of thermodynamics, that is, I could make no assertions about the priority of authors without referring to their actual works....Any contributions I could make to that study are in the way of altering the point of view here and there for clearness or variety, and picking holes here and there to ensure strength and stability.

"As for instance I think that you might make something of the theory of the absolute scale of temperature by reasoning pretty loud about it and paying it due honour at its entrance. To pick a hole—say in the 2nd law of $\Theta\Delta$cs., that if two things are in contact the hotter cannot take heat from the colder without external agency.

"Now let A and B be two vessels divided by a diaphragm and let them contain elastic molecules in a state of agitation which strike each other and the sides.

"Let the number of particles be equal in A and B but let those in A have the greatest energy of motion. Then even if all the molecules in A have equal velocities, if oblique collisions occur between them their velocities will become unequal, and I have shown that there will be velocities of all magnitudes in A and the same in B, only the sum of the squares of the velocities is greater in A than in B.

"When a molecule is reflected from the fixed diaphragm CD no work is lost or gained.

"If the molecule instead of being reflected were allowed to go through a hole in CD no work would be lost or gained, only its energy would be transferred from the one vessel to the other.

"Now conceive a finite being who knows the paths and velocities of all the molecules by simple inspection but who can do no work except open and close a hole in the diaphragm by means of a slide without mass.

"Let him first observe the molecules in A and when he sees one coming the square of whose velocity is less than the mean sq. vel. of the molecules in B let him open the hole and let it go into B. Next let him watch for a molecule of B, the square of whose velocity is greater than the mean sq. vel. in A, and when it comes to the hole let him draw the slide and let it go into A, keeping the slide shut for all other molecules.

"Then the number of molecules in A and B are the same as at first, but the energy in A is increased and that in B diminished, that is, the hot system has got hotter and the cold colder and yet no work has been done, only the intelligence of a very observant and neat-fingered being has been employed.

"Or in short if heat is the motion of finite portions of matter and if we can apply tools to such portions of matter so as to deal with them separately, then we can take advantage of the different motion of different proportions to restore a uniformly hot system to unequal temperatures or to motions of large masses.

"Only we can't, not being clever enough."

To this is appended a pencilled annotation by Thomson:

"Very good. Another way is to reverse the motion of every particle of the Universe and to preside over the unstable motion thus produced."

In an undated letter, which must have been written about this time, Maxwell constructed the following Catechism:

"Concerning Demons.

"1. Who gave them this name? Thomson.

"2. What were they by nature? Very small BUT lively beings incapable of doing work but able to open and shut valves which move without friction or inertia.

"3. What was their chief end? To show that the 2nd Law of Thermodynamics has only a statistical certainty.

"4. Is the production of an inequality of temperature their only occupation? No, for less intelligent demons can produce a difference in pressure as well as temperature by merely allowing all particles going in one direction while stopping all those going the other way. This reduces the demon to a valve. As such value him. Call him no more a demon but a valve like that of the hydraulic ram, suppose."

Maxwell, writing on Dec. 23, 1867, acknowledged receipt of the two-chaptered pamphlet in these words:

"I have received your histories of Thermodynamics and Energetics, and will examine them, along with Robertson on the Unconditioned who holds that our ultimate hope of sanity lies in sticking to metaphysics and letting physics go down the wind. I have read some metaphysics of various kinds and find it more or less ignorant discussion of mathematical and physical principles, jumbled with a little physiology of the senses. The value of the metaphysics is equal to the mathematical and physical knowledge of the author divided by his confidence in reasoning from the names of things.

"You have also some remarks on the sensational system of philosophising (sensation in the American not the psychological sense). Beware also of the hierophantic or mystagogic style. The sensationalist says, 'I am now going to grapple with the Forces of the Universe, and if I succeed in this extremely delicate experiment you will see for yourselves exactly how the world is kept going.' The hierophant says, 'I do not expect to make you or the like of you understand a word of what I say, but you may see for yourselves in what a mass of absurdity the subject is involved.'

"Your statement however seems tolerably complete considering the number of pages. One or two ideas should be brought in with greater pomp of entry perhaps....

"There is a difference between a vortex theory ascribed to Maxwell at page 57, and a dynamical theory of magnetism by the same author in Phil. Trans. 1865. The former is built up to show that the phenomena are such as can be explained by mechanism. The nature of the mechanism is to the true mechanism what an orrery is to the Solar System. The latter is built on Lagrange's Dynamical Equations and is not wise about vortices. Examine the first part which treats of the mutual action of currents before you decide that Weber's is the only hypothesis on the subject....

"You wrote me about experiments in the Laboratory. There is one which is of a high order but yet I think within the means and powers of students, namely, the determination of Joule's Coefft. by means of mercury. Mercury is ($13\cdot57/0\cdot033$) times better than water so that about 9 feet would give 10° F....[Plan described for obtaining a vertical fall of mercury and measuring temperatures above and below]... I think it a plan free from many mechanical difficulties, and in a lofty room with plenty of mercury and strong ironwork, and a cherub aloft to read the level and the thermometer and a monkey to carry up mercury to him (called Quicksilver Jack), the thing might go on for hours, the coefficient meanwhile converging to a value to be appreciated only by the naturalist."

Tait also sent copies of the two-chaptered pamphlet to Helmholtz and Clausius for their criticisms before the publication of the complete work. In his letter to Helmholtz of date Feb. 2, 1867, he said:

Herewith I send copies of the first two chapters of a little work which I intend soon to publish. Its main object is to serve as a text-book for students till Thomson and I complete our work on Natural Philosophy....My object in sending this to you at present is to ask you and through you Prof. Kirchhoff, whether in attempting to do justice to Joule and Thomson I have done injustice to you or your colleague.

Helmholtz replied at considerable length on Feb. 23, 1867. The greater part of this letter was quoted by Tait in the preface to the book, and was also reproduced in Helmholtz's collected papers. In another portion of the letter, not quoted by Tait, Helmholtz said that he did not think it quite fair to Kirchhoff to be mentioned simply in one line of print with his predecessors in the field of radiation and absorption. On March 1, Tait, after thanking Helmholtz for his "frank and friendly letter" continued:

With regard to Kirchhoff my object was to ascertain whether his paper on what is called, I think, *Wirkungsfunction* and which had reference to the solution of gases in liquids, should have been referred to....The spectrum analysis question is referred to very briefly in my pamphlet which accounts for my not having given his remarkable researches more prominence. But, with reference to your letter, I was under the impression that Stewart had established his priority in giving a complete proof of the equality of Radiation and Absorption....What I recollect is that Stewart answered Kirchhoff's paper in the Philosophical Magazine, and that Kirchhoff did not reply to that answer in which Stewart gave the details of his (supposed) prior proof....

As to Mayer I had no idea that his illness was due to the cold way in which his papers were received; nor, had I known this, would I have written so strongly against his claims to the *establishment* of the Conservation of Energy....I have always given him full credit for the developments and consequences which he drew from his premises, but at the same time I have held that his premises (though now known to be true) had no basis better than a piece of bad Latin....

In his letter of March 19, 1867, Helmholtz made the following remark regarding Kirchhoff:

"He enters very unwillingly into controversies, and he told me that he had regarded as sufficient what stood in his paper (in the Phil. Mag. 4, XXV, 259), and had not found that any substantially new arguments had been brought forward by his opponents[1]."

At the end of a long letter of date 27 March, 1867, mainly occupied with a discussion of his and Stewart's experiments on the rotating disk in vacuo, Tait asked Helmholtz:

Is it fair to ask you whether you think with Clausius that my little pamphlet will only do me harm—or with Thomson and Joule (who, of course, are interested parties) as well as Stewart, who have reported favourably on it? I wish to avoid strife and to produce a useful little text-book; but, if Clausius is right, I had better burn it at once.

In his reply of April 30, 1867, Helmholtz considered some of the difficulties in the proposed experiments on the rotating disk, then thanked Tait for his offer to publish his translation of the paper on " Vortex Motion," and finished with the following wise words:

"In regard to the question of the publication of your sketch of the history of the mechanical theory of heat it is very difficult to give advice. For my part I must say that I have a great aversion to all priority quarrels and have indeed never myself protested against the greatest misappropriation (*eingriffe*); and I find that I have in this way really come well off, and that in the end my Own has been adjudged again. But I know that my best friends think differently on this matter, and that I stand pretty much alone in my opinion. Further as regards the services of Joule and Thomson in the matter under discussion they appear to me to be so completely and generally recognised by all intelligent people with whom I have spoken that a polemic in their interests is hardly needed.

"If then you divest your writing of its polemical garb it will in my opinion be thankfully received and will have more influence than with this polemic.

"This is my opinion, since you have wished to hear it; naturally I shall not take it ill if you do not follow it, since I do not know enough of the personal conditions which may be moving you."

Tait however thought otherwise, and in the interests of his friends and for what he regarded as the truth he sent forth his book in all its individuality.

The general character of the three chapters may be inferred from their titles, namely, (i) Historical Sketch of the Dynamical Theory of Heat, (ii) Historical Sketch of the Science of Energy, (iii) Sketch of the Fundamental Principles of Thermodynamics. The book opens with the arresting question, What is Heat? The gradual undermining of the old Caloric Theory

[1] It may be remarked here that Lord Rayleigh, in his paper "On Balfour Stewart's Theory of the Connexion between Radiation and Absorption" (*Phil. Mag.*, Jan. 1901), agrees with Tait that Stewart had made out his case.

by the discoveries of Black, Davy, and Rumford, and the ultimate triumph of the Energy Theory are sketched in a racy interesting manner; and for the first time in an elementary text-book Carnot's Cycle of Operations and his notion of the Perfect Reversible Engine are expounded in detail. The early and practically contemporaneous researches in the dynamical theory of heat of Clausius, Rankine, and Thomson are explained, and the whole subject brought into relation with the laws of Radiation.

In the second chapter Tait takes a much wider sweep, and passes in review a series of striking examples of the transformations and the conservation of energy. These are taken from the recognised branches of physics, dynamics, sound, electricity and magnetism, in all their aspects, solar radiation, gravitational energy, physiological activity, and tidal retardation and other illustrations of the dissipation of energy.

The third chapter is an exposition of the dynamical theory of heat as it was developed by the pioneer workers on the subject. When engaged in putting it together Tait wrote the following three letters to Thomson:

6 G. G., E., 4/1/68.

Dear T,

The compts. and best wishes of the season from all here to you and Lady Thomson, of whom we hope to hear good accounts....

I don't understand Macmillan at all.... He says everything is arranged, and doesn't tell me how[1]. But I'll find out before I take any further step. Meanwhile, how about the remaining sheets of the pamphlet which you took with you?

I have been writing at the third chapter of my D. T.[2] and have undergone some very laborious reading at Clausius' *Abhandlungen*. I find that he calls $\frac{H}{t}$ (he calls it T, not having begun by defining temp^{re} properly) the equivalence-value of a quantity H of heat at temp^{re} t. Then the second law becomes the assertion of equivalences

$$\frac{H}{t} + \frac{H_0}{t_0} = 0,$$

where H_0 is negative. Then he goes to your

$$\Sigma \frac{H_t}{t} = 0$$

for reversible processes (in your proofs of which, by the way, some lines and steps

---

[1] This has reference to the 2nd edition of "*T and T'*," which however was not seriously entered upon till 1875.

[2] Dynamical Theory of Heat.

are altogether left out, so that it is no proof at all, a fact I only discovered on copying it out into my MSS), and he generalises it into

$$\int \frac{dH}{t},$$

which he calls the "Entropy." This is the equivalence-value of the actual heat and of the potential of internal work—the latter of which he calls "Disgregation." All the rest is mere playing with diff^1 $=^{ns}$. Have you done no more about dissipation than is to be found in the Phil. Mag.? The entropy and equivalence-values, etc., are merely dissipation enunciated in other language. I have my third Chap. nearly finished, and shall probably have both ready for your perusal when you return, as I wish your "polishing" to be done before I go to press. I have kept a few of Rankine's geometrical things at the commencement, and have then bagged freely from you. In the 4th Chap. I shall do some geometrical things about Leyden jars and then bag freely from H^2 [i.e., H. Helmholtz's paper, " Ueber die Erhaltung der Kraft "].

Macmillan has sent me an interleaved copy of Vol. I^1, stitched up in 4 equal parts. This looks like speedy work for the new edition.

I got hold of Maxwell lately, and have managed to extract from him a promise to send constant contributions to the R. S. E. He has also undertaken to write accounts of his own, and some other people's work for my D. T.

Snow, sleet, and hail alternately—while you are among the oranges and myrtles !!!

Yrs.   T'.

6 G. G., E.
13/1/68.

Dear T,

  I have written to Bertrand, and sent him papers as you counselled—no answer.

My laboratory is getting on capitally but it will take some time to dry before we can safely work in it.

When two soap-bubbles unite into one, how much work is done? Note, that the common surface *diminishes*, but the whole bulk *increases*.

I have finished Chap. III of my Sketch of Thermodynamics, and as you are not available I have sent it to Maxwell to look over. You will be rather surprised when you see the quiet way in which I have bagged from you and Rankine....

Yrs.   T'.

6 G. G., E.
18/1/68.

Dr T,

  Have you got a note I sent you a week or two ago—giving you Forbes' address?...

---

[1] That is, of the *Natural Philosophy*; compare footnote (1) on previous page.

My third chapter is ready, and I had thought of sending it to press—but on second thoughts, I refrained—knowing that you would make far more serious alterations on *printed* pages (especially if formed into *sheets*) than you would care to do on a mere MS. Once you have looked over it in MSS you will not have the face to protest against it in type.

You are getting imbued with a little of Pecksniff—rather as regard motives and actions than as regards style; still you *have* caught some of the style also.

But it is simply true, as I told you, that your *printed* proof of

$$\Sigma \frac{H_t}{t} = 0$$

is no proof at all—not even a *chain* of reasoning, merely a set of detached links! How you let it be printed in such a state I can't imagine. Everybody sees you had the proof in your eye, but whether you or the printers omitted a leading step I can't of course tell....

When do you return? U must come to I, or I 2 U, as soon as possible,—for there is *very* much pressing work.

Yrs.   T'.

In July, 1868, Maxwell acknowledged receipt of the complete work in these words:

"I will write you about your treatise at earliest but (1) I, personally, am satisfied with the book as a development of T' and as an account of a subject when the ideas are new and as I well know almost *unknown* to the most eminent scientific men. It is a great thing to get this expressed anyhow and I think you have done it intelligibly as well as accurately. But with respect to the bits of matter I sent you, do you not think there are breaches of continuity between some, e.g., the statement about dynamical theories and the context, if they do not actually contradict the context, at least the N. B. Review part of it. If you disagree with anything of mine, out with it, for it is better to go into print having one opinion than with two opinions to throw the reader into perplexity.

"(2) I shall see what case Clausius has.

"(3) Who is Charles that I might believe on him?"

In a review of the second edition of Tait's *Thermodynamics* in *Nature*, Vol. XVII (1877), Maxwell wrote:

"In the popular treatise, whatever threads of science are allowed to appear, are exhibited in an exceedingly diffuse and attenuated form, apparently with the hope that the mental faculties of the reader, though they would reject any stronger food, may insensibly become saturated with scientific phraseology, provided it is diluted with a sufficient quantity of more familiar language....In this way by simple reading the student may become possessed of the phrases of the science without having been put to the trouble of thinking a single thought about it....

"The technical treatises do less harm, for no one ever reads them except under compulsion....

"Prof. Tait has not adopted either of these methods. He serves up his strong meat for grown men at the beginning of the book, without thinking it necessary to employ the language either of the nursery or of the school; while for younger students he has carefully boiled down the mathematical elements into the most concentrated form, and has placed the result at the end as a *bonne bouche*, so that the beginner may take it in all at once and ruminate upon it at his leisure.

"A considerable part of the book is devoted to the history of thermodynamics, and here it is evident that with Prof. Tait the names of the founders of his science call up the ideas, not so much of the scientific documents they have left behind them in our libraries, as of the men themselves, whether he recommends them to our reverence as masters in science, or bids us beware of them as tainted with error. There is no need of a garnish of anecdotes to enliven the dryness of science, for science has enough to do to restrain the strong human nature of the author, who is at no pains to conceal his own idiosyncrasies, or to smooth down the obtrusive antinomies of a vigorous mind into the featureless consistency of a conventional philosopher."

The succeeding paragraphs contained a masterly account of the scientific methods of Rankine, Clausius, and Thomson.

In this *bonne bouche* of a third chapter, as Maxwell humorously called it, Tait gives an extremely compact and instructive sketch of the mathematical elements of the subject.   Beginning with Watt's energy diagram he develops Carnot's cycle in its modern form, and then, possibly following Maxwell's advice quoted above, discusses with great clearness Thomson's scale of absolute temperature. With this in hand and with the further assumption based on experiment that Carnot's function is inversely as the absolute temperature, Tait is able to present Thomson's original treatment in a simplified form. When, however, Tait explicitly referred to $\int \frac{dq}{t}$ as

"Thomson's expression for the amount of heat dissipated during the cycle" Clausius found cause of complaint, claiming the above integral as his.   In the preface to the second edition Tait showed very clearly that Thomson had the whole thing formulated as early as 1851 ; but not until he and Joule had experimental evidence of the value of Carnot's function would Thomson use any other than the unintegrated form with the symbol μ for Carnot's function.

In a postcard of date Feb. 12, 1872, Maxwell remarked :

"As for C., though I imbibed my $\Theta\Delta$cs from other sources, I know that he is a prime source, and have in my work for Longman been unconsciously acted on by the motive not to speak about what I don't know. In my spare moments I mean to take such draughts of Clausiustical Ergon as to place me in that state of disgregation in which one becomes conscious of the increase of the general sum of Entropy. Meanwhile till

> Ergal and Virial from their throne be cast
> And end their strife with suicidal yell

I remain, Yrs. $\frac{dp}{dt}$."

In a letter of date Oct. 13, 1876, Maxwell made clearer references to the point at issue between Tait and Clausius, and gave at the same time some interesting confessions as to his own knowledge or, rather, ignorance of the subject. He wrote:

"When you wrote the Sketch your knowledge of Clausius was somewhat defective. Mine is still, though I have spent much labour upon him and have occasionally been rewarded, e.g., *earlier* papers, molecular stotting[1], electrolysis, entropy, and concentration of rays....

"N.B. In the latter paper the name of Hamilton does not occur. When you are a-trouncing him, trounce him for that. Only perhaps Kirchhoff ignored Hamilton first and Clausius followed him unwittingly not being a constant reader of the R. I. A. Transactions, and knowing nothing of H. except (lately) his Princip., which he and others try to degrade into the 2nd law of $\Theta\Delta$ as if any pure dynamical statement would submit to such an indignity.

"With respect to your citation of Thomson it would need to be more explicit. The likest thing I find to what you give is in the 1st paper on D. T. of H. (17 March, 1851, p. 272 & 273), but I do not find dq divided by anything like t.

"I think Rankine, by introducing his thermodynamic function ϕ which is $\int dq/t$, made a great hit, because ϕ is a real quantity whereas q is not, only $dq = td\phi$. There are many things in T which are equivalent to this because T has worked at the same subject and worked correctly and all mathematical truth is one, but you cannot expect Clausius to see this unless it stands very plain in print. In short Rankine's statements are identical with those of C, but T's are only equivalent....

"With respect to our knowledge of the condition of energy within a body, both Rankine and Clausius pretend to know something about it. We certainly know how much goes in and comes out and we know whether at entrance or exit it is in the form of heat or work, but what disguise it assumes when in the privacy of bodies, or, as Torricelli says, "nell' intima corpulenza de' solide naturali," is known only to R., C. and Co."

The paper mentioned by Maxwell was not, however, the paper referred to by Tait, as will appear immediately.

Among Tait's correspondence an interesting letter from Thomson (Kelvin)

---

[1] *Stot*, a Scottish word meaning to impinge and rebound, still in constant use among school children of all classes, e.g., to stot a ball. Compare German *stossen*.

was found bearing on this controversy.   It is valuable as showing that Tait's views were fully endorsed by his friend.   The letter begins abruptly with a quotation from Thomson's 1852 paper "On a Universal Tendency in Nature to the Dissipation of Mechanical Energy" (*Proc. R. S. E.* III; also *Phil. Mag.* IV, Oct. 1852; *Math. and Phys. Papers*, Vol. I, No. LIX). To make the quotation quite intelligible, a preliminary sentence seems to be necessary (see *Math. and Phys. Papers*, Vol. I, p. 512).

"'Let S denote the temperature of the steam...; T the temperature of the condenser; μ the value of Carnot's function, for any temperature t; and R the value of

$$\epsilon^{-\frac{1}{J}\int_{T}^{S}\mu dt.}\text{'}\text{''}$$

The letter begins with the integral, and then continues the quotation:

"'Then $(1-R)w$ expresses the greatest amount of mechanical effect that can be economised in the circumstances from a quantity w/J of heat produced by the expenditure of a quantity w of work in friction, whether of the steam in the pipes and entrance ports, or of any solids or fluids in motion in any part of the engine; and the remainder, Rw, is absolutely and irrecoverably wasted, unless some use is made of the heat discharged from the condenser.'   The whole thing is included in this illustration and the preceding 'universal' generalisation of it, of which this is a particular illustration. I don't believe Clausius yet to this day understands as much of the *fact* of dissipation of energy as is stated in that first paper in which the theory is propounded and the name given, and it does not appear that he has ever made any acknowledgment whatever of T in the matter.   This *must* be because he does not understand it; *not* because he would consciously appropriate what is not his own.

"As for the very letters of the formula, T in the same article says, 'If the system of thermometry adopted be such that

$$\mu = \frac{J}{t+\alpha},\text{'}$$

Accepting Clausius' statement that 'neither the expression $\left(t_0\int\frac{dq}{t}\right)$ nor anything of like meaning can be found in the article referred to by T',' the only conclusion is that he is ignorant of the fact that

$$\epsilon^{-\frac{1}{J}\int_{T}^{S}\mu dt} = \frac{T+\alpha}{S+\alpha},$$

and so had his eyes closed to the fact that Rw/J means the same as

$$(T+\alpha)\frac{dq}{S+\alpha},$$

or $t_0\dfrac{dq}{t}$ according to the notation of T'.

"In that same article occurs the expression[1]

$$\iiint \epsilon^{-\frac{1}{J}\int_0^t \mu dt}\, c\, dx\, dy\, dz,$$

which (considering that there is absolutely no limitation of the body to which the $\iiint$ may be applied) supplies with tolerable completeness the $\int$ of the bone over which Clausius snarls, and triumphantly justifies T"'s [Tait's] § 178.

"Lastly remark that the very formula for the 'part of it (the heat) rejected as waste into the refrigerator at the temperature T' in the other article[2] referred to by T' (§ 179) is

$$dx\, dy\, dz\, c\, dt \cdot \epsilon^{-\frac{1}{J}\int_T^t \mu dt.}$$

The $dx\, dy\, dz\, c\, dt$ here is T"'s dq and $\epsilon^{-\frac{1}{J}\int_T^t \mu dt}$ is $1/t$ when the thermodynamic thermometry is used.

"Last lastly remark that while T was keeping the notation μ he was working along with Joule (*Phil. Mag.* 1852, second half-year) to find whether $t = J/\mu - \alpha$ agreed approximately enough with air thermometer ordinary reckoning to be a convenient assumption, and (*Phil. Mag.* 1853 first half-year) intimated that it was so (and set forth the same more fully afterwards with Joule, Trans. R. S.): and from 1851 (Dynamical Theory of H., Part I forward) T had the formula

$$\mu = \frac{J}{\frac{1}{E}+t},$$

and kept putting forward in all his papers till he finally adopted [? J/T], leaving absolutely no room for Clausius' pretensions. Cl. in fact never showed any right whatever to

$$\mu = \frac{J}{\frac{1}{E}+t},$$

and till this day has not put it on its right foundation.

(Signed) T."

The argument in this letter is practically identical with what Thomson himself allowed Tait to publish in the *Philosophical Magazine* for May 1879 as a note to Tait's own communication "On the Dissipation of Energy"; but the tone of it is more personal. The statements in the last paragraph

---

[1] [Marginal note by Thomson himself.] The misprints corrected in *Phil. Mag.*, Jan. 1853, bemuddle the formula for final uniform temperature but *not* the meaning of the dissipation and the formula $t_0 \int \frac{dq}{t}$ for it.

[2] "On the Restoration of Mechanical Energy from an unequally heated Space," *Phil. Mag.* v, 1853, *Math. and Phys. Papers*, Vol. I, No. LXIII, p. 555.

can easily be verified by referring to the papers mentioned, which are now conveniently collected together as Articles XLVIII, XLIX, LIX, and LXIII in Volume 1 of the *Mathematical and Physical Papers*. All of these except the later parts of XLVIII preceded the publication of Clausius' Fourth Memoir, which appeared in Poggendorf's *Annalen* in December 1854, and in which the Entropy integral is given by Clausius for the first time.

The second edition of Tait's *Thermodynamics* was published in 1877. In it he makes more emphatic his criticism of the original form of the axiom which Clausius used as the basis of the Second Law of Thermodynamics, and is less eulogistic in his references to Clausius' thermodynamic work in general. The facts are all given in due order; but Clausius was not satisfied with the manner in which his work was presented, and criticised strongly the general "Tendency" of Tait's historical sketch of the dynamical theory of heat.

Tait has by some writers been accused of Chauvinism in his treatment of scientific history. It seems to me that the charge is ill-founded. His championship of Joule and Thomson as two of the real founders of Thermodynamics and of Balfour Stewart as having established, in relation to the laws of radiation, certain truths that were almost universally ascribed to Kirchhoff, is probably what is in the mind of those who make the charge. But all are agreed as to the eminence of Joule and Thomson, and nothing that Tait wrote could ever be interpreted as detraction of Kirchhoff. Nevertheless Balfour Stewart's work was not then appreciated at its true value. Even Lord Rayleigh's more recent championship[1], which is quite as strong as Tait's, has not yet had its full impression on the scientific world. Probably the charge of Chauvinism against Tait may be attributed in some measure to the vigour of his onslaught on anything which he regarded as bad history, and to the glee with which he exposed it. Except in France, Boyle's Law is the name now universally given to what used to be even in this country called Marriotte's Law[2]; but it needed Tait to discover evidence in Newton's *Principia* and in Marriotte's own writings that Marriotte had a skilful way of

---

[1] See his paper "On Balfour Stewart's Theory of the Connexion between Radiation and Convection," *Phil. Mag.*, 1, January 1901,—regarding which Lord Rayleigh says, "Kirchhoff's independent investigation of a year and a half later [Dec. 1859] is more formal and elaborate but scarcely more convincing."

[2] The name occurs even in the First Edition of "T and T'," § 597 !

expounding other people's discoveries as if they were his own. Boyle, no doubt, was an Englishman; but it cannot be claimed that Marriotte preceded him. There is no Chauvinism here on Tait's part. On the other hand it was Tait, who, accepting the statement of Gay-Lussac, secured for "le Citoyen Charles" the recognition of his rights in relation to the law of gases, named after Dalton in this country and after Gay-Lussac on the Continent. It was Tait more than any other individual writer who popularised Carnot's Cycle of Operations and the Perfect Engine, which are expounded not only in his purely scientific works but also in *The Unseen Universe*. Again, Tait, by translating Helmholtz's paper on Vortex Motion, gave a new direction to hydrodynamical study in this country. No doubt he felt warmly any attempt, conscious or unconscious, to credit to others discoveries made by any of his own countrymen, and in this he was not peculiar; but I know of no case in which he claimed for a fellow countryman anything which could be demonstrably associated with the name of another at an earlier date. He used to say that, if laws are to be named after their first discoverers, then Ohm's Law should be called after Fourier, and Doppler's Principle after Römer. In these instances there is nothing Chauvinistic.

There are now many books on Thermodynamics of various standards, each having its own merit. But as an account of the fundamental principles in their historic setting Tait's *Sketch* cannot be surpassed. The prominence given to Carnot's Principle, the simplicity and directness of the mathematical methods introduced into the third chapter, the beautiful illustrations of the transformation of energy given in the second chapter, and the clear account of the manner in which Thomson seized hold of the original conception of absolute temperature,—all give the book a character peculiarly its own. Abbé Moigno, the well-known mathematician and editor of *Les Mondes*, saw at once the value of the work, and with the help of M. Alfred Le Cyre, published a French translation in 1870. The preface opens with these sentences:

"Lorsque je lus pour la première fois l'Esquisse historique de la théorie dynamique de la chaleur, trois choses me frappèrent vivement: 1°, l'auteur résume rapidement et complètement les travaux accomplis dans cette branche aujourd'hui si étendue de la physique mathématique; 2°, il rend parfaitement à chacun la justice qui lui est due; 3°, il établit en quelques pages trés-nettes et trés-élégantes synthétiquement d'abord, analytiquement ensuite, les lois fondamentales de la dynamique de la chaleur."

The next work by Tait which calls for notice is his *Recent Advances in Physical Science* (Macmillan & Co., 1876, 2nd Edition, 1876), the published form of a course of lectures which Tait gave by request to a company of some ninety Edinburgh citizens, mostly professional men. The lectures were delivered in his usual style from the briefest notes, and the book was compiled from the verbatim shorthand report. Of all Tait's published works it gives the best idea of his method as a lecturer. One of its greatest merits to a real student of the subject is the exposition of Carnot's Principle. The name of Carnot was first introduced in Lecture IV on the Transformation of Energy, and occurred again and again throughout the succeeding chapter on The Transformation of Heat into Work. The story goes that when Tait began the Sixth Lecture with the words " I shall commence this afternoon by taking a few further consequences of the grand ideas of Carnot," an elderly pupil sitting towards the back was heard to protest vehemently against the name of Carnot.

The published book contains thirteen lectures, but some of the lectures delivered were not published. I remember for example being one of a few undergraduates who were allowed to join the class on two of the occasions on which it met in the University. This change of meeting-place was for the sake of the experimental illustrations, which could not well be performed in an ordinary hall. These two lectures on the Polarisation of Light and Radiant Heat do not appear in the volume, probably because much of the subject matter could not be regarded as recent in the sense in which the doctrine of energy was recent.

In addition to the clear exposition of the foundations of the modern theory of energy, Tait gave in these lectures an admirable account of the physical basis of spectrum analysis and the first great discoveries made by Kirchhoff and Bunsen, and by Huggins, Lockyer, Young and others. Astrophysics is now a branch of astronomy claiming its own specialists and possessing its own literature ; but, in the seventies, solar and stellar spectroscopy was but a particular illustration of the broad principle of spectrum analysis. Another important section of *Recent Advances* was devoted to the discussion of the atom and molecule, their magnitudes and masses, and even their ultimate constitution.

The book was reviewed in all our best papers and journals at considerable length, in general with high commendation. The following quotation from an article in the *Quarterly Review*, Vol. 142, entitled " Modern Philosophers on the Probable Age of the Earth " may be taken as a good type of the appreciative notices which abounded.

" His lectures now before us, from their nature, belong to the class of composition for which we avow our predilection. They were delivered extempore to a scientific audience, and printed from short-hand notes. They lose nothing of their vigour, to use an expression of Lord Macaulay, by translation out of English into Johnsonese. We are allowed to seize the thought in the making, and if it loses anything in grace, the loss is more than counterbalanced by power.

" Those who wish thoroughly to understand the subject of this paper should study Professor Tait's lectures on the sources of energy, and the transformation of one sort of energy into another. Matthew Arnold's phrase, 'let the mind play freely round' any set of facts of which you may become possessed, often recurs to the mind on reading these papers. There is a rugged strength about Professor Tait's extempore addresses which, taken together with their encyclopaedic range, and the grim humour in which the professor delights, makes them very fascinating. They have another advantage. Men not professionally scientific find themselves constantly at a loss how to keep up with the rapid advance which has characterised recent years. One has hardly mastered a theory when it becomes obsolete. But in Professor Tait we have a reporter of the very newest and freshest additions to scientific thought in England and on the Continent, with the additional advantage of annotations and explanations by one of the most trustworthy guides of our time."

The second edition of *Recent Advances* was translated into German by G. Wertheim (Braunschweig, F. Vieweg und Sohn, 1877); into French by Krouchkoll (Paris, Gauthier Villars, 1887); and into Italian by D' Angelo Emo (Fano, Tipografia Sonciniana, 1887).

After the publication of *Recent Advances* Tait became occupied with the preparation of the second edition of " T and T'." In the preface to the second volume which appeared in 1883 it is stated that the continuation of the great work had been abandoned. Tait accordingly turned his attention to the production of a series of elementary text-books, more in the line of what he originally intended before Thomson joined him in 1861.

In 1884 and 1885 Tait brought out three books on *Heat, Light*, and *Properties of Matter*.

What gives the book on *Heat* its distinguishing features are the introductory chapters, especially Chapter IV. After a rapid historic survey of the growth of the modern conception of heat, Tait introduces the First Law of Thermodynamics. Typical examples are given of the effects and production of heat, leading up to the great principle of Transformation and to the Second Law of Thermodynamics. Then follows Kelvin's definition of absolute temperature. By thus early introducing the true conception of temperature he is able to discuss all the familiar thermal changes in volume and state in terms of the absolute temperature.   A German translation by Dr Ernst Lecher was published in 1885 (Wien, Toeplitz und Deuticke).

The book on *Light* (second edition 1889; third edition 1900) was based on the article "Light" which he supplied to the ninth edition of the *Encyclopaedia Britannica*.   Many paragraphs are identical in the two publications; but the article contains a sketch of Hamilton's Characteristic Function which does not appear in the book; while the book contains an able discussion of Radiation and Spectrum Analysis, which are done under separate headings in the *Encyclopaedia*.   The mathematical discussions are of a higher order than in *Heat*, the geometrical theorem on which he finally builds the explanation of the rainbow being especially worthy of note.   Particularly interesting are the quotations from Newton, Huyghens, and Laplace with reference to the undulatory and emission theories of light.

Of these text-books written by Tait on different branches of natural philosophy perhaps the most characteristic is the *Properties of Matter* (1885, successive editions, 1890, 1894, 1899 and 1907, the last under the able editorship of Professor W. Peddie).   A German translation by G. Siebert was published in 1888 (Wien, A. Pichler's Witwe und Sohn).   The *Properties of Matter* is the book which will best recall to his former students the personality of Tait as a lecturer.   It embodies much of the earlier half of the course of study through which Tait gave his many students a "common sense view of the world we live in."

The headings of the chapters show the scope of the book, concerning which Lord Rayleigh in his review (see *Nature*, August 6, 1885, Vol. xxxii) remarked that it was not easy to give a reason why electric and thermal properties of matter should be excluded.   The reason is undoubtedly historic, the phrase "Properties of Matter" dating from the time when the mechanical *ponderable* matter was distinguished from the *imponderables* heat, light, electricity and magnetism.   The first three chapters are devoted to a discussion of what matter is, and contains lively criticism of the metaphysicians. Then come Time and Space; Impenetrability, Porosity, Divisibility; Inertia, Mobility, Centrifugal Force; Gravitation; Deformability and Elasticity; Compressibility of Gases and Vapours; of Liquids; Compressibility and Rigidity of Solids; Cohesion and Capillarity; Diffusion, Osmose, Transpiration, Viscosity; Aggregation of Particles.

Lord Rayleigh in his review specially referred to the treatment of elasticity, remarking that the Chapters on Deformation and Compression "are perhaps the most valuable part of the work, and will convey a much needed precision of ideas to many students of physics whose want of mathematical training deters them from consulting the rather formidable writings of the original workers in this field. The connection of Young's modulus of elasticity...with the more fundamental elastic constants...is demonstrated in full.... In his treatment of the compression of solids and liquids the author is able to make valuable contributions derived from his own experimental work.

"In the chapter on 'Gases' a long extract is given from Boyle's 'Defence of the Doctrine Touching the Spring and Weight of Air,' in order to show how completely the writer had established his case in 1662. As to this there can hardly be two opinions; and Professor Tait is fully justified in insisting upon his objections to 'Marriotte's Law.' In Appendix IV a curious passage from Newton is discussed, in which the illustrious author appears to speak of Marriotte sarcastically. It is proper that these matters should be put right...."

A paragraph from Balfour Stewart's review of Tait's *Heat* (*Nature*, June 26, 1884, Vol. xxx) seems to be worthy of quotation as an interesting description of Tait's method and style in all his books.

"A treatise on heat by one so eminent, both as physicist and teacher of Physics, needs no apology, and yet no doubt the author is right in stating that his work is adapted to the lecture room rather than to the study or the laboratory. Freshness and vigour of treatment are its characteristics, and the intelligent student who reads it conscientiously will rise from it not merely with a knowledge of heat but of a good many other things beside.

"'If science,' says our author, 'were all reduced to a matter of certainty, it could be embodied in one gigantic encyclopaedia, and too many of its parts would then have... little more than the comparatively tranquil or rather languid interest which we feel in looking up in a good gazetteer such places as Bangkok, Akhissar, or Tortuga.' Not a few text-books of science are precisely of the nature of such a guide without its completeness, and while they carry the student successfully to the end of his journey, the way before him is made so utterly deficient in human interest that he reaches his goal with a sigh of relief, and looks back upon his journey with anything but satisfaction —as a task accomplished rather than a holiday enjoyed. Now the presence of such a human interest is the great charm of the work before us. It may be a fancy on our part, but we cannot help likening our author to the well-known guide of Christiana and her family. Both have been equally successful in the slaughter of those giants whom the older generation of pilgrims had to find out for themselves and encounter alone. But here the likeness ends, for it is quite certain that those who place themselves under the scientific guidance of our author will not be treated like women or children, but they will be taught to fight like men. And surely to combat error is an essential part of the education of the true man of science, for, if not trained up as a good soldier of the truth to defend the king's highway, he will be only too apt to turn freebooter and gain his livelihood by preying on the possessions of others."

These text-books, especially the *Heat* and *Properties of Matter*, were of course very useful helps to the students of the general class of Natural Philosophy. In the earlier days "Little T and T'" and Tait's *Thermodynamics* were the only books which were serviceable in supplementing the lectures. The former was a sealed book to the majority of those studying for the ordinary M.A. degree; and the latter in its first chapter covered a limited ground, while most of the second chapter was too condensed food for the ordinary mind to assimilate. We had, accordingly, to trust largely to the lectures, for the mode of treatment and the illustrations given were peculiarly Tait's own.

The article "Mechanics" which Tait contributed to the *Encyclopaedia Britannica* in 1883 formed the foundation of an advanced text-book on *Dynamics*, which was published in 1895 (A. and C. Black). Having used paper-bound copies of the article as a text-book in his Honours Class for the twelve intermediate years Tait was able, when its publication in usual book form was determined on, to modify and improve along lines which experience had indicated. As explained by Tait in the letter to Cayley quoted above (p. 155), the *Britannica* article was originally planned by Maxwell; but the details had to be arranged by Robertson Smith, the editor, so as not to overlap other articles. The book accordingly, although largely a reprint, contains sections on Attraction, Hydrodynamics, and Waves which were not in the original article.

If from the point of view of the student the book has a fault, it is that of brevity and conciseness. But there can be only one opinion as to its thoroughness and accuracy. The ground covered is greater than in any other book on the subject, for it includes not only what is ordinarily understood by Dynamics of particles and rigid bodies but also the more important parts of elasticity and motion of fluids. The foundations are Newton's Laws of Motion; for although Tait had himself, in scientific papers and otherwise, tried to devise a system free from the explicit assumption of Force in the Newtonian sense, yet to the end he regarded Newton's Laws of Motion as the most practical way of introducing the student to a study of the subject.

Naturally there are strong resemblances between Tait's *Dynamics* and "T and T'," especially in certain modes of proof; but in his own book Tait restrains himself from treating developments which make a great demand upon the mathematical knowledge of the reader.

Occasionally the extreme brevity of a statement is such that the student on a first reading fails to see immediately all that is implied; but a critical examination of such statements shows that they are complete without being redundant. Among the parts which are particularly characteristic of Tait's methods the following may be mentioned : discussion of Fourier's series, of strains, of Attractions and Potential, of Action (under which is included the flow of electricity in a surface), of the strength of tubes under internal and external pressures, of the bending and vibration of rods, of vortex motion, and of surface waves on fluids. Perhaps the practical nature of the book is best indicated by the way in which Lagrange's generalised coordinates are introduced. Having established in ordinary Cartesian symbolism Hamilton's principle of Varying Action, Tait then uses this principle to deduce the usual Lagrangian equations of motion. The demonstration is not general or exhaustive, but it is sufficient for the kind of problems which most naturally present themselves to a student beginning the study of higher dynamics.

Tait's demonstrations, whether geometrical or analytical, are characterised by neatness and elegance. He used to say that he could always improve a demonstration given by some one else. When reading a newly published paper he was able very rapidly to come to an opinion as to its originality and accuracy. Thus, as already noticed (p. 113), he was very critical of certain of the mathematical processes used in investigations regarding the kinetic theory of gases. If a theorem could not be proved without a prodigious array of symbols covering pages, he had a feeling that the theorem was not worth the proving. His attitude of mind towards much of mathematical literature is well brought out in the answer he gave to one of his sons about the year 1878. He was turning over the pages of a mathematical journal which had just come by post. When asked if he was going to read the journal right through, he remarked: "Certainly not. I am not such a flat as to read other people's mathematics. I look to see what result the beggar brings out, and then if he's right I can usually find a shorter cut."

About 1892 Tait formed the project of printing a small pamphlet of concise paragraphs to take the place of lecture notes for his students, who would thus be able to pay undivided attention to the explanations and amplifications given in the lectures. Some twenty or thirty pages were put in type, but pressure of other work, more particularly the editing of the reprint of Scientific Papers, prevented the project being carried to completion. When reminded by the publishers that these pages had been lying in type for nearly six years Tait felt that he was not able to carry out fully the original intention, and compromised the matter by confining these notes to a highly condensed discussion of Newton's Laws of Motion, in other words, the foundations of dynamics. A small book of fifty-two pages, and entitled *Newton's Laws of Motion*, was finally published in 1899 by A. and C. Black.

The book contained a brief introduction on Matter and Energy and then two chapters on Kinematics and Dynamics respectively. In a review by A. E. H. L. in the columns of *Nature* (Vol. LXI, January 18, 1900) the book was commended as being

"for the most part excellent, the geometrical methods employed being especially elegant. Room is found for an elementary discussion of strain, of compounded simple harmonic motions, of attractions, including the distribution of electricity on a sphere under influence, and of the velocity of waves along a stretched cord, in addition to interesting and unhackneyed accounts of the matters which are the stock in trade of books on the elements of mechanics. The book on the whole is thoughtful, in many parts it is much better than the current text-books on the subject, and the parts that call for criticism are no worse than the corresponding parts of most other books on the subject; but they are the most important parts, and they might have been so much better. There was a great opportunity, but it has been missed."

Part of the criticism virtually amounted to a complaint that certain sections were not sufficiently expanded. Tait's own preface may be regarded as an answer to this kind of objection; for the book is explicitly stated not " to be a text-book" but " a short and pointed summary of the more important features of ... the basis of the subject." For example " to explain " (as was desired by the critic) " the mathematical notion of a limit " requires not " some space " but a good deal of space, if the explanation is to be complete. Nevertheless, the following brief paragraphs show that the conception of physical and dynamical continuity, on which fundamentally the notion of the limit rests, was explicitly recognised by Tait:

10. When we pass from the consideration of displacement to that of motion, the idea of time necessarily comes in. For motion essentially consists in *continued* displacement. In the kinematics of a point, all sorts of motion are conceivable: but we limit ourselves to such as are possible in the case of a *particle of matter*.

11. These limitations are simple, but very important.

(*a*) The path of a material particle must be a *continuous* line. [A gap in it would imply that a particle could be annihilated at one place and reproduced at another.]

(β) There can be no instantaneous finite change in the direction, or in the speed, of the motion. [*Inertia* prevents these, unless we introduce the idea of finite transformations of energy for infinitely small displacements, or (in the Newtonian system) infinite forces.]....

14. If the speed be variable its value, during any period, must sometimes exceed

and sometimes fall short of the average value. But (by 11 (β), above, and therefore *solely* in consequence of inertia) the shorter the period considered, the more closely will the actual speed of a material particle agree with the average value: and that without limit.

Again, very early in the book, Tait warned the reader against the inevitable anthropomorphism which clings to our words and phrases; yet he was attacked for using Newton's anthropomorphic definition of force as a *cause* and at the same time pointing out its true nature as simply a *rate of change* of a quantity in time. As regards the general criticism that Tait followed too slavishly Newton's presentation of the foundations of dynamics, there is a great deal to be said on both sides. Tait's experience had convinced him that for junior men Newton's method was the best, dealing as it did with immediate sensations and perceptions. For that reason he called the book *Newton's Laws of Motion*.

But, although in this small pamphlet Tait felt himself compelled to adhere to Newton's method, every one interested in the subject knew that he had in one published paper attempted to establish the laws of motion on a wider basis free from the explicit use of the word Force. This paper "On the Laws of Motion, Part I," was printed (but only in Abstract) in the *Proceedings* of the Royal Society of Edinburgh, 1882; and a German translation appeared in a German mathematical journal. The Second Part was never written out in a form suitable for publication. When busy with the preparation of the 1882 paper, Tait wrote to Cayley on Nov. 20, 1882, in these words:

Do you know of any attempt to construct the whole system of *Mechanics* (for it would, under the circumstances, be absurd to call it *Dynamics*) from general principles, such as Conservation and Transformation of Energy, Least Action, etc., without introducing either Force, Momentum or Impulse? I have worked out a scheme of the kind having been led to it by writing a long article for the *Encyc. Brit.* Not that it goes in there, of course, but because in speaking of the anthropomorphic terms in which Newton's Laws are expressed—(e.g., a body *compelled* by force to *do* so and so; a body *persevering* in its state of etc. etc.)—I tried to find out some simple mode of getting rid of what I find Maxwell has called *Personation*.

Of course, Force constantly **comes in**, but not in any sense as an *agent*, merely as the space-rate of transformation of energy. It **plays a part** in some sense akin to that of temperature-gradient in heat-conduction. But I see, by the words I have doubly underlined, how very difficult it is to avoid anthropomorphism. I suppose it must always be so, unless scientific men protest effectively against "the sun rises," "the wind blows," etc. etc.

If any such scheme has appeared, I should like to consider it before bringing my notions before the R.S.E.

I have said in my article that no one who has ever rolled a pea on the table under the tips of his index and middle fingers, crossed, will afterwards believe anything whatever on the testimony of his "muscular sense" alone. Yet what other ground have we, for believing in the objectivity of force, than the impression on our muscular sense?

On January 20, 1883, Cayley replied:

"Dear Tait,

I ought to have written ever so long ago in answer to your question as to the construction of a system of mechanics from general principles without Force, Momentum, or Impulse—but it could only have been to say that I did not know of any attempt at such a construction—the idea was quite new to me, and I have not taken it in enough to see anything about it myself—so that you will have lost nothing by the delay. I hope your proposed communication to the R.S.E. will be published......"

On February 26, 1883, Cayley acknowledged receipt of the Paper in these words:

"Dear Tait,

The whole discussion is beyond me—I understand force—I do not understand energy. I am willing to believe that Newton's Action = Reaction potentially includes d'Alembert's principle—but I never saw my way with the former, and do see my way with the latter—and I accept Virtual Velocities + d'Alembert's principle as the foundation of Mechanics. In this position of outer darkness, it would be quite useless to attempt any remark on your paper.

"I send herewith a paper from the A.M.J.; please look at the statement pp. 2–4 of Abel's theorem in its most simple form......"

Tait replied as follows:

38 George Square,
Edinburgh.
28/2/83.

My dear Cayley,

Many thanks for your paper, which I have already looked at and will read. It seems to me that this work may, with a little trouble, be brought to bear on the very important and difficult question of Kinetic Stability. If so, I hope you will develop it largely. I suppose you know Boole's paper in *Phil. Trans.* It was from it that I first got a notion of what Abel's Theorem really means.

Your disclaimer in reference to my *Abstract* is really a vote in my favour. For *Virtual Velocities* is merely the principle of Energy in a mathematical guise; and d'Alembert's principle is **either** the first or second interpretation of Newton's Lex III; and you say that you adhere to *them*.

I say advisedly, *either* the first or second, for there are two quite distinct things which go by the name of d'Alembert's principle:

30—2

I. Some people say *this* is d'Alembert:

Let *m* at *x, y, z,* be a particle the *applied* forces being *X, Y, Z,* and the *internal* forces ξ, η, ζ.

Then $$m\ddot{x} = X + \xi, \text{ &c.,}$$

whence $$\Sigma(m\ddot{x}) = \Sigma X, \text{ &c.,}$$

ξ &c. going out. And the same sort of thing when factors δx, &c. are used.

II. Others say *this* is d'Alembert:

Let the notation be as before. Then the *statical* conditions are

$$\Sigma(X + \xi) = 0, \text{ &c.}$$

whence, introducing the *reversed effective forces*, you get for the *kinetical* conditions

$$\Sigma(X - m\ddot{x} + \xi) = 0, \text{ &c.}$$

And the same sort of thing with any *permissible* displacements as factors. I is merely Lex III direct. II is amply met by Newton's second interpretation of Lex III, where he points out the *Reactiones*, "ex acceleratione oriundis," as forces to be taken into account.

Which is *your* view of d'Alembert?

But there is a point in my paper which may interest you, where I show that the hitherto puzzling *Least Action* merely expresses the inertia condition, so far as the component motion parallel to an equipotential surface is concerned......

In the winter of 1874, a few months after the delivery by Tyndall of his famous presidential address before the British Association at Belfast, it began to be whispered among the students of Edinburgh University that Tait was engaged on a book which was to overthrow materialism by a purely scientific argument. When, in the succeeding spring, *The Unseen Universe*[1] appeared it was at once accepted as the fulfilment of this rumour. The title page of the book contained the words, "THE UNSEEN UNIVERSE, or **Physical Speculations on a Future State**. The things which are seen are temporal, but the things which are not seen are eternal. London, Macmillan and Co. 1875"; and at the top was a trefoil knot, the symbol of the Vortex Atom imagined by Thomson and discussed at considerable length by the authors of the book. In spite of its anonymous publication it seemed to be known from the beginning that the work was written by Balfour Stewart and P. G. Tait. Anyone at all familiar with Tait's scientific style and with his views of the historic development of the modern theory of energy could not fail to see that his hand must have been mainly responsible

[1] Tait greatly enjoyed Gustav Wiedemann's punning criticism that the book should be called the "Unsinn Univers."

for Chapters III and IV, on the Present Physical Universe and on Matter and Ether. Whatever may be thought of the argument of the book, one merit was that, by means of these physical chapters, the great ideas associated with the names of Carnot and Joule were presented to the minds of vast numbers of readers who would never otherwise have come into touch with them.

The book was heralded in a curious old-world fashion by means of an anagram, which was published in *Nature*, October 15, 1874, and signed West, that is, according to Tait's elucidation, We S(tewart) T(ait). This anagram spelled out the sentence

"Thought conceived to affect the matter of another universe simultaneously with this may explain a future state."

This sentence may therefore be regarded as one of the central doctrines of *The Unseen Universe*. It occurs at the end of paragraph 199 in Chapter VII.

The book created a great sensation. It was at once recognised as the work of a scientific author or authors. The fourth edition, which was published in April 1876, exactly a year after the first publication, appeared with the authors' names on the title-page, and subsequent reprints did not differ materially from this edition. The one conspicuously new feature was an introduction setting forth succinctly the motive of the book, which had been strangely misunderstood by some of the earlier critics. Also a few important changes were made throughout, but on the whole the book was essentially the same through all the editions.

One addition to the original form of the text is well worth attention, being a fine example of the kind of humour which Tait occasionally delighted in. The end of paragraph 103 originally ended with the sentence :

"The one (i.e., matter) is like the eternal unchangeable Fate or Necessitas of the ancients ; the other (i.e., energy) is Proteus himself in the variety and rapidity of its transformations."

In the later editions this sentence is followed by six lines of Greek verse, namely :

φύσις, διαδοχαῖς σχημάτων τρισμυρίοις,
ἀλλάσσεται τύπωμα, Πρωτέως δίκην,
πάντων ὅσ' ἔστι ποικιλώτατον τέρας·
τῆς δ' αὐτ' Ἀνάγκης ἐστ' ἀκίνητον σθένος,
μόνη δ' ἀπάντων ταὐτὸ διαμένουσ' ἀεὶ
βροτῶν τε καὶ θεῶν πάντ' ἀποτρύει γένη.

A footnote states:

Thus paraphrased for us:

> Nature, bewildering in diversity,
> Of marvels Marvel most inscrutable,
> Like Proteus, altereth her shape and mould;
> But Fate remaineth ever immovable,
> And, changeless in persistency, outwears
> The Time of men, the gods' Eternity.

Recalling that Professor D'Arcy W. Thompson had once remarked to me that he believed his father Professor D'Arcy Thompson of Galway had supplied Tait with some Greek verses, I drew his attention to the lines, and obtained the following reply of date June 4, 1908:

"Many thanks indeed for your letter of the 31st May, which, with its enclosures, interests me very much indeed.

" I cannot of course absolutely testify that the verses are my father's, but everything points that way:

(1) I know that my father did some verses for *The Unseen Universe*, and, as far as my recollection goes, they were on just such lines as these you send;

(2) The Greek is extremely like Euripides, the author whom my father told me he had imitated;

(3) The English paraphrase strikes me as being exactly in my father's style.

" My father certainly told me that Tait had *asked* him to make those verses for the book, so that little piece of waggery of inserting them for the admiration of the reader and the mystification of the scholar was Tait's doing......

(Signed) D'ARCY W. THOMPSON."

This is an example of the way in which Tait prevailed upon his friends to help in adding interest to the pages of *The Unseen*.

Robertson Smith, the eminent Semitic scholar and theologian, seems to have given valuable hints throughout, as may be inferred from the following letter written by Tait on June 5, 1875.

My dear Smith,

Macmillan gives me private information that in a few weeks a second edition of the *U. U.* will be wanted. He deprecates any material change, partly on its own merits, mainly on the inevitable delay it would involve.

Now, while I still most strongly hold to your kind promise to (some day soon) rewrite the first chapter for us, I think Mac. is right that there should be no material change in the second edition—especially as but few of the great critics have yet spoken out, and we must not at once abandon our first essay as if afraid of what may be ultimately said of it. We must be *at first* a Lucretian Atom not a vortex ring, strong in solid singleness, not wriggling meanly away from the knife! Will you therefore, by little instalments, as it suits you, give me *soon* all the more vital improvements which occur to you as possible without much altering the pages, etc. (the type having been kept up, so as to save expense)?

You have of course seen Clifford's painful essay in the *Fortnightly*.......

An advanced ritualist, MacColl, has cracked us up in a letter to the *Guardian* last week. This week the other ritualist paper *The Church Herald* says our book is *infidel*. Last week the *Spiritualist* said that with a few slight changes the book would be an excellent text-book for *its* clients. The *Edinburgh Daily Review* says we are subtle and dangerous materialists. Hanna (late of Free St John's here) says the work is the most important defence of religion that has appeared for a long time! Which of these is nearest the truth?......

The *Church Herald* is down on us for *your* suggestion about "*for a little while lower than the angels.*"

Truly the reviews and critiques of *The Unseen Universe* were as varied as the religious and irreligious views of the critics who wrote them. To one it was a "masterly treatise," to another it was full of "the most hardened and impenitent nonsense that ever called itself original speculation." Some sneered at the authors for their ignorance of philosophical thought and phraseology; others were captivated by the "acute analytical faculty," the "broad logical candid turn of mind" displayed. Many of the early appreciations of the book were certainly crude, hastily conceived, and hurriedly presented before their readers. On various sides the intention of the argument was not clearly apprehended. There was a novelty in the mode of presenting it, with an appeal to the profoundest truths of modern physics, which rather confused the mind of the ordinary critic unskilled in Carnot cycles and reversible engines. One critic there was, the versatile and brilliant Clifford, who knowing these truths in all their purely physical significance, gave the authors a terrible trouncing in the *Fortnightly Review* (June 1, 1877). The critique is reprinted in his *Lectures and Essays*, but with some of the liveliest passages deleted. The most important omission is the opening paragraph, which in its original form presented Clifford in the guise of a clever debater, who burlesques the argument he intends to demolish. The final paragraph sufficiently shows Clifford's point of view and is of interest here from its incidental description of Tait as a "wide-eyed hero," between whom and Clifford there existed indeed a warm affection, divergent though their views were on questions of religion. Scoffing at the attempt to preserve the Christian faith in an enlightened scientific age, Clifford wrote:

"'Only for another half-century let us keep our hells and heavens and gods.' It is a piteous plea; and it has soiled the hearts of these prophets, great ones and blessed, giving light to their generation, and dear in particular to our mind and heart. These sickly dreams of hysterical women and half-starved men, what have they to do with the sturdy strength of a wide-eyed hero who fears no foe with pen or club? This sleepless vengeance of fire upon them that have not seen and have not believed, what has it to do with the gentle patience of the investigator that shines through every page of this book, that will ask only consideration and not belief for anything that has not with infinite pains been wholly established? That which you keep in your hearts, my brothers, is the slender remnant of a system which has made its red mark in history, and still lives to threaten mankind. The grotesque forms of its intellectual belief have survived the discredit of its moral teaching. Of this what the kings could bear with, the nations have cut down; and what the nations left, the right hand of man by man revolts against day by day. You have stretched out your hands to save the dregs of the sifted sediment of a residuum. Take heed lest you have given soil and shelter to the seed of that awful plague which has destroyed two civilisations, and but barely failed to slay such promise of good as is now struggling to live among men."

Racy and instructive though it was, Clifford's review did not really touch the central doctrines of Stewart and Tait's speculations. One of their aims was to show that there was nothing in physical science which denied the possibility of our intelligences existing after death in another universe. They also argued that certain aspects of the modern theory of energy suggested, if they did not demonstrate, the probability of such an Unseen Universe. The reasonings could not satisfy either the extreme right or the extreme left. It was little wonder then that the prophet and the agnostic alike fell foul of the book,—the prophet, because the authors strove to bring under the Law of Continuity certain mysteries of his faith, the agnostic, because starting from the known they endeavoured to cross the fringe of the unknown.

Many of the ideas and speculations put forward by Stewart and Tait were novelties to the vast majority of their readers. These ideas are now familiar as the sunshine. It would be impossible to say, however, to what extent the authors of *The Unseen Universe* impressed some of their views upon the world, or to what degree they were simply the earliest exponents of thoughts which were gradually taking shape in the human mind.

The tenth edition of *The Unseen Universe* was translated into French by a naval Lieutenant A.-B., with a preface to French readers by Professor D. de St-P. (Paris Libraire Germer Baillière et Cie, 1883).

In 1878 Stewart and Tait published a sequel to *The Unseen Universe* under the name of *Paradoxical Philosophy*. The book was cast into the form of a dialogue, the purpose of it being to convert Dr Hermann Stoffkraft, a German materialist, to a belief in the doctrines of *The Unseen Universe*. The deed is done; but of course, as in the orthodox novel, the end is obvious from the beginning. A delicately humorous and yet scientifically critical review was written for *Nature* (Dec. 19, 1878) by Clerk Maxwell. Certain paragraphs from that review hit off with such remarkable clearness the whole bearing of the two books that no apology is needed for their reproduction here.

"We cannot accuse the authors of leading us through the mazy paths of science only to entrap us into some peculiar form of theological belief. On the contrary, they avail themselves of the general interest in theological dogma to imbue their readers unawares with the newest doctrines of energy. There must be many who would never have heard of Carnot's reversible engine, if they had not been led through its cycle of operations while endeavouring to explore the Unseen Universe. No book containing so much thoroughly scientific matter would have passed through seven editions in so short a time without the allurement of some more human interest.......

"The words on the title-page: 'In te, Domine, speravi, non confundar in aeternum' may recall to an ordinary reader the aspiration of the Hebrew Psalmist, the closing prayer of the 'Te Deum' or the dying words of Francis Xavier; and men of science, as such, are not to be supposed incapable either of the nobler hopes or of the nobler fears to which their fellow men have attained. Here, however, we find these venerable words employed to express a conviction of the perpetual validity of the 'Principle of Continuity,' enforced by the tremendous sanction, that if at any place or at any time a single exception to that principle were to occur, a general collapse of every intellect in the universe would be the inevitable result.

"There are other well known words in which St Paul contrasts things seen with things unseen. These also are put in a prominent place by the authors of *The Unseen Universe*. What, then, is the Unseen to which they raise their thoughts?

"In the first place the luminiferous aether, the tremors of which are the dynamical equivalent of all the energy which has been lost by radiation from the various systems of grosser matter which it surrounds. In the second place a still more subtle medium, imagined by Sir William Thomson as possibly capable of furnishing an explanation of the properties of sensible bodies; on the hypothesis that they are built up of ring vortices set in motion by some supernatural power in a frictionless liquid: beyond which we are to suppose an indefinite succession of media, not hitherto imagined by anyone, each manifoldly more subtle than any of those preceding it. To exercise the mind in speculations on such media may be a most delightful employment for those who are intellectually fitted to indulge in it, though we cannot see why they should on that account appropriate the words of St Paul."

After a playful discussion of some of the theories of the origin of consciousness and of the meaning of personality, Clerk Maxwell summed up thus:

"The progress of science, therefore, so far as we have been able to follow it, has added nothing of importance to what has always been known about the physical consequences of death, but has rather tended to deepen the distinction between the visible part, which perished before our eyes, and that which we are ourselves, and to show that this personality, with respect to its nature as well as to its destiny, lies quite beyond the range of science."

In his letters to Tait, Maxwell let his humour play round the curious speculations and metaphysics of the authors of *The Unseen Universe* and of *The Paradoxical Philosophy*. For example, at the end of the letter of Sept. 7, 1878, from which quotations have already been given (pp. 151–2), Maxwell remarked:

"It is said in *Nature* that *U.U.* is germinating into some higher form. If you think of extending the collection of hymns given in the original work, do not forget to insert 'How happy could I be with Ether.'"

After the publication of *Paradoxical Philosophy*, he sent Tait "A Paradoxical Ode" consisting of three stanzas, said to be "after Shelley." The movement of the verses, the rhythm and the rhyming, strongly suggest portions of *Prometheus Unbound*, although the imitation is not quite accurate as to form. With marvellous ingenuity has Maxwell woven into his verses much that characterised not only the speculations of *The Unseen Universe* but also certain features of Tait's scientific work in relation to the classification of knots. The verses are given here as they appear in the original draft which was pasted into Tait's Scrap Book. It differs in slight details from the version published in Maxwell's *Life*.

TO HERMANN STOFFKRAFT, PH.D.

A PARADOXICAL ODE.

(*After Shelley.*)

I.

My soul's an amphicheiral knot[1]
Upon a liquid vortex wrought
By Intellect in the Unseen residing,
While thou dost like a convict sit
With marlinspike untwisting it
Only to find my knottiness abiding,

---

[1] An amphicheiral knot is a knot which can be changed into its own mirror reflexion—amphicheiral similarity is the similarity between a right and a left hand. See Tait "On Knots" (*Trans. R.S.E.*, 1876–7; *Sci. Pap.*, Vol. I, pp. 288, 314, etc.).

Since all the tools for my untying
In four-dimensioned space are lying[1],
Where playful fancy intersperses
Whole avenues of universes[2],
Where Klein and Clifford fill the void
With one unbounded, finite homoloid,
Whereby the infinite is hopelessly destroyed.

II.

But when thy Science lifts her pinions
In Speculation's wild dominions
I treasure every dictum thou emittest;
While down the stream of Evolution
We drift, and look for no solution
But that of the survival of the fittest.
Till in that twilight of the gods
When earth and sun are frozen clods,
When, all its energy degraded,
Matter in aether shall have faded,
We, that is, all the work we've done
As waves in aether, shall for ever run
In swift expanding spheres, through heavens beyond the sun[3].

III.

Great Principle of all we see
Thou endless Continuity[4]
By thee are all our angles gently rounded;
Our misfits are by thee adjusted,
And as I still in thee have trusted,
So let my methods never be confounded!
O never may direct Creation
Break in upon my contemplation
Still may the causal chain ascending,
Appear unbroken and unending,
And where that chain is lost to sight,
Let viewless fancies guide my darkling flight
Through Aeon-haunted worlds, in order infinite.

$$\frac{dp}{dt}.$$

[1] A tri-dimensional knot cannot exist in four-dimensional space.
[2] See *The Unseen Universe*, 4th edit., § 220.
[3] See *The Unseen Universe*, § 196.
[4] See *The Unseen Universe*, Chapter II.

In a later letter Maxwell wrote:

"Last three lines of Ode to Stoffkraft should be as follows:

> While Residents in the Unseen—
> Aeons or Emanations—intervene,
> And from my shrinking soul the Unconditioned screen."

On Aug. 28, 1879, ten weeks before his death, Maxwell sent Tait a curious composition purporting to be a soliloquy or self-communion by Tait himself. In spite of the rapid advance of the fatal illness to which he succumbed Maxwell's quaint humour still found expression to his life-long friend. In this last of many letters, the speculations in *The Unseen Universe* and the quaternion operator Nabla which Tait used with so great effect are mingled together in a fashion most strange and fanciful. The jest lurks in the closing sentence, pathetic though this is in its confession of physical weakness.

"HEADSTONE IN SEARCH OF A NEW SENSATION."

"While meditating, as is my wont on a Saturday afternoon, on the enjoyments and employments which might serve to occupy one or two of the aeonian aetherial phases of existence to which I am looking forward, I began to be painfully conscious of the essentially finite variety of the sensations which can be elicited by the combined action of a finite number of nerves, whether these nerves are of protoplasmic or eschatoplasmic structure. When all the changes have been rung in the triple bob major of experience, must the same chime be repeated with intolerable iteration through the dreary eternities of paradoxical existence? The horror of a somewhat similar consideration had as I well knew driven the late J. S. Mill to the very verge of despair till he discovered a remedy for his woes in the perusal of Wordsworth's Poems.

"But it was not to Wordsworth that my mind now turned, but to the noble Viscount the founder of the inductive philosophy and to the Roman city whence he was proud to draw his title, consecrated as it is to the memory of the Protomartyr of Britain.

"Might not I, too, under the invocation of the holy ALBAN become inspired with some germinating idea, some age-making notion by which I might burst the shell of circumstance and hatch myself something for which we have not even a name, freed for ever from the sickening round of possible activities and exulting in a life every action of which would be a practical refutation of the arithmetic of the present world.

"Hastily turning the page on which I had inscribed these meditations, I noticed just opposite the name of the saint another name which I did not recollect having written.   Here it is

ИAB⅃A

"Here then was the indication, impressed by the saint himself, of the way out of all my troubles.   But what could the symbol mean?   I had heard that the harp from which Heman or Ethan drew those modulations from the plaintive to the triumphant which modern music with its fetters of tonality may ignore but can never equal—I had heard that this harp had been called by a name like this.   But not in all Wales could such a harp be found, nor yet the lordly music which has not been able to come down through the illimitable years.

"Here I was interrupted by a visitor from Dresden who had come all the way with his Erkenntniss-Theorie under his arm, showing that space must have three dimensions, and that there's not a villain living in all Denmark but he's an arrant knave.   Peruse his last epistle and see whether he could be transformed from a blower of his own trumpet into a Nabladist.

"I have been so seedy that I could not read anything however profound without going to sleep over it.

$$\frac{dp}{dt} \cdot \text{"}$$

CHAPTER VII

ADDRESSES, REVIEWS, AND CORRESPONDENCE

In general Tait did not, like some of his colleagues, begin each session at Edinburgh University with a special introductory lecture upon some chosen subject. Occasionally, however, recent discoveries or new ways of looking at physical problems attracted his attention, and gave an unusual character to the opening lecture. The following two examples will illustrate what is meant.

In the opening lecture of November 1869 he gave an account of the first great results in solar spectroscopy, and discussed to some extent the nature of nebulae and comets. This was the occasion on which for the first time he gave an explanation of some of the phenomena of comets' tails. Tait's "beautiful sea-bird analogy"—as Kelvin called it—was also given in a series of interesting articles on Cosmical Astronomy which appeared in *Good Words* in 1875. The following quotation is taken from a full abstract of the opening lecture just referred to, which was published in *Nature*, Dec. 16, 1869, and was translated in *La Revue des Cours Scientifiques*, 1870.

Finally let us consider what we have recently learned about comets—bodies which have hitherto puzzled the astronomer quite as much as the nebulae....There seems to be good grounds for imagining that a comet is a mere shower of stones (meteorites and fragments of iron). This at least is certain that such a shower would behave, in its revolution about the sun, very much as comets are seen to do....

Such small comets as have been observed have given continuous spectra from their tails, so far as could be judged with regard to an object so feebly illuminated. This, then, it would appear, is simply reflected solar light. The heads, however, give spectra somewhat resembling those of the nebulae I have just mentioned—the spectra of incandescent gases. This is quite consistent with the descriptions given by Hevelius and others of some of the grander comets; which presented no peculiarities of colour in the tail, but where the head was blueish or greenish. Now these appearances are easily reconciled with the shower-of-stones hypothesis. The nucleus, or head, of a comet is that portion of the shower where the stones are most numerous, where their relative velocities are greatest, and where, therefore, mutual impacts, giving off incandescent gases, are the most frequent and the most violent. This simple hypothesis explains easily many very striking facts about comets, such as their sometimes appearing to send off in a few hours a tail many hundreds of millions of miles in length. Wild notions of repulsive forces vastly more powerful than the sun's gravity have been entertained; bold speculations as to decomposition (by solar light) of gaseous matter left behind it in space by the comet have also been propounded; but it would seem that the shower-of-stones hypothesis accounts very simply for such an appearance. For, just as a distant flock of seabirds comes suddenly into view as a dark line when the eye is brought by their evolutions into the plane in which they fly, so the scattered masses which have lost velocity by impact, while they formed part of the head, or those which have been quickened by the same action, as well as those which lag behind the others in virtue of the somewhat larger orbits which they describe, show themselves by reflected solar light as a long bright streak whenever the earth moves into any tangent plane to the surface in which they are for the time mainly gathered.

A year later Tait found occasion, after the usual exposition of the significance of the Conservation of Energy, to warn his students against looseness of language. He illustrated his point by quoting from Bain's logic. It would seem that he was put on the track of this book by W. Robertson Smith, who had been carrying on his theological studies at the Free Church College, Edinburgh, during 1868–70, and at the same time acting as Tait's Assistant. In an undated letter, the main subject matter of which fixes the date as April of 1870, Robertson Smith asked "Have you seen Bain's Logic? Full of rubbish about conservation of force, by which he means momentum!!!" Tait's own lively criticisms of Bain's inaccuracies in scientific statement will be found in *Nature*, Dec. 1, 1870.

It is the custom in the Scottish Universities for the Arts Professors in rotation to address the graduates at the annual graduation ceremonial. In the old days when the Arts Chairs were limited by statute to seven, each Professor was called upon at intervals of about seven years to act as "Promotor" and give an address. Tait was Promotor in 1866, 1874, 1881, and 1888. Before his turn came round again, the new regulations had come into force, and the Arts Faculty had been widened out to embrace nearly a dozen other chairs in literature, history, education, science and art.

Tait's first address on the value of the Edinburgh degree of M.A. was published by the Senatus as a pamphlet. It contained a strong protest against the proposition to amalgamate the Scottish Universities as one grand National University with a central Examining Board. A quotation from this address was given above (p. 11) in reference to cramming or coaching for examinations. The question of central boards of examiners and their necessary concomitant cram received many a hard hit from Tait in his graduation addresses, which present in somewhat whimsical guise his horror of the examiner who is not at the same time a teacher. The same views are expressed in an article on " Artificial Selection " which appeared in *Macmillan's Magazine* for 1872 and which contains some racy illustrations of how not to examine. The following quotation from this article indicates the ideal University which Tait at that time pictured to himself:

A combination of the Scottish and English University systems, to the exclusion of what is manifestly bad in each, is the thing really wanted. England's superiority consists in very great measure of money and lands—that of Scotland in making the University Professors the actual teachers. Let us have in the great English Universities Professors teaching the many, to take the place of the all-pervading Coach—in addition of course to the almost unequalled body of Professors they now possess....In Scottish Universities let many of the chairs be doubled or even trebled; let there be, for instance, a Professor of Experimental Physics in each, and a Professor of Applied Mathematics, in the place of the present solitary Professor of the enormous subject Natural Philosophy; let us have a Professor of Chemistry and Medicine, and a Professor of the Theory of Chemistry, etc....Let the multifarious duties now discharged by one over-burdened man be distributed among two, three or four; let their salaries not depend for so much as half the whole amount on the numbers attending their classes, so that there shall be no possible incitement to lower their standard to attract more listeners. But also let us take every care that they be kept rigorously to their work, and at once laid aside whenever they have ceased to be working teachers.

This unfortunately is not likely to be done. The extreme poverty of the Scottish Universities, more especially of the Metropolitan one, prevents their doing much. And Scotland's share of the Imperial Revenue has always been insignificant compared with her contributions to it. Still it is surely possible that a few annual thousands might be obtained from Parliament to furnish her universities properly with laboratories; and the overworked and underpaid professors with adequate remuneration and with additional assistants, from whom in turn their successors might be chosen. Then the country, having done something to deserve success, cannot fail to attain it.

Recent developments have in some respects, although not in all, been along the very lines here sketched by one who, because of his conservative political sympathies, was believed by many to be averse to progress of any kind. In one particular, however, we have not worked towards the ideal imagined by Tait. The exaltation of the examination still continues; and some of his strong characteristic statements are quite to the point in these days. Take the following extracts from the address to the graduates of 1874 (see *Nature*, April 30, 1874), and from the address of 1881.

It is a mere common-place to say that examination, or, as I have elsewhere called it, artificial selection is, as too often conducted, about the most imperfect of human institutions; and that in too many cases it is not only misleading, but directly destructive, especially when proper precautions are not taken to annihilate absolutely the chances of a candidate who is merely crammed, not in any sense educated. Not long ago I saw an advertisement to the effect:—' History in an hour, by a Cambridge *Coach*." How much must this author have thought of the ability of the examiners before whom his readers were to appear? There is one, but so far as I can see, only one, way of entirely extirpating cram as a system, it may be costly—well, let the candidates bear the expense, if the country (which will be ultimately the gainer) should refuse. Take your candidates, when fully primed for examination, and send them off to sea—without books, without even pen and ink; attend assiduously to their physical health, but let their minds lie fallow. Continue this treatment for a few months, and then turn them suddenly into the Examination Hall. Even six months would not be wasted in such a process if it really enabled us to cure the grand inherent defect of all modern examinations. It is amusing to think what an outcry would be everywhere raised if there were a possibility of such a scheme being actually tried—say in Civil Service Examinations. But the certainty of such an outcry, under the conditions supposed, is of itself a complete proof of the utter abomination of the cramming system. I shall probably be told, by upholders of the present methods, that I know nothing about them, that I am prejudiced, bigoted, and what not. That, of course, is the natural cry of those whose "craft is in danger"—and it is preserved for all time in the historic words, "Thou wert altogether born in sin, and dost thou teach us?" I venture now to state, without the least fear of contradiction, a proposition which (whether new or not) I consider to be of inestimable value to the country at large:—Wherever the examiners are not in great part the teachers also, there will cram to a great extent supersede education.......

This Chinese passion for excessive examination threatens to become as great a nuisance here as the Celestials themselves have proved in San Francisco. I had better not tell you any more of my own experiences; I have already said quite enough about myself. But I can tell you what occurred to a friend of my own, a professor in a neighbouring State. My friend belongs to the well-known kingdom of Yvetot— that happy land in which (as Thackeray tells us) the good king

> Each year called out his fighting men,
> And marched a league from home; and then—
> Marched back again.

The Professor had once a favourite student, of much more than ordinary promise, who sought a post under government—a post for which he was in all respects singularly fitted. Now, none but the very highest of such posts can be obtained in Yvetot except after a strict examination; so the youth had to submit to the ordeal.   One of the chief subjects on which he was examined was the Mandingo language.   The attempt proved unsuccessful; the student was "remitted"—that is the correct phrase; they are very dainty and delicate in their phrases in Yvetot. Yet, as it happened, that youth had been for some years in an office in the capital of Mandingo itself; and had acquired, not, indeed, a pedantic knowledge, but, what is far better, a thorough working knowledge of the language!   He was remitted *because his examiner was unable to test that kind of knowledge.*

Again, the Professor had an old serving-man who desired Government employment as a door-porter and messenger.   He had done that kind of work extremely well for years in private houses.   But even posts like these are never given by Government (in Yvetot) except after examination.   No one has yet found out how to examine in the art of door-keeping, so the would-be porter had to be examined in physical geography and continued fractions!   Of course he also was remitted, and he died of shame a few days later.   *But* (in Yvetot) *all Government door-keepers must know physical geography!*

My friend the Professor has an almost morbid hatred of cram, and his *Répétiteur* thinks with him.   But one day there came to him an unfortunate man who had been remitted in a Government examination, and who desired to try his fate once more.   There were but a few days before the awful tribunal of examiners was again to meet.   To teach the victim in the available time was impossible; so my friend thought that for once he would try whether the art of cramming comes by nature, or has to be painfully acquired.   The *Répétiteur* entered into the scheme with hearty goodwill, and conducted the process.   When the examination was over the lately remitted one was almost at the head of the list.   The Professor's only remark was,—"*I should like to examine these examiners!*"

To return to our own land.   While the present state of things continues, the universities have no option.   They must give degrees, and in consequence they must examine.   With human beings, as with guns and girders, testing is a very delicate process.   You may double-shot your gun, or load your girder far beyond what it will thenceforth be required to bear, and both may stand the test; yet the very testing may have produced a flaw in the metal, some day to finish its career by what will then be called (euphemistically, of course) a terrible *accident.*

In presence of these painful realities, it is your duty, alike to the University and to your fellow-men, to endeavour, as far as you have opportunity, to extend the inestimable blessings and privileges of education.   But, in doing so, never for a moment forget what education really is.   It is not Latin, nor Greek, nor philosophy, whether natural or unnatural[1].   It is not even the three R's.   These are indispensable preliminaries to education, but preliminaries only.   He who possesses them has been taught; it does not at all follow that he has been educated.   The confusion of teaching with education is a common but monstrous fallacy.   You may know Liddell and Scott by heart without becoming Greek scholars.   You may be able to differentiate,

---

[1] I remember that when Tait uttered these words with characteristic emphasis he turned with triumphant glee toward his colleague, Campbell Fraser, professor of Logic and Metaphysics, as much as to say, "I've got you this time—there is no reply!"

and integrate, and solve all manner of regulation problems in mathematics, and yet be no mathematicians. Machinery can be made to do all that sort of thing better than any human being can. You must have risen far above the mere efforts of memory, and of "rule of thumb," before you can consider yourselves educated. Your minds must be able to do something which no machine can do for them; otherwise they sink below the level of a machine, for it is absolutely free from human liabilities to error.

But there is no cause for dread less extended education should promote sameness in opinions or pursuits—a thing in itself undesirable. Unless human knowledge were complete, there could not but be serious differences of opinion, even amongst the most highly educated, and even on subjects of the gravest importance. Yet, though extensive differences of opinion cannot but exist, you must not on that account think it of little moment what opinions you hold. Every human being who has received the priceless gift of reason is righteously responsible for its employment to the uttermost in all the varied circumstances of life. Opinions held lightly or on insufficient grounds will never be of much use in inciting to, or in directing, action. "Let every man be fully persuaded *in his own mind.*"

On May 24, 1873, Tait delivered the Rede Lecture in the Senate House, Cambridge, choosing for his subject the Thermoelectric Diagram. A fairly full abstract was published in *Nature*, Vol. VIII, and is reprinted as No. XXVIII in the *Scientific Papers*. An enlarged diagram showing the positions of the lines of the more interesting metals was specially prepared, being drawn in such a way as to give at a glance the Centimetre-gramme-second values of the various quantities involved—electromotive power, Peltier Effect, Thomson Effect, etc. This was subsequently shown at the South Kensington Loan Exhibition of Scientific Apparatus.

In the preparations for the Rede Lecture, Maxwell gave valuable help, as may be gathered from the following letter, written in reply to questions from Tait.

11 SCROOPE TER.
10 *March*, 1873.

O T',

Θαγξ φορ Αλλες.

(1) I have no Assistant. If I can do you any service, well and good, if not, why not?

(2) Prof. Liveing will lend you his bags, give you his gases, and furnish you with lime light. If you are particular about your lantern bring it yourself, like Guy Fawkes or the man in the Moon. The gases will go for half an hour. If you want them for longer, say so. Bring your own galvanometer.

(3) Thermopylae exist, but Peltier only in the form of a repulsive electrometer, and the effet Thomson is an " effet defective."

(4) The Senate house is a place to write in, to graduate in, and to vote in. The Public Orator I believe can speak in it provided he employs the Latin tongue. What those venerable walls would say if the vernacular were sounded within them I dare not even think. If you have a good audience there will not be much echo from Geo. II or Pitt, and if you erect a lofty platform, the light spot on the screen and the under side of your table may be seen by all.

(5) If you do your ΘH as you did your Quaternions to the British Asses you will do very well, always remembering that to speak familiarly of a 2nd Law, as of a thing known for some years, to men of culture who have never even heard of a 1st Law, may arouse sentiments unfavourable to patient attention....

Both Moral and Intellectual Entropy are noble subjects, though the dictum of Pecksniff concerning the idea of Todgers be unknown to me and not easily verified.

I do not know much about reversible operations in morals. The science or practice depends chiefly on the existence of singular points in the curve of existence at which influences, physically insensible, produce great results. The man of tact says the right word at the right time, and a word spoken in due season how good is it? The man of no tact is like vinegar upon natron when he sings his songs to a heavy heart. The ill timed admonition only hardens the conscience, and the good resolution, made just when it is sure to be broken, becomes macadamized into pavement for the abyss.

$$\text{Yrs.} \quad \frac{dp}{dt}.$$

In the early seventies the Director of the Museum of Science and Art in Edinburgh, now the Royal Scottish Museum, arranged courses of scientific lectures to the Industrial Classes. Courses were given by Dr Buchan, Professor Tait, Professor Crum Brown, Dr (afterwards Professor) McKendrick on special branches of their respective sciences. Tait's lectures on Cosmical Astronomy were delivered during January 1874, the titles of the successive lectures being (1) Our sources of information as to bodies non-terrestrial, (2) Their dimensions and distances, (3) Their masses and rates of motion, (4) Their composition and modes of aggregation, (5) Their mutual action, (6) Their ultimate state. When preparing these lectures, Tait took the opportunity of fulfilling a promise to Dr Norman Macleod, the editor of *Good Words*, and contributed a corresponding series of articles to that popular magazine.

At the British Association Meeting in Glasgow in 1876, Professor Andrews was President; and Tait out of a feeling of loyalty to his old friend and colleague agreed to give one of the evening lectures. The subject was " Force," and its main scientific features were a strong demand for accuracy in scientific language, and a demonstration that force in the strictly Newtonian sense of the word has no real objective existence but is a mere space-variation of energy.   The lecture abounded in illustrations from all sides of human experience and was severely critical on laxity of thought and of expression on the part, not only of journalists essaying to speak of scientific things, but even of recognised writers of scientific books.   The lecture was published in *Nature* and subsequently reprinted as an appendix to the second edition of *Recent Advances in Physical Science*.   It appears as No. xxxvii in the *Scientific Papers*.   The raciest and most critical passages were, however, omitted.   In these Tait let himself go to the intense amusement of many of his audience and to the horror of some who did not quite appreciate the form Tait's humour occasionally assumed.   Lord Brougham and Professor Tyndall, though not explicitly named, were singled out as having been guilty of carelessness of diction in the expression of scientific truth ; and the audience were startled when Tait capped his exposure of the recent President of the British Association by the question, "Are these thy gods, Oh Israel ? "

Tait used to tell how he early noticed in the audience one alert listener who seemed almost to anticipate the points, so quickly did he respond to the humour and sarcasm of the lecturer.   His expectant and eager expression was a delightful inspiration to Tait.

The real fun of the lecture is well shown forth in the humorous verses which Maxwell sent to Tait a few days later, with the heading " For P. G. Tait but *not* for Ebony"—meaning *Blackwood's Magazine*.   The following version is taken from the original draft, which was pasted into Tait's Scrap Book.

REPORT OF TAIT'S LECTURE ON FORCE :—B.A. 1876.

Ye British Asses, who expect to hear
    Ever some new thing,
I've nothing new to tell, but what, I fear,
    May be a true thing,
For Tait comes with his plummet and his line
    Quick to detect your
Old bosh new dressed, in what you call a fine
    Popular lecture.

Whence comes that most peculiar smattering
    Heard in our section ?
Pure nonsense, to a scientific swing
    Drilled to perfection ?

That small word "Force" is made a barber's block
 Ready to put on
Meanings most strange and various, fit to shock
 Pupils of Newton.

Ancient and foreign ignorance they throw
 Into the bargain ;
The Sage of Leipzig mutters from below
 Horrible jargon.
The phrases of last century in this
 Linger to play tricks—
Vis viva and *Vis Mortua* and *Vis*
 Acceleratrix.

These long-nebbed words that to our text-books still
 Cling by their titles,
And from them creep, as entozoa will,
 Into our vitals.
But see! Tait writes in lucid symbols clear
 One small equation ;
And Force becomes of Energy a mere
 Space-Variation.

Force, then, is force, but mark you! not a thing,
 Only a Vector ;
Thy barbéd arrows now have lost their sting
 Impotent spectre!
Thy reign, O Force! is over. Now no more
 Heed we thine action ;
Repulsion leaves us where we were before,
 So does attraction.

Both Action and Reaction now are gone.
 Just ere they vanished,
Stress joined their hands in peace, and made them one ;
 Then they were banished.
The Universe is free from pole to pole
 Free from all forces.
Rejoice! ye stars—like blessed gods ye roll
 On in your courses.

No more the arrows of the Wrangler race,
 Piercing, shall wound you.
Forces no more, those symbols of disgrace
 Dare to surround you.
But those whose statements baffle all attacks,
 Safe by evasion,—
Whose definitions, like a nose of wax,
 Suit each occasion,

Whose unreflected rainbow far surpassed
    All our inventions,
Whose very energy appears at last
    Scant of dimensions :—
Are these the gods in whom ye put your trust,
    Lordlings and ladies?
The "secret potency of cosmic dust"
    Drives them to Hades.

While you, brave Tait! who know so well the way
    Forces to scatter,
Calmly await the slow but sure decay
    Even of matter.

On January 29, 1880, in the City Hall, Glasgow, under the auspices of the Glasgow Science Lecture Association, Tait gave a lecture on Thunderstorms, for which he collected a vast amount of curious information.   At one time he intended to include this lecture in the first volume of the *Scientific Papers*; but gave up the idea on the ground no doubt that the lecture did not contain any distinct addition of his own to our scientific knowledge.   Nevertheless it touches in an interesting way on many of the features of thunderstorms. It was reported in full in the columns of *Nature* and it has been thought well to reprint it in this volume as an admirable specimen of the popular scientific lecture.

Had Tait devoted himself to popular lecturing, there is no doubt he would have impressed himself strongly on the community. He had a full command of terse vigorous language, a pleasant resonant voice, the power of speaking deliberately and emphatically, a clear utterance, and a strong personality behind it all.   His humour could always be counted upon as adding a sparkle to the physical arguments and descriptions.   Finally, his honesty of mind would never lead him to gloss over difficulties, or give a doubtful lead on the applications of some broad principle.

Tait acted as Reviewer and Critic of many scientific works—chiefly in the columns of *Nature* and occasionally in the *Philosophical Magazine*.

It may be said emphatically that Tait never wrote for the mere sake of writing. His desire always was to bring out what he believed to be the truth, and this he did in many cases by exposing the errors. He had no patience with rhetorical writing in a book claiming to be scientific; and it went hard with an author who indulged in such verbiage. Tait had also a keen eye for faults of expression, for looseness of phrase, and for lack of precision in the ideas which it was intended to communicate.

As examples of the severely critical vein we may refer to his two articles on Sensation and Science in *Nature*, Vol. IV, July 6, 1871, and Vol. VI, July 4, 1872. The first is devoted to an exposure of the extraordinary misconception on the part of Professor Haughton as to the physical significance of the Principle of Least Action. The criticism is deservedly severe. In a writer of Haughton's standing and reputation the misconception was inexcusable, for the simple reason that his words would carry weight and be accepted as scientific truth by very many of his hearers and readers. Haughton's aim was to apply to the animal kingdom this principle of least action, which appears sometimes in various more or less irreconcileable guises as "the minimum of effort," "the least quantity of material," "a wonderful economy of force," "a performing its allotted task (by a muscle) with the least amount of trouble to itself," "minimum amount of muscular tissue," and so on.

"A very Proteus is this so-called principle," wrote Tait. "There is no knowing where to have it....It is a minimum, an economy, a least quantity, and what not; sometimes of effort, sometimes of material, then of trouble, and anon of muscular tissue, or of force of the same kind as that with which the bee constructs its cell! But the most curious feature about it is that in none of its metamorphoses does it in the slightest degree resemble the Least Action of Maupertuis, with which it would seem throughout to be held as identical."

The second article on Sensation and Science dealt with a book on Comets and things in general by Professor Zöllner of Leipzig, an extraordinary man of brilliant but unequal parts. The work, as Tait described it, "deals not alone with the nature of Comets, the inferiority of British to German physicists, and the grave offence of which a German is guilty when he sees anything to admire except at home; but also with the errors of Thomas Buckle, the relations of Science to Labour and Manufacture, and the analogies of development in Languages and Religion." Zöllner was specially wrath with Helmholtz for sanctioning the German translation of Thomson and Tait's *Natural Philosophy*. Tait could not bring himself to take the man and his writings seriously; but Helmholtz thought it necessary in his Preface to the Second Part of the German edition to reply at considerable length to Zöllner's attacks. A translation (by Crum Brown) of this reply is given in *Nature*, Vol. x, 1874.

When reading Zöllner's book Tait called Tyndall's attention to the terrible onslaught the author had made on Tyndall's theory of comets. In his reply Tyndall wrote : " I have glanced over it (Zöllner's book) not read it, myself. I can see that he means to mangle me—kill me first and chop me into mince-meat afterwards. But whether it is that the fire of my life has fallen to a cinder, the book has produced very little disturbance in my feelings....Ten years ago I should have been at the throat of Zöllner, but not now. I would rather see you and Clausius friends than Zöllner and myself. Trust me C. is through and through an honest high-minded man."

The reference to Clausius had to do with the controversy then going on between Tait and Clausius in regard to the second law of thermodynamics.

In many of his reviews Tait found occasion not only to hit off the character of the writer but also to descant on the true way and the false in the teaching of science. A few examples may still be of interest. The following extracts are from a review which appeared in *Nature* on January 30, *1873*, of De Morgan's inimitable *Budget of Paradoxes*.

This work is absolutely unique. Nothing in the slightest degree approaching it in its wonderful combinations has ever, to our knowledge, been produced. True and false science, theological, logical, metaphysical, physical, mathematical, etc., are interwoven in its pages in the most fantastic manner : and the author himself mingles with his puppets, showing off their peculiarities, posing them, helping them when diffident, restraining them when noisy, and even occasionally presenting himself as one of their number. All is done in the most perfect good-humour, so that the only incongruities we are sensible of are the sometimes savage remarks which several of his pet bears make about their dancing master.

De Morgan was a man of extraordinary information. We use the word advisedly as including all that is meant by the several terms knowledge, science, erudition, etc. Everywhere he was thoroughly at home. An old edition and its value-giving peculiarities or defects, a complex mathematical formula with its proof and its congeners, a debated point in theology or logic, a quotation from some almost-unheard-of author, all came naturally to him, and from him. With a lively and ready wit, and singularly happy style, and admirable temper, he was exactly fitted to write a work like this. And every page of it shows that he thoroughly enjoyed his task.

De Morgan was a very dangerous antagonist. Ever ready, almost always thoroughly informed, gifted with admirable powers of sarcasm which varied their method according to the temperament of his adversary, he was ready for all comers, gaily tilted against many so-called celebrities ; and—upset them. It is unfortunate that the issue of his grand contest with Sir William Hamilton (the great Scottish Oxford Philosopher) is but in part indicated in this volume—it is softened down, in fact, till one can hardly recognise the features of the extraordinary *Athenaeum* correspondence of 1847. There the ungovernable rage of the philosopher contrasts most strongly with the calm sarcasm of the mathematician, who was at every point his master, and who "played" him with the dexterity and the tenderness of old Isaak himself! But it is characteristic of De Morgan that, though he was grievously insulted throughout the greater part of this discussion, no trace of annoyance seems to have remained with him after the death of his antagonist; for none would gather from the *Budget* more than the faintest inkling of the amount of provocation he received.

Take again the following introductory paragraphs of a very full and instructive review of Clerk Maxwell's great work on *Electricity and Magnetism*. The review appeared in *Nature*, April 24, 1873.

In his deservedly celebrated treatise on "Sound," the late Sir John Herschel felt himself justified in saying, "It is vain to conceal the melancholy truth. We are fast dropping behind. In Mathematics we have long since drawn the rein and given over a hopeless race." Thanks to Herschel himself, and others, the reproach, if perhaps *then* just, did not long remain so. Even in pure mathematics, a subject which till lately has not been much attended to in Britain, except by a few scattered specialists, we stand at this moment at the very least on a par with the élite of the enormously disproportionate remainder of the world. The discoveries of Boole and Hamilton, of Cayley and Sylvester, extend into limitless regions of abstract thought, of which they are as yet the sole explorers. In applied mathematics no living men stand higher than Adams, Stokes, and W. Thomson. Any one of these names alone would assure our position in the face of the world as regards triumphs already won in the grandest struggles of the human intellect. But the men of the next generation—the successors of these long-proved Knights—are beginning to win their spurs, and among them there is none of greater promise than Clerk Maxwell. He has already, as the first holder of the new chair of Experimental Science in Cambridge, given the post a name which requires only the stamp of antiquity to raise it almost to the level of that of Newton. And among the numerous services he has done to science, even taking account of his exceedingly remarkable treatise on "Heat," the present volumes must be regarded as preeminent.

We meet with three sharply-defined classes of writers on scientific subjects (and the classification extends to all such subjects, whether mathematical or not). There are, of course, various less-defined classes, occupying intermediate positions.

First, and most easily disposed of, are the men of calm, serene, Olympian self-consciousness of power, those upon whom argument produces no effect, and whose grandeur cannot stoop to the degradation of experiment! These are the *à priori* reasoners, the metaphysicians, and the *Paradoxers* of De Morgan.

Then there is the large class, of comparatively modern growth, with a certain amount of knowledge and ability, diluted copiously with self-esteem—haunted, however, by a dim consciousness that they are only popularly famous—and consequently straining every nerve to keep themselves in the focus of the public gaze. These, also, are usually, men of "paper" science, kid-gloved and black-coated—with no speck but of ink.

Finally, the man of real power, though (to all seeming) perfectly unconscious of it—who goes straight to his mark with irresistible force, but neither fuss nor hurry—reminding one of some gigantic but noiseless "crocodile," or punching engine, rather than of a mere human being.

The treatise we have undertaken to review shows us, from the very first pages, that it is the work of a typical specimen of the third of these classes. Nothing is asserted without the reasons for its reception as truth being fully supplied—there is no parade of the immense value of even the really great steps the author has made—no attempt at sensational writing when a difficulty has to be met; when necessary, there is a plain confession of ignorance without the too common accompaniment of a sickening mock-modesty....

The main object of the work, besides teaching the experimental facts of electricity and magnetism, is everywhere clearly indicated—it is simply to upset completely the notion of *action at a distance*. Everyone knows, or at least ought to know, that Newton considered that no one who was capable of reasoning at all on physical subjects could admit such an absurdity: and that he very vigorously expressed this opinion. The same negation appears prominently as the guiding consideration in the whole of Faraday's splendid electrical researches, to which Maxwell throughout his work expresses his great obligations. The ordinary form of statement of Newton's law of gravitation seems directly to imply this action at a distance; and thus it was natural that Coulomb, in stating his experimental results as to the laws of electric and magnetic action which he discovered, as well as Ampère in describing those of his electrodynamic action, should state them in a form as nearly as possible analogous to that commonly employed for gravitation.

The researches of Poisson, Gauss, etc., contributed to strengthen the tendency to such modes of representing the phenomena; and this tendency may be said to have culminated with the exceedingly remarkable theory of electric action proposed by Weber.

All these very splendid investigations were, however, rapidly leading philosophers away towards what we cannot possibly admit to be even a bare representation of the truth. It is mainly to Faraday and W. Thomson that we owe our recall to more physically sound, and mathematically more complex, at least, if not more beautiful, representations. The analogy pointed out by Thomson between a stationary distribution of temperature in a conducting solid, and a statical distribution of electric potential in a non-conductor, showed at once how results absolutely identical in law and in numerical relations, could be deduced alike from the assumed distance-action of electric particles, and from the contact-passage of heat from element to element of the same conductor.

After quoting Maxwell's own frank and ample acknowledgement of his debt to these two men, Tait continued:

It certainly appears, at least at first sight, and in comparison with the excessively simple distance action, a very formidable problem indeed to investigate the laws of the propagation of electric or magnetic disturbance in a medium. And Maxwell did not soon, or easily, arrive at the solution he now gives us. It is well-nigh twenty years since he first gave to the Cambridge Philosophical Society his paper on *Faraday's Lines of Force*, in which he used (instead of Thomson's heat-analogy) the analogy of an imaginary incompressible liquid, without either inertia or internal friction, subject, however, to friction against space, and to creation and annihilation at certain sources and sinks. The velocity-potential in such an imaginary fluid is subject to exactly the same conditions as the temperature in a conducting solid, or the potential in space outside an electrified system. In fact the so-called equation of continuity coincides in form with what is usually called Laplace's equation. In this paper Maxwell gave, we believe for the first time, the mathematical expression of Faraday's Electro-tonic state, and greatly simplified the solution of many important electrical problems. Since that time he has been gradually developing a still firmer hold of the subject, and he now gives us, in a carefully methodised form, the results of his long-continued study....

It is quite impossible in such a brief notice as this to enumerate more than a very few of the many grand and valuable additions to our knowledge which these volumes contain. Their author has, as it were, flown at everything ;—and, with immense spread of wing and power of beak, he has hunted down his victims in all quarters, and from each has extracted something new and invigorating—for the intellectual nourishment of us, his readers.

In his review of Maxwell's remarkable little book *Matter and Motion* (*Nature*, Vol. xvi, June 14, 1877) Tait was led into an interesting discussion of the necessity for accuracy and for paying attention to the things which count. He pointed his moral by quoting some sentences from recent text-books on Natural Philosophy (which it had been his intention to review along with Maxwell's book), and then proceeded to contrast them with Maxwell's unpretentious volume.

Clerk Maxwell's book is not very easy reading. No genuine scientific book can be. But the peculiar characteristic of it is that (while anyone with ordinary abilities can read, understand, and profit by it) it is the more suggestive the more one already knows. We may boldly say that there is no one now living who would not feel his conceptions of physical science at once enlarged, and rendered more definite by the perusal of it....

Clerk Maxwell's work, then, is simply Nature itself, so far as we understand it. The peaks, precipices, and crevasses are all there in their native majesty and beauty. Whoso wishes to view them more closely is free to roam where he pleases. When he comes to what he may fear will prove a dangerous or impassable place, he will find the requisite steps cut, or the needful rope attached, sufficiently but not obtrusively, by the skilful hand of one who has made his own roads in all directions, and has thus established a claim to show others how to follow.

In the rival elementary works the precipice and the crevasse are not to be seen: there are, however, many pools and ditches; for the most part shallow, but *very* dirty. You are confined to the more easily accessible portions of the region. In the better class of such books these are trimly levelled—the shrubs and trees are clipped into forms of geometrical (i.e. unnatural) symmetry like a Dutch hedge. Smooth straight walks are laid down leading to old well-known "points of view,"—and, as in Trinity of former days, undergraduates are warned against walking on the grass-plats.

These "royal roads" to knowledge have ever been the main cause of the stagnation of science in a country. He would be a bold man indeed who would venture to assert that the country which, in times all but within the memory of many of us, produced such mighty master-minds as Lagrange, Fourier, Ampère, and Laplace, does not now contain many who might well have rivalled the achievements even of men like these. But they have no chance of doing so; they are taught, not by their own struggles against natural obstacles, with occasional slight assistance at a point of unexpected difficulty, but by being started off in groups, "eyes front" and in heavy marching order, at hours and at a pace determined for all alike by an Official of the Central Government, along those straight and level (though perhaps sometimes rough) roads which have been laid down for them! Can we wonder that, whatever their natural fitness, they don't now become mountaineers?

It seems appropriate at this point to reproduce parts of the account which Tait gave of the life and work of his life-long friend James Clerk Maxwell. Schoolboys at the same school, contemporaries at Cambridge, profoundly interested in the same great branch of science, and constant correspondents throughout their busy lives, they were the truest of friends knit heart to heart by bonds which only death could sever. Tait had an unstinted admiration for the genius of Maxwell, a deep love for the man, and a keen appreciation of his oddities and humour. In their correspondence they were always brimming over with fun and frolic, and puzzling each other with far-fetched puns, and literary allusions of the most extraordinary kind. I have been able throughout this memoir to give a good deal from Maxwell's letters to Tait. Unfortunately the other side of the correspondence has disappeared. Some lines written to Tait on a half sheet of note-paper whose contents referred to proof corrections are worth preserving as a neat example of Maxwell's power of moralising on physical truth :

> "The polar magnet in his heart of steel
> Earth's gentle influence appears to feel;
> But trust him not! he's biassed at the core
> Force will but complicate that bias more.
> No Power but that of all-dissolving Fire
> Can quite demagnetize the hardened wire."

The following extracts are from Tait's account of Maxwell's work in *Nature*, January 29, 1880:

At the instance of Sir W. Thomson, Mr Lockyer, and others I proceed to give an account of Clerk Maxwell's work, necessarily brief, but I hope sufficient to let even the non-mathematical reader see how very great were his contributions to modern science. I have the less hesitation in undertaking this work that I have been intimately acquainted with him since we were schoolboys together.

If the title of mathematician be restricted (as it too commonly is) to those who possess peculiarly ready mastery over symbols, whether they try to understand the significance of each step or no, Clerk Maxwell was not, and certainly never attempted to be, in the foremost rank of mathematicians. He was slow in "writing out," and avoided as far as he could the intricacies of analysis. He preferred always to have before him a geometrical or physical representation of the problem in which he was engaged, and to take all his steps with the aid of this : afterwards, when necessary, translating them into symbols. In the comparative paucity of symbols in many of his great papers, and in the way in which, when wanted, they seem to grow full-blown from pages of ordinary text, his writings resemble much those of Sir William Thomson, which in early life he had with great wisdom chosen as a model.

There can be no doubt that in this habit, of constructing a mental representation of every problem, lay one of the chief secrets of his wonderful success as an investigator. To this were added an extraordinary power of penetration, and an altogether unusual amount of patient determination. The clearness of his mental vision was quite on a par with that of Faraday; and in this (the true) sense of the word he was a mathematician of the highest order.

But the rapidity of his thinking, which he could not control, was such as to destroy, except for the very highest class of students, the value of his lectures. His books and his written addresses (always gone over twice in MS) are models of clear and precise exposition; but his *extempore* lectures exhibited in a manner most aggravating to the listener the extraordinary fertility of his imagination.

Clerk Maxwell spent the years 1847–50 at the University of Edinburgh, without keeping the regular course for a degree. He was allowed to work during this period, without assistance or supervision, in the Laboratories of Natural Philosophy and of Chemistry: and he thus experimentally taught himself much which other men have to learn with great difficulty from lectures or books. His reading was very extensive. The records of the University Library show that he carried home for study, during these years, such books as Fourier's *Théorie de la Chaleur*, Monge's *Géométrie Descriptive*, Newton's *Optics*, Willis' *Principles of Mechanism*, Cauchy's *Calcul Différentiel*, Taylor's *Scientific Memoirs*, and others of a very high order. These were read *through*, not merely consulted. Unfortunately no list is kept of the books consulted in the Library. One result of this period of steady work consists in two elaborate papers, printed in the *Transactions of the Royal Society of Edinburgh*. The first (dated 1849), "On the Theory of Rolling Curves," is a purely mathematical treatise, supplied with an immense collection of very elegant particular examples. The second (1850) is "On the Equilibrium of Elastic Solids." Considering the age of the writer at the time, this is one of the most remarkable of his investigations. Maxwell reproduces in it, by means of a special set of assumptions, the equations already given by Stokes. He applies them to a number of very interesting cases, such as the torsion of a cylinder, the formation of the large mirror of a reflecting telescope by means of a partial vacuum at the back of a glass plate, and the Theory of Örsted's apparatus for the compression of water. But he also applies his equations to the calculation of the strains produced in a transparent plate by applying couples to cylinders which pass through it at right angles, and the study (by polarised light) of the doubly-refracting structure thus produced. He expresses himself as unable to explain the permanence of this structure when once produced in isinglass, gutta percha, and other bodies. He recurred to the subject twenty years later, and in 1873 communicated to the Royal Society his very beautiful discovery of the *temporary* double refraction produced by shearing in viscous liquids.

During his undergraduateship in Cambridge he developed the germs of his future great work on "Electricity and Magnetism" (1873) in the form of a paper "On Faraday's Lines of Force," which was ultimately printed in 1856 in the "Trans. of the Camb. Phil. Society." He showed me the MS of the greater part of it in 1853. It is a paper of great interest in itself, but extremely important as indicating the first steps to such a splendid result. His idea of a fluid, incompressible and without mass, but subject to a species of friction in space, was confessedly adopted from the analogy pointed out by Thomson in 1843 between the steady flow of heat and the phenomena of statical electricity.

After a fairly exhaustive account of Maxwell's principal contributions to scientific literature, Tait continued :

Maxwell has published in later years several additional papers on the Kinetic Theory, generally of a more abstruse character than the majority of those just described. His two latest papers (in the *Phil. Trans.* and *Camb. Phil. Trans.* of last year) are on this subject : one is an extension and simplification of some of Boltzmann's valuable additions to the Kinetic Theory. The other is devoted to the explanation of the motion of the radiometer by means of this theory. Several years ago (*Nature*, Vol. XII, p. 217), Prof. Dewar and the writer pointed out, and demonstrated experimentally, that the action of Mr Crookes' very beautiful instrument was to be explained by taking account of the increased length of the mean free path in rarefied gases, while the then received opinions ascribed it either to evaporation or to a quasi-corpuscular theory of radiation. Stokes extended the explanation to the behaviour of disks with concave and convex surfaces, but the subject was not at all fully investigated from the theoretical point of view till Maxwell took it up. During the last ten years of his life he had no rival to claim concurrence with him in the whole wide domain of molecular forces, and but two or three in the still more recondite subject of electricity.

"Everyone must have observed that when a slip of paper falls through the air, its motion, though undecided and wavering at first, sometimes becomes regular. Its general path is not in the vertical direction, but inclined to it at an angle which remains nearly constant, and its fluttering appearance will be found to be due to a rapid rotation round a horizontal axis. The direction of deviation from the vertical depends on the direction of rotation....These effects are commonly attributed to some accidental peculiarity in the form of the paper..." So writes Maxwell in the *Cam. and Dub. Math. Jour.* (May, 1854) and proceeds to give an exceedingly simple and beautiful explanation of the phenomenon. The explanation is, of course, of a very general character, for the complete working out of such a problem appears to be, even yet, hopeless; but it is thoroughly characteristic of the man, that his mind could never bear to pass by any phenomenon without satisfying itself of at least its general nature and causes.

Similar in character to the quotations just given are the following culled from a series of articles and reviews which appeared in *Nature* between 1875 and 1887, all dealing with the life and work of Sir George Stokes. The earliest article formed the fifth of the *Nature* series of Scientific Worthies (*Nature*, July 15, 1875).

GEORGE GABRIEL STOKES.

A great experimental philosopher, of the age just past, is reported to have said, " Show me the scientific man who never made a mistake, and I will show you one who never made a discovery." The implied inference is all but universally correct, but now and then there occur splendid exceptions (such as are commonly said to be requisite to prove a rule), and among these there has been none more notable than the present holder of Newton's Chair in Cambridge, George Gabriel Stokes, Secretary of the Royal Society.

To us, who were mere undergraduates when he was elected to the Lucasian Professorship, but who had with mysterious awe speculated on the relative merits of the men of European fame whom we expected to find competing for so high an honour, the election of a young and (*to us*) unknown candidate was a very startling phenomenon. But we were still more startled, a few months afterwards, when the new Professor gave public notice that he considered it part of the duties of his office to assist any member of the University in difficulties he might encounter in his mathematical studies. Here was, we thought (in the language which Scott puts into the mouth of Richard Cœur de Lion), "a single knight, fighting against the whole *mêlée* of the tournament." But we soon discovered our mistake, and felt that the undertaking was the effect of an earnest sense of duty on the conscience of a singularly modest, but exceptionally able, and learned man. And, as our own knowledge gradually increased, and we became able to understand his numerous original investigations, we saw more and more clearly that the electors had indeed consulted the best interests of the University; and that the proffer of assistance was something whose benefits were as certain to be tangible and real as any that mere human power and knowledge could guarantee.

And so it has proved. Prof. Stokes may justly be looked upon as in a sense one of the intellectual parents of the present splendid school of Natural Philosophers whom

Cambridge has nurtured—the school which numbers in its ranks Sir William Thomson and Prof. Clerk Maxwell.

All of these, and Stokes also, undoubtedly owe much (more perhaps than they can tell) to the late William Hopkins. He was, indeed, one whose memory will ever be cherished with filial affection by all who were fortunate enough to be his pupils.

But when they were able, as it were, to walk without assistance, they all (more or less wittingly) took Stokes as a model. And the model could not but be a good one: it is all but that of Newton himself. Newton's wonderful combination of mathematical power with experimental skill, without which the Natural Philosopher is but a fragment of what he should be, lives again in his successor. Stokes has attacked many questions of the gravest order of difficulty in pure mathematics, and has carried out delicate and complex experimental researches of the highest originality, alike with splendid success. But several of his greatest triumphs have been won in fields where progress demands that these distinct and rarely associated powers be brought simultaneously into action. For there the mathematician has not merely to save the experimenter from the fruitless labour of pushing his enquiries in directions where he can be sure that (by the processes employed) nothing new is to be learned; he has also to guide him to the exact place at which new knowledge is felt to be necessary and attainable. It is on this account that few men have ever had so small a percentage of *barren* work, whether mathematical or experimental, as Stokes.

The following review by Tait of Stokes' *Mathematical and Physical Papers* (Vols. I and II) appeared in *Nature*, December 13, 1883:

There can be but one opinion as to the value of the collection before us, and (sad to say) also as to the absolute *necessity* for it. The author, by common consent of all entitled to judge, takes front rank among living scientific men as experimenter as well as mathematician. But the greater part of his best work has hitherto been buried in the almost inaccessible volumes of the *Cambridge Philosophical Transactions*, in company with many other papers which deserve a much wider circulation than they have yet obtained. Stokes' well-deserved fame was thus practically secured by means of a mere fraction of his best work....

The present publication will effect a very remarkable amount of transference of credit to the real author, from those who (without the possibility of suspicion of *mala fides*) are at present all but universally regarded as having won it. Two or three years ago, only, the subject for a Prize Essay in a Continental scientific society was *The nature of unpolarized, as distinguished from polarized, light.* But all that science is even yet in a position to say, on this extremely curious subject, had been said by Stokes *thirty years ago* in the *Cambridge Philosophical Transactions....*

Prof. Stokes has wisely chosen the chronological order, in arranging the contents of the volumes. Such a course involves, now and then, a little inconvenience to the reader; but this is much more than compensated for by the insight gained into the working of an original mind, which seems all along to have preferred a bold attack upon each more pressing scientific difficulty of the present, to attempts at smoothing the beginner's road into regions already well explored. When, however, Prof. Stokes does write an elementary article, he does it admirably. Witness his *Notes on Hydrodynamics*, especially that entitled *On Waves*.

Before that article appeared, an article as comprehensive as it is lucid, the subject was almost a forbidden one even to the best student, unless he were qualified to attack the formidable works of Laplace and Airy, or the still more formidable memoirs of Cauchy and Poisson. Here he finds at least the main points of this beautiful theory, disencumbered of all unnecessary complications, and put in a form intelligible to all who have acquired any right to meddle with it. It is quite impossible to tell how much real good may be done by even *one* article like this. Would there were more such! There are few, even of the most gifted men, who do not occasionally require extraneous assistance after the earlier stages of their progress: all are the better for it, even in their maturer years.

The contents of these two volumes consist mainly, almost exclusively, of papers connected with the *Undulatory Theory of Light* or with *Hydrodynamics*. On the former subject at least, Stokes stands, without a living rival, the great authority. From the *Aberration of Light*, the *Constitution of the Luminiferous Ether*, the full explanation of the singular difficulties presented by *Newton's Rings*, to the grand theoretical and experimental treatise on the *Dynamical Theory of Diffraction*, we have a series of contributions to this branch of optics which, even allowing for improved modern surroundings, will bear comparison with the very best work of Newton, Huyghens, Young, or Fresnel in the same department.

Specially remarkable among the Hydrodynamical papers is that on *Oscillatory Waves*, to which a very important addition has been made in the reprint. The investigation of the " profile " of such a wave is here carried to a degree of approximation never before attempted.

Besides these *classes* of papers we have the very valuable treatise on *Friction of Fluids in Motion, and on the Equilibrium and Motion of Elastic Solids*. This was Stokes' early masterpiece, and it may truly be said to have revolutionized our knowledge on the subjects it treats. To mention only one point, though an exceedingly important one, it was here that for the first time was clearly shown the error of assuming any *necessary* relation between the rigidity and the compressibility of an elastic solid, such as had been arrived at from various points of view by the great Continental mathematicians of the earlier part of the present century.

Of the few purely mathematical papers in the present volumes the most important is the well-known examination of the *Critical Values of the Sums of Periodic Series*, a subject constantly forced on the physicist whenever he has to treat a case of discontinuity....

Tait contributed to *Nature* three reviews on Stokes' *Burnett Lectures*, which were delivered in Aberdeen in three Courses and were published in three corresponding volumes about a year apart. The review of the First Course, *On the Nature of Light*, appeared on April 10, 1884 (Vol. xxix). The Second Course, *On Light as a Means of Investigation*, was reviewed on August 20, 1885 (Vol. xxxii). The following extracts are of special interest:

The interest raised by the first series of these lectures is fully sustained by this second instalment, though the subject-matter is of a very different order. *Then*, the main question was the nature of light itself; *now*, we are led to deal chiefly with the uses of light as an instrument for indirect exploration. It is one of the most amazing results of modern science that the nature of mechanisms, too minute or too distant to be studied directly with the help of the microscope or the telescope, can be thus in part at least, revealed to reason. This depends on the fact that a ray of light, like a human being, bears about with it indications alike of its origin and of its history; and can be made to tell whence it sprang and through what vicissitudes it has passed.

The lecturer begins by pointing out that this indirect use of light already forms an extensive subject; and he then specially selects for discussion half-a-dozen important branches of it....

The first of these is Absorption. Here we have the explanation of the colours of bodies; the testing ray having gone in, and come out "shorn." This leads to the application of the prism in the immediate discrimination of various solutions which, to the unaided eye, appear to have the same colour. It is shown how, by a mere glance, the chemist may often be saved from fruitless toil, occasionally from grave error.

From the study of what rays are absorbed, the transition is an easy and natural one to the study of *what becomes of them* when they are absorbed. Here we have heating, chemical changes, phosphorescence, etc. The remainder of the lecture is devoted to an exceedingly interesting treatment of the beautiful subject of fluorescence.

The second lecture begins with Rotation of the Plane of Polarisation of light by various liquids, with its important application to saccharimetry. Then we have Faraday's discovery of the corresponding phenomenon produced in the magnetic field, with its application in the discrimination of various classes of isometric compounds.......

Then comes the "still vexed" question of the history of Spectrum Analysis. The present view of it must, of course, be carefully read: it is much too long to be here extracted in full, and to condense would be to mutilate it. Of course the claims of the author himself are the only ones to which scant justice is done. But the President of the *British Association* of 1871 fortunately gave, in his opening address, the means of filling this *lacuna*. Just as the Gravitation-theory of an early Lucasian Professor was publicly taught in Edinburgh University before it became familiar among scientific men, so the present Lucasian Professor's suggestions for the analysis of the solar atmosphere, by means of the dark lines in the spectrum, were publicly explained in the University of Glasgow for *eight successive years* before the subject became generally known through the prompt and widespread publicity given to the papers of Bunsen and Kirchhoff!

The following are Sir William Thomson's words of 1871: "It is much to be regretted that this great generalisation was not published to the world twenty years ago...because we might now be (*sic*) in possession of the inconceivable riches of astronomical results which we expect from the next ten years' investigation by spectrum analysis, had Stokes given his theory to the world when it first occurred to him."

The third lecture is devoted to the information which spectrum analysis affords as to the chemical composition of the sun's atmosphere, and its physical condition; the classification of stars, the constitution of nebulae, and the nature of comets....

The remarks on the nebulae and on comets will be read with great avidity; and, by the majority of readers, with some surprise. For it is stated that the planetary nebulae, " making abstraction of the stellar points, consist of glowing gas." And of comets we find: " There can no longer be any doubt that the nucleus consists, in its inner portions at least, of vapour of some kind, and we must add incandescent vapour..." An ingenious suggestion as to the source of this incandescence is introduced as the "green-house theory." The nucleus is supposed to be surrounded by an envelope of some kind, transparent to the higher but opaque to the lower forms of radiation. Thus solar heat can get freely at the nucleus, but cannot escape until it has raised the nucleus (in part at least) to incandescence. The coma and tail are formed by the condensation of small quantities of this vapour, so that they are mere mists of excessive tenuity. Herschel's suggestion, that the development of the tail is due to electric repulsion exerted by a charge on the sun, is spoken of with approval; and the production of the requisite charge of the mist-particles is regarded as a concomitant of condensation. Nothing, however, is said as to the opposite charge which the comet itself must receive, nor of the peculiar effects which would arise from this cause: whether in the form of a modification of the shape of the comet's head, or of a modification of its orbit and period due to a constantly increasing attraction exerted by the sun upon a constantly diminishing mass.

Of course, if this novel theory can stand the test of a full comparison with facts, it will have established its claim to become part of science. But it is hard to take leave of the simple old ideal comet: the swarm of cosmical brickbats: something imposing because formidable: and to see it replaced by what is, in comparison, a mere phantom, owing its singular appearance to the complexity of the physical properties it possesses and the recondite transformations perpetually taking place in its interior. The old idea of a comet's constitution was not only formidable, but was capable of explaining so much, and of effecting this by means so simple and so natural, that one almost felt it deserved to be well-founded! The new idea makes it resemble the huge but barely palpable 'Efreet of the *Arabian Nights*, who could condense himself so as to enter the bottle of brass with the seal of Suleymán the son of Dáood!......

The following sentences are from Tait's review of the Third Course of Stokes' *Burnett Lectures*, namely, *On the Beneficial Effects of Light* (see *Nature*, June 2, 1887, Vol. XXXVI):

This volume completes the course of the First Burnett Lecturer on the New Foundation. We have already (Vol. XXIX, p. 545, and Vol. XXXII, p. 361) noticed the first two volumes; and we are now in a position to judge of the work as a whole. But we must first speak of the contents of the present volume.

The author commences by extending the term "Light" to radiation in general....

Next comes a curious suggestion of analogy between the behaviour of fluo-rescent bodies (which always *degrade* the refrangibility of the light they give off) and the heat-radiation from bodies which have been exposed to sun-light. Sun-light, as it reaches us after passing through the atmosphere, is less rich in ultra-red rays than is the radiation from the majority of terrestrial sources; while the radiation from bodies which have been heated by direct sun-light is entirely ultra-red. Here we have, for the terrestrial atmosphere, the "green-house theory" which, in the second course, was applied to explain some of the singular phenomena exhibited by comets.

This is followed by an extremely interesting discussion of the functions of the colouring matters of blood and of green leaves: with the contrasted effects, upon plants, of total deprivation of light, and of continuously maintained illumination. A particularly valuable speculation, as to the probable nature of the behaviour of chlorophyll, is unfortunately too long for extraction.

So far, radiation has been treated without any special reference to vision. But the author proceeds to describe the physical functions and adaptations of the eye: with particular reference to the arrangements for obviating such of the theoretical defects as, while involved in its general plan, *would also tend to diminish its practical usefulness*. The introduction of this obviously natural proviso, one which we do not recollect having seen prominently put forward till now, exhibits in a quite new light the intrinsic value of those objections to the "argument from design" which have been based upon the alleged imperfection of the eye as an optical instrument.

The analogy of fluorescence is once more introduced, but now for the purpose of suggesting a mechanical explanation of the mode in which the sense of vision is produced. This is brought forward after the modern photo-chemical theory of vision has been discussed.......The triplicity of the colour-sense, and the mechanism of single vision with two eyes, are treated at some length. But throughout this part of the work it is frankly confessed that there are many elementary questions, some of funda-mental importance, which we are still unable even approximately to answer....

No higher praise need be bestowed on the scientific part of this third volume than is involved in saying that it is a worthy successor to the other two. Together, they form a singularly instructive, and yet (in the best sense) popular, treatise on a fascinating branch of natural philosophy. Were this their only aim, no one could deny that it has been thoroughly attained.

But their aim is of a loftier character. Here and there throughout the work there have been occasional references to the main purpose which has determined the author's mode of arranging his facts and his deductions from them. In the few closing pages this purpose is fully developed, and a brief but exceedingly clear statement shows at once how much in one sense, and yet how little in another, can be gathered as to the personality and the character of the Creator from a close and reverent study of His works.

Tait contributed important reviews on two works by W. K. Clifford. The first of these, which appeared in *Nature* (Vol. XVIII) on May 23, 1878, referred to the *Elements of Dynamic*, Part I, *Kinematic*, which was particularly interesting to Tait because of the use the author made throughout of quaternion methods. I give the review in full:

Though this preliminary volume contains only a small instalment of the subject, the mode of treatment to be adopted by Prof. Clifford is made quite obvious. It is a sign of these times of real advance, and will cause not only much fear and trembling among the crammers but also perhaps very legitimate trepidation among the august body of Mathematical Moderators and Examiners. For, although (so far as we have seen) the word quaternion is not once mentioned in the book, the analysis is in great part purely quaternionic, and it is not easy to see what arguments could now be brought forward to justify the rejection of examination-answers given in the language of quaternions—especially since in Cambridge (which may claim to lay down the law on such matters) Trilinear Coördinates, Determinants, and other similar methods were long allowed to pass unchallenged before they obtained formal recognition from the Board of Mathematical Studies.

Everyone who has even a slight knowledge of quaternions must allow their wonderful special fitness for application to Mathematical Physics (unfortunately we cannot yet say Mathematical *Physic*!): but there is a long step from such semi-tacit admissions to the full triumph of public recognition in Text-Books. Perhaps the first attempt to obtain this step (in a book not ostensibly quaternionic) was made by Clerk Maxwell. In his great work on *Electricity* all the more important Electrodynamic expressions are given in their simple quaternion form—though the quaternion *analysis* itself is not employed: and in his little tract on *Matter and Motion* (*Nature*, Vol. XVI, p. 119) the laws of composition of vectors are employed throughout. Prof. Clifford carries the good work a great deal farther, and (if for this reason alone) we hope his book will be widely welcomed.

To show the general reader how much is gained by employing the calculus of Hamilton we may take a couple of very simple instances, selecting them not because they are specially favourable to quaternions but because they are familiar in their Cartesian form to most students. Every one who has read *Dynamics of a Particle* knows the equations of non-acceleration of moment of momentum of a particle, under the action of a single centre of force, in the form

$$\left.\begin{aligned} x\ddot{y} - y\ddot{x} &= 0 \\ y\ddot{z} - z\ddot{y} &= 0 \\ z\ddot{x} - x\ddot{z} &= 0 \end{aligned}\right\}$$

with their first integrals, which express the facts that the orbit is in a plane passing through the centre, and that the radius-vector describes equal areas in equal times. But how vastly simpler as well as more intelligible is it not to have these *three* equations written as *one* in the form

$$V\rho\ddot{\rho} = 0$$

and the three first integrals above referred to as the immediate deduction from this in the form

$$V\rho\dot{\rho} = \alpha.$$

Take again Gauss's expression for the work done in carrying a unit magnetic pole round any closed curve under the action of a unit current in any other closed circuit. As originally given, it was [a long unwieldy expression in x, y, z, x', y', z']. With the aid of the quaternion symbols this unwieldy expression takes the compact form

$$\iint \frac{S \cdot \rho d\rho d\rho'}{T\rho^3}.$$

The meanings of the two expressions are identical, and the comparative simplicity of the second is due solely to the fact that it takes space of three dimensions as it finds it ; and does not introduce the cumbrous artificiality of the Cartesian coördinates in questions such as this where we can do much better without them.

In most cases at all analogous to those we have just brought forward, Prof. Clifford avails himself fully of the simplification afforded by quaternions. It is to be regretted, therefore, that in somewhat higher cases, where even greater simplification is attainable by the help of quaternions, he has reproduced the old and cumbrous notations. Having gone so far, why not adopt the whole ?

Perhaps the most valuable (so far at least as physics is concerned) of all the quaternion novelties of notation is the symbol

$$\nabla = i\frac{\partial}{\partial x} + j\frac{\partial}{\partial y} + k\frac{\partial}{\partial z},$$

whose square is the negative of Laplace's operator : i.e.

$$\nabla^2 = -\left[\left(\frac{\partial}{\partial x}\right)^2 + \left(\frac{\partial}{\partial y}\right)^2 + \left(\frac{\partial}{\partial z}\right)^2\right].$$

A glance at it is sufficient to show of what extraordinary value it cannot fail to be in the theories of Heat, Electricity, and Fluid Motion. Yet, though Prof. Clifford discusses Vortex Motion, the Equation of Continuity, etc., we have not observed in his book a single ∇. There seems to be a strange want of consistency here, in coming back to such " beggarly elements " as

$$\delta_x u + \delta_y v + \delta_z w$$

instead of $\qquad -S\nabla\sigma,$

especially when, throughout the investigation, we have σ used for

$$ui + vj + wk,$$

and when, in dealing with strains, the Linear and Vector Function is quite freely used. Again, for the vector axis of instantaneous rotation of the element at x, y, z (ρ), when the displacement at that point is $\sigma = iu + jv + kw$, we have the cumbersome form

$$\tfrac{1}{2}\{(\delta_y u - \delta_z v)\,i + (\delta_z u - \delta_x w)\,j + (\delta_x v - \delta_y u)\,k\}$$

instead of the vastly simpler and more expressive

$$\tfrac{1}{2}V\nabla\sigma.$$

It may be, however, that this apparent inconsistency is in reality dictated by skill and prudence. The suspicious reader, already put on his guard by Clerk Maxwell's first cautious introduction of the evil thing, has to be treated with anxious care and nicety of handling : lest he should refuse altogether to bite again. If he rises to the present cast we shall probably find that Prof. Clifford has ∇, in the form as it were of a gaff, ready to fix him irrevocably. That he will profit by the process, in the long run, admits of no doubt : so the sooner he is operated on the better. What is now urgently wanted, for the progress of some of the most important branches of mathematical physics, is a "coming" race of intelligent students brought up, as it were, at the feet of Hamilton ; and with as little as may be of their freshness wasted on the artificialities of x, y, z. Till this is procured, quaternions cannot have fair play. Nut-cracking, though occasionally successful for a moment, is the most wasteful and destructive of all methods of sharpening the teeth.

What we have at some length discussed is the most prominent feature of the present work, but by no means its only distinctive one. No writer, who has any claim upon his readers at all, can treat even the most hackneyed subject without giving a new and useful turn to many a long-known truth. Many of Prof. Clifford's proofs are exceedingly neat, and several useful novelties (e.g. Three-bar Motion) are introduced. We have to complain, however, of a great deal of unnecessary new and very strange nomenclature : for a large part of which the author is not responsible, his error (for such we cannot help considering it) consisting in giving this stuff a place of honour in his book. One does not require to be very violently conservative to feel dismayed at an apparently endless array of such new-fangled terms as Pedals, Rotors, Cylindroids, Centrodes, Kites, Whirls, and Squirts! Yet these are but a few gleaned at random from the book. Something, it seems, *must* be hard in a text-book—simplify the Mathematic, and the Anglic (i.e. the English) immediately becomes perplexing.

In *Nature* of June 11, 1885 (Vol. XXXII), Tait reviewed Clifford's *The Common Sense of the Exact Sciences*, a book which was published six years after the author's death. The following sentences seem of sufficient permanent interest to deserve quotation :

Once more a characteristic record of the work of a most remarkable, but too brief, life lies before us. In rapidity of accurate thinking, even on abstruse matters, Clifford had few equals ; in clearness of exposition, on subjects which suited the peculiar bent of his genius and on which he could be persuaded to bestow sufficient attention, still fewer. But the ease with which he mastered the more prominent features of a subject often led him to dispense with important steps which had been taken by some of his less agile concurrents. These steps, however, he was obliged to take when he was engaged in exposition ; and he consequently gave them (of course in perfect good faith) without indicating that they were not his own. Thus, especially in matters connected with the development of recent mathematical and kinematical methods, his statements were by no means satisfactory (from the historical point of view) to those who recognised, as their own, some of the best "nuggets" that shine here and there in his pages. His *Kinematic* was, throughout, specially open to this objection: and it applies, though by no means to the same extent, to the present work. On the other hand, the specially important and distinctive features of this work, viz. the homely, yet apt and often complete, illustrations of matters intrinsically difficult, are entirely due to the author himself....

The chief good of this book, and in many respects it is very good, lies in the fact that the versatility of its gifted author has enabled him to present to his readers many trite things, simple as well as complex, from so novel a point of view that they acquire a perfectly fresh and unexpected interest in the eyes of those to whom they had become commonplace. Surely this was an object worthy of attainment! But it is altogether thrown away on the non-mathematical, to whom neither new nor old points of view are accessible.

Tait's review of Poincaré's *Thermodynamique* appeared in *Nature*, Jan. 14, 1892 (Vol. XLV). It is a good example of honesty in criticism; for in spite of the great and deserved fame of the author, Tait could only condemn the book as a *physical* treatise.

The great expectations with which, on account of the well-won fame of its author, we took up this book have unfortunately not been realised. The main reason is not far to seek, and requires no lengthened exposition. Its nature will be obvious from the following examples....

Some forty years ago, in a certain mathematical circle at Cambridge, men were wont to deplore the necessity of introducing words *at all* in a physico-mathematical text-book: the unattainable, though closely approachable, Ideal being regarded as a work devoid of aught but formulae!

But one learns something in forty years, and accordingly the surviving members of that circle now take a very different view of the matter. They have been taught, alike by experience and by example, to regard mathematics, so far at least as physical enquiries are concerned, as a mere auxiliary to thought....This is one of the great truths which were enforced by Faraday's splendid career.

And the consequence, in this country at least, has been that we find in the majority of the higher class of physical text-books few except the absolutely indispensable formulae. Take, for instance, that profound yet homely and unpretentious work, Clerk Maxwell's *Theory of Heat*. Even his great work, *Electricity*, though it seems to bristle with formulae, contains but few which are altogether unnecessary. Compare it, in this respect, with the best of more recent works on the same advanced portions of the subject.

In M. Poincaré's work, however, all this is changed. Over and over again, in the frankest manner (see, for instance, pp. xvi, 176), he confesses that he lays himself open to the charge of introducing unnecessary mathematics: and there are many other places where, probably thinking such a confession would be too palpably superfluous, he does not feel constrained to make it.

T.   35

M. Poincaré not only ranks very high indeed among pure mathematicians but has done much excellent and singularly original work in applied mathematics: all the more therefore should he be warned to bear in mind the words of Shakespeare:

> "Oh! it is excellent
> To have a giant's strength; but it is tyrannous
> To use it like a giant."

From the physical point of view, however, there is much more than this to be said. For mathematical analysis, like arithmetic, should never be appealed to in a physical enquiry till unaided thought has done its utmost. Then, and not till then, is the investigator in a position rightly to embody his difficulty in the language of symbols, with a clear understanding of what is demanded from their potent assistance. The violation of this rule is very frequent in M. Poincaré's work, and is one main cause of its quite unnecessary bulk. Solutions of important problems, which are avowedly imperfect because based on untenable hypotneses (see, for instance, §§ 284—286), are not useful to a student, even as a warning: they are much more likely to create confusion, especially when a complete solution, based upon full experimental data and careful thought, can be immediately afterwards placed before him. If something is really desired, in addition to the complete solution of any problem, the proper course is to prefix to the complete treatment one or more exact solutions of simple cases. This course is almost certain to be useful to the student. The whole of M. Poincaré's work savours of the consciousness of mathematical power: and exhibits a lavish, almost a reckless, use of it.

One test of the soundness of an author, writing on Thermodynamics, is his treatment of temperature, and his introduction of absolute temperature. M. Poincaré gets over this part of his work very expeditiously. In §§ 15—17 temperature, t, is conventionally defined as in the Centigrade thermometer by means of the volume of a given quantity of mercury; or by any continuous function of that volume which increases along with it. Next (§ 22) absolute temperature, T, is defined, provisionally and with a caution, as $273 + t$; from the (so-called) laws of Marriotte and Gay-Lussac. Then, finally (§ 118), absolute temperature is virtually defined afresh as the reciprocal of Carnot's function. (We say *virtually*, as we use the term in the sense defined by Thomson. M. Poincaré's *Fonction de Carnot* is a different thing.) But there seems to be no hint given as to the results of experiments made expressly to compare these two definitions. Nothing, for instance, in this connection at all events, is said about the long-continued early experimental work of Joule and Thomson, which justified them in basing the measurement of absolute temperature on Carnot's function.

In saying this, however, we must most explicitly disclaim any intention of charging M. Poincaré with even a trace of that sometimes merely invidious, sometimes purely Chauvinistic, spirit which has done so much to embitter discussions of the history of the subject. On the contrary, we consider that he gives far too little prominence to the really extraordinary merits of his own countryman Sadi Carnot. He writes not as a partisan but rather as one to whom the history of the subject is a matter of all but complete indifference. So far, in fact, does he carry this that the name of Mayer, which frequently occurs, seems to be spelled incorrectly on by far the greater number of these occasions! He makes, however, one very striking historical statement (§ 95):

"Clausius...lui donna le nom de *Principe de Carnot*, bien qu'il l'eût énoncé sans avoir connaissance des travaux de Sadi Carnot."

Still, one naturally expects to find, in a Treatise such as this, some little allusion at least to Thermodynamic Motivity; to its waste, the Dissipation of Energy; and to the rest of those important early results of Sir W. Thomson, which have had such immense influence on the development of the subject. We look in vain for any mention of Rankine or of his Thermodynamic Function; though we have enough, and to spare, of it under its later *alias* of Entropy. The word dissipation does indeed occur, for we are told in the Introduction that the *Principe de Carnot* is "*la dissipation de l'entropie.*"

We find Bunsen and Mousson cited, with regard to the effect of pressure upon melting points, almost before a word is said of James Thomson; and, when that word does come, it wholly fails to exhibit the real nature or value of the great advance he made.

Andrews again, *à propos* of the critical point, and his splendid work on the isothermals of carbonic acid, comes in for the barest mention only *after* a long discussion of those very curves, and of the equations suggested for them by Van der Waals, Clausius, and Sarrau: though his work was the acknowledged origin of their attempts.

The reason for all this is, as before hinted, that M. Poincaré has, in this work, chosen to play almost exclusively the part of the pure technical analyst; instead of that of the profound thinker, though he is perfectly competent to do that also when he pleases. And, in his assumed capacity, he quite naturally looks with indifference, if not with absolute contempt, on the work of the lowly experimenter. Yet, in strange contradiction to this, and still more in contradiction to his ascription of the Conservation of Energy to Mayer, he says of that principle: "personne n'ignore que c'est un fait expérimental."

But the most unsatisfactory part of the whole work is, it seems to us, the entire ignoration of the true (i.e. the statistical) basis of the second Law of Thermodynamics. According to Clerk Maxwell (*Nature*, XVII, 278)

"The touch-stone of a treatise on Thermodynamics is what is called the second law."

We need not quote the very clear statement which follows this, as it is probably accessible to all our readers. It certainly has not much resemblance to what will be found on the point in M. Poincaré's work: so little, indeed, that if we were to judge by these two writings alone it would appear that, with the exception of the portion treated in the recent investigations of v. Helmholtz, the science had been retrograding, certainly not advancing, for the last twenty years.

In his reply (*Nature*, March 3, 1892), Poincaré practically confined his attention to the discussion of the Thomson Effect, offering fuller explanations of his meaning. This, however, did not touch on Tait's chief objections which were epitomised as follows (March 10, 1892):

1.   The work is far too much a mere display of mathematical skill.   It soars above such trifles as historical details, while overlooking in great measure the experimental bases of the theory; and it leaves absolutely unnoticed some of the most important branches of the subject.

(Thus, for instance, Sadi Carnot gets far less than his due, Rankine is not alluded to, and neither Thermodynamic Motivity nor the Dissipation of Energy is even mentioned!)

2.   It gives an altogether imperfect notion of the true foundation for the reckoning of absolute temperature.

3.   It completely ignores the real (i.e. the statistical) basis of the Second Law of Thermodynamics.

On March 24, Poincaré gave his reasons why, in the cases cited, he did not discuss the questions in the way desired by Tait.   A final letter from Tait on April 7 ended the discussion.   As was customary in such discussions, Tait sent a proof of this letter to Kelvin.   The proof has been preserved, and is interesting inasmuch as it shows that Kelvin was in entire agreement with Tait.   In his first letter Poincairé had referred to the distinction between true and apparent electromotive forces; and Tait replied:

It is necessary to add that I made *no reference whatever* to M. Poincaré's distinctions between "true" and "apparent" electromotive force;—simply because I regard these, along with many other celebrated terms such as "disgregation" etc., as mere empty names employed to conceal our present ignorance.

Kelvin underlined the words "regard these" and wrote in the margin "So do I, K."   Other annotations were even more definite in their expression, and Kelvin's general acquiescence in the position taken by Tait was indicated by the brief note appended to the sheet, "Netherhall, Mar. 28/92 OT OK K."

Tait's review of A. McAulay's *Utility of Quaternions in Physics* appeared in *Nature*, Dec. 28, 1893 (Vol. XLIX) under the title "Quaternions as an Instrument in Physical Research."   The following are some of the characteristic passages:

Just as "one shove of the bayonet" was truly said to be more effective than any number of learned discussions on the art of war: this really practical work, giving genuine quaternion solutions of new problems as well as largely extended developments of old ones, is of incomparably greater interest and usefulness than the recently renewed, but necessarily futile, attempts to *prove* that a unit vector cannot possibly be a quadrantal versor....

Here, at last, we exclaim, is a man who has caught the full spirit of the quaternion system: "the real *aestus*, the *awen* of the Welsh bards, the *divinus afflatus* that transports the poet beyond the limits of sublunary things!"...Intuitively recognising its power, he snatches up the magnificent weapon which Hamilton tenders to all, and at once dashes off to the jungle on the quest of big game. Others, more cautious or perhaps more captious, meanwhile sit pondering gravely on the fancied imperfections of the arm; and endeavour to convince a bewildered public (if they cannot convince themselves) that, like the Highlander's musket, it requires to be treated to a brand-new stock, lock, and barrel, *of their own devising*, before it can be safely regarded as fit for service. "Non *his* juventas orta parentibus...." What could be looked for from the pupils of a school like that?

Mr McAulay himself has introduced one or two rather startling innovations. But he retains intact all the exquisitely designed Hamiltonian machinery, while sedulously oiling it, and here and there substituting a rolling for a sliding contact, or introducing a *lignum vitae* bearing....

The "startling innovations," however, as we called them above, are unquestionably Mr McAulay's own—and he has certainly gone far to justify their introduction. He has employed the sure tests of ready applicability and extreme utility, and these have been well borne. Objections based upon mere unwontedness or even awkwardness of appearance must of course yield when such important advantages as these (if they be otherwise unattainable) are secured; but it certainly requires a considerable mental wrench to accustom ourselves to the use of

$$X_1 \frac{d}{dx_1}$$

as an equivalent for the familiar expression

$$\frac{dX}{dx}.$$

If this be conceded, however, it is virtually *all* that Mr McAulay demands of us, and we are free to adopt his system....... A single example, of a very simple character, must suffice. Thus in the strain of a homogeneous isotropic solid, due to external potential u, we have for the strain-function ϕ (when there are no molecular couples) the equation

$$S . \nabla \phi \alpha + S \alpha \nabla u = 0$$

which (in virtue of the property of α, already spoken of) is equivalent to three independent scalar conditions. Suppose we wish to express these, without the α, in the form of one vector condition. Mr McAulay boldly writes the first term as

$$S . \alpha \phi'_1 \nabla_1 \quad \text{or rather as} \quad S . \alpha \phi \nabla,$$

for in so simple a case the suffixes are not required, and the strain-function is self-conjugate under the restriction above. Then, at once, the property of α shows us that

$$\phi \nabla + \nabla u = 0,$$

which is the vector equation required. Here it is obvious that, in the usual order of writing,

$$\phi \nabla = \frac{d}{dx} \phi i + \frac{d}{dy} \phi j + \frac{d}{dz} \phi k.$$

This simple example shows the *nature* of the gain which Mr McAulay's method secures. Those who wish to know its *extent* must read the work itself. They will soon be introduced to novel forms of concentrated operators, with regard to which, as I have not yet formed a very definite opinion, I shall content myself by hazarding the remark that, while they are certainly powerful and eminently useful, they must at present be regarded as singularly uncouth.

Some of Tait's contributions to *Nature* were signed G. H., i.e. Guthrie Headstone = Guthrie Tête Peter. These were always short notes, at times quizzical, bearing on some passing interest. One of serious import was the reference to the Tay Bridge disaster of December 28, 1879. The letter to *Nature* (July 22, 1880) is worth quoting from.

There are two interesting scientific questions, apart from engineering proper, which are suggested by the late enquiry, although no reference seems to have been made to them in the reports.

The first is the origin of the extraordinary flash seen at the moment of the downfall of the bridge by many spectators several miles away. It is scarcely doubtful that an impact was the only possible cause.

The second is the important question of the amount of wind-pressure which would suffice to force a train bodily off from the top of the bridge.......

The flash seems to prove that the train had been blown off the rails, and had come into violent contact with the sides of the high girders. Then, and not sooner, the piers were subjected to a strain they were unable to bear.

Tait's strong objection to acting as a scientific witness prevented him from appearing at the Enquiry, but this in no way interfered with his making suggestions to his old school friend Allan D. Stewart, M.Inst.C.E., who had been consulted by the Engineer, Sir Thomas Bouch, in regard to the design of the continuous girders, though not of the columns whose failure caused the catastrophe. Several years later Tait's attention was called by his son (then an Engineering Apprentice)[1] to an official diagram showing how the engine, carriages, etc. and the girders were found relative to the proper centre line of the bridge, and he agreed that the upsetting or derailing of a second-class carriage near the rear of the train was the immediate cause of the collapse of the bridge. This view seemed to be substantiated by the marks found on the girders, the smashing of the second-class carriage, and the distance separating this carriage from the front portion of the train.

In October 1873 there appeared in the *British Quarterly Review* a trenchant criticism of Herbert Spencer's *First Principles* (Second Edition).

[1] W. A. Tait, of Leslie and Reid, Engineers to the Edinburgh and District Water Trust.

The writer of the Review was the Senior Wrangler of 1868, who is now the Right Honourable Sir John Fletcher Moulton, one of the Lord Justices of England. His strictures were of course replied to by Spencer, who, as is usual in such controversies, passed by without remark the most condemnatory parts. It is interesting to note, however, that in the next edition of the *First Principles* readers will look in vain for certain of the most striking of the quotations introduced by the Reviewer in his exposure of Spencer's ignorance of physical principles. These sentences were silently removed without a word of explanation, presumably on the principle elsewhere enunciated by Spencer when in reply to other criticisms he remarked[1] "Though still regarding the statement I had actually made as valid, I concluded it would be best to remove the stumbling block out of the way of future readers."

At the close of his "Replies to Criticisms" in the *Fortnightly Review*, Herbert Spencer endeavoured to turn the keen edge of the British Quarterly Reviewer's damaging attack; and in February 1874 this portion with additions was issued as a pamphlet entitled "Mr Herbert Spencer and the British Quarterly Review." To anyone acquainted with physical principles the reply is a revelation of how a man with a marvellous power of assimilating knowledge and a unique gift of exposition can fail in understanding not merely the criticism but also the *facts* on which the criticism is based. For example Herbert Spencer's conception of what is meant by the adiabatic condensation and rarefaction of air through which sound is passing is not merely crude and incomplete—it is profoundly erroneous.

One of Spencer's main positions was that "our cognition of the Persistence of Force is *à priori*," against which the Reviewer quoted from Tait's *Thermodynamics* to the effect that

"Natural Philosophy is an experimental, and not an intuitive science. No *à priori* reasoning can conduct us demonstratively to a single physical truth."

Spencer believed he found a discrepancy between this statement and the following sentence from Thomson and Tait's *Treatise on Natural Philosophy* :

"As we shall show in our chapter on Experience, physical axioms are axiomatic to those only who have sufficient knowledge of the action of physical causes to enable them to see at once their necessary truth."

[1] See *Appendix to First Principles dealing with Criticisms*, p. 586, issued as a separate pamphlet in July 1880.

From this he argued that these physical axioms were *à priori* and maintained that Newton's Laws of Motion were also in this sense axiomatic. He wrote (page 316 in *Replies to Criticisms*):

"Not a little remarkable, indeed, is the oversight made by Professor Tait, in asserting that 'no *à priori* reasoning can conduct us demonstratively to a single physical truth,' when he has before him the fact that the system of physical truths constituting Newton's *Principia*, which he has joined[1] Sir William Thomson in editing, is established by *à priori* reasoning."

Unfortunately for Herbert Spencer's argument his quotation from Thomson and Tait's *Treatise* was incomplete, and the British Quarterly Reviewer shattered the support, which Spencer imagined he had found in the sentence quoted, by simply continuing the quotation. Referring to the passage quoted by Spencer, the Reviewer (in a note to *The British Quarterly*, January, 1874, pp. 215–8) remarked:

"Had Mr Spencer, however, read the sentence that follows it, we doubt whether we should have heard aught of this quotation. It is: 'Without further remark we shall give Newton's Three Laws; it being remembered that, as the properties of matter *might* have been such as to render a totally different set of laws axiomatic, these laws must be considered as resting on convictions drawn from observation and experiment, *not* on intuitive perception.' This not only shows that the term 'axiomatic' is used in the previous sentence in a sense that does not exclude an inductive origin, but it leaves us indebted to Mr Spencer for the discovery of the clearest and most authoritative expression of disapproval of his views respecting the nature of the Laws of Motion."

This awkward accusation of ignorance of *ipsissima verba* of the authors he was quoting Spencer did not condescend to answer. Deprived of their support, he turned his battery of words upon the position taken by Thomson and Tait, and proceeded to propound this dilemma:

Consider, he says, what is implied by framing the thought that "the properties of matter might have been such as to render a totally different set of laws axiomatic"... Does it express an experimentally ascertained truth? If so, I invite Professor Tait to describe the experiments. Is it an intuition? If so, then along with doubt of an intuitive belief concerning things as they are, there goes confidence in an intuitive belief concerning things as they are not. Is it an hypothesis? If so, the implication is that a cognition of which the negation is inconceivable (for an axiom is such) may be discredited by inference from that which is not a cognition at all, but simply a supposition.

[1] This is inaccurate: it was Professor Blackburn who was joined with Thomson in editing the *Principia*.

This argument cleverly evades the real question by attacking a somewhat infelicitous way of stating the important principle that, apart from experiment and observation, the human mind cannot formulate the laws of nature. In its assumption of the meaning of the word axiom, it disregards the obvious meaning of Thomson and Tait's statements in the very paragraph quoted ("T and T'," § 243). Spencer seemed to have recognised this, for he no longer contended that Thomson and Tait gave any support to his view; but he still continued to assert that Newton must himself have regarded the Laws of Motion as *à priori* principles (*Replies to Criticisms*, p. 326). In response to Spencer's challenge Tait now entered the field and penned the following short letter to *Nature*, March 26, 1874.

HERBERT SPENCER *versus* THOMSON AND TAIT.

A friend has lent me a copy of a pamphlet recently published by Mr Herbert Spencer, in which certain statements of mine are most unsparingly dealt with, especially in the way of attempted contrast with others made by Sir W. Thomson and myself. I am too busy at the present season to do more than request you to reprint one of the passages objected to (leaving it to your readers to divine to what possible objections it is open), and to illustrate by a brief record of my college days something closely akin to the mental attitude of the objector.

"Natural Philosophy is an experimental, and not an intuitive science. No *à priori* reasoning can conduct us demonstratively to a single physical truth" (Tait, *Thermodynamics*, p. 1).

One of my most intimate friends in Cambridge, who had been an ardent disciple of the late Sir W. Hamilton, Bart. and had adopted the preposterous notions about mathematics inculcated by that master, was consequently in great danger of being plucked. His college tutor took much interest in him, and for a long time gave him private instruction in elementary algebra in addition to the college lectures. After hard labour on the part of each, some success seemed to have been obtained, as my friend had at last for once been enabled to follow the steps of the solution of a question involving a simple equation. A flush of joy mantled his cheek, he felt his degree assured, and he warmly thanked his devoted instructor. Alas, this happy phase had but a brief duration; my friend's early mental bias too soon recovered its sway, and he cried in an agony of doubt and despair, "But what if x should turn out, after all, *not* to be the unknown quantity?"

Compare this with the following extract from Mr Spencer's pamphlet:

"... if I examine the nature of this proposition that 'the properties of matter *might have been*' other than they are. Does it express an experimentally-ascertained truth? If so, I invite Prof. Tait to describe the experiments!"

P. G. TAIT.

In his reply published in *Nature*, April 2, Spencer complained that

Tait had torn a sentence from its context (no doubt in imitation of Spencer's own method of quotation), and maintained that the unknown quantity was the application of Tait's story. It should be remembered of course that Spencer had been arguing in his *First Principles* and in his *Replies to Criticisms* that Newton's Laws of Motion were known *à priori*, whereas Tait regarded the Properties of Matter, including the Laws of Motion, as unknown until they were discovered by the legitimate methods of experience, and by these alone. (See also Tait's reply, p. 285 below.)

This passage at arms excited a great deal of interest among the students of the Physical Laboratory. One of our number was R. B. Haldane[1], second to none in knowledge of philosophy and in power of debate. We fought the Spencer-Tait controversy over and over again. I remember that W. K. Clifford visited Edinburgh about that time, and in the Tea Room at one of the April Meetings of the Royal Society of Edinburgh much lively talk went on regarding the controversy. Tait was in great spirits and said to Clifford, " There is not a man in England I suppose, other than Herbert Spencer, who does not see the point of my story." Clifford responded with a hearty laugh, " No doubt, it is a very good story."

In the same reply of April 2, Spencer practically reproduced the argument as given in his pamphlet, and made the remark that Tait himself,

" by saying of physical axioms that the appropriately-cultivated intelligence sees at once their necessary truth, tacitly classes them with mathematical axioms, of which this self-evidence is also the recognised character."

Writing to the same number of *Nature* the British Quarterly Reviewer disposed of Spencer's claim that he knew what Newton thought by quoting from two letters in which Newton wrote to Cotes in these words:

" In experimental philosophy it (i.e. hypothesis) is not to be taken in so large a sense as to include the first Principles or Axiomes which I call the Laws of Motion. These Principles are deduced from phenomena and made general by Induction, which is the highest evidence that a Proposition can have in this Philosophy"...

" On Saturday last I wrote you representing that Experimental philosophy proceeds only upon phenomena and deduces general Propositions from them only by Induction. And such is the proof of mutual attraction. And the arguments for the impenetrability, mobility, and force of all bodies, and for the laws of motion are no better."

On April 16, Herbert Spencer so far admitted his imperfect knowledge, and withdrew his contention that Newton regarded the Laws of Motion as axioms in the limited sense for which he had been all along arguing; but he

---

[1] Secretary of State for War since 1905.

went on to maintain that, in the sense in which he understood it, *à priori* intuition preceded experimental verification.

He illustrated his meaning by deducing the second law of motion from the *à priori* assumption that definite quantitative relations exist between cause and effect.   He evidently thought that " definite quantitative relations " meant proportionality[1].   But whatever final meaning Spencer attached to the phrase " *à priori* intuition," there was no getting away from the obvious meaning of one of the passages quoted by the British Quarterly Reviewer :—

"Deeper than demonstration—deeper even than definite cognition—deep as the very nature of mind is the postulate at which we have arrived (i.e. the Persistence of Force).   Its authority transcends all other whatever; for not only is it given in the constitution of our own consciousness, but it is impossible to imagine a consciousness so constituted as not to give it" (*First Principles*, p. 192).

This was one of the stumbling blocks which Spencer out of consideration for the future reader removed from his later editions.

In reply to the criticism that the phrase Persistence of Force was used in various quite distinct senses, Spencer remarked (*Replies to Criticisms*, p. 311) that had "he (the Reviewer) not been in so great a hurry to find inconsistencies, he would have seen why, for the purposes of my argument, I intentionally use the word Force: Force being the generic word, including both that species known as Energy, and that species by which Matter occupies space and maintains its integrity."

This recalls Maxwell's metrical Report on Tait's Lecture on Force.

> That small word "Force," is made a barber's block,
>   Ready to put on
> Meanings most strange and various, fit to shock
>   Pupils of Newton.
>   ....
> But those whose statements baffle all attacks,
>   Safe by evasion,—
> Whose definitions, like a nose of wax,
>   Suit each occasion, etc.

(See above, p. 254.)

[1] E.g., there is a definite quantitative relation between speed of projection and height reached, between strength of electric current and heat generated; but there is no simple proportionality. The whole discussion in *Nature* (Vols. IX and X) is well worth reading.   See especially the British Quarterly Reviewer's letters on April 2, April 16, and June 11.   It is not an exaggeration to say that in each succeeding letter Spencer takes up a different position, having been driven from one after another of his fancied strongholds.   The discussion was important as showing to what extent Herbert Spencer's *First Principles* could be relied on as an exposition of physical fact and theory.

Towards the close of his letter to Tait of date 27 August, 1874, reproduced in full in the Appendix to Chapter IV, pp. 171–5 above, Maxwell indulged in some exquisite fooling, in which Spencer's utterances are humorously parodied.

Another sly hit at the synthetic philosopher was given on a post card of date July 27, 1876, when Maxwell asked Tait

"Have you (read) Willard Gibbs on Equilibrium of Heterogeneous Substances? If not, read him. Refreshing after H. Spencer on the Instability of the Homogeneous."

In *Nature*, July 17, 1879, Tait reviewed Sir Edmund Becket's book *On the Origin of the Laws of Nature*. It opened with these words :

This is a very clever little book and deserves to be widely read. Its subject, however, is scarcely one for our columns. For it is essentially "apologetic," and its strong point is not so much accurate science as keen and searching logic. It dissects with merciless rigour some of the more sweeping assertions of the modern materialistic schools, reducing them (when that is possible) to plain English so as to make patent their shallow assumptions....He follows out in fact, in his own way, the hint given by a great mathematician (Kirkman) who made the following exquisite translation of a well-known definition :—

"Evolution is a change from an indefinite, incoherent, homogeneity, to a definite, coherent, heterogeneity, through continuous differentiations and integrations."

(Translation into plain English.)

"Evolution is a change from a no-howish, untalkaboutable, all-alikeness, to a some-howish and in-general-talkaboutable not-all-alikeness by continuous something-elsifications and sticktogetherations."

Some quotations were then given of the method of Sir Edmund Becket in dealing with certain modes of argument, and Tait concluded

When the purposely vague statements of the materialists and agnostics are thus stripped of the tinsel of high-flown and unintelligible language, the eyes of the thoughtless who have accepted them on authority (!) are at last opened, and they are ready to exclaim with Titania

"Methinks I was enamoured of an ass."

As the touch of Ithuriel's spear at once happily revealed the deceiver, these frank and clear exposures of the pretensions of pseudo-science cannot fail of producing great ultimate good.

In his appendix to *First Principles*, dealing with Criticisms, Spencer replied at considerable length to the criticism which seemed to be implied in these quotations and statements. He pointed out (p. 566) that a "formula expressing all orders of change in their general course...could not possibly be framed in any other than words of the highest abstractness"; and by way of a general enquiry into mental idiosyncrasies proceeded to put together in one group the two mathematicians Kirkman and Tait and two literary men, a North American Reviewer and Matthew Arnold. We are told (p. 570) that

"men of letters, dieted in their early days on grammars and lexicons and in their later days occupied with belles lettres, Biography and a History made up mainly of personalities, are by their education and course of life left almost without scientific ideas of a definite kind.......The mathematician too and the mathematical physicist, occupied exclusively with the phenomena of number space and time, or, in dealing with forces, dealing with them in the abstract, carry on their researches in such ways as may, and often do, leave them quite unconscious of the traits exhibited by the general transformations which things, individually and in their totality, undergo."

These exhibit "certain defects of judgment...to which the analytical habit, unqualified by the synthetical habit, leads." Much of which seems to be beside the mark. Tait certainly was not occupied exclusively with the phenomena of number, space and time. His was a mind intuitively physical. He was in experimental touch with things as they are in a way to which Spencer was an absolute stranger. As for lack of scientific ideas of a definite kind, no better examples can be found than in *First Principles*, especially in the chapter on the Instability of the Homogeneous. Enough has been given of Herbert Spencer's own words to enable us to appreciate Tait's rejoinder. This was given as part of his opening lecture in 1880, and appeared in *Nature*, Nov. 25, the same year; but it will suffice to reproduce only those parts which are not a repetition of previous letters.

Mr Spencer has quite recently published a species of analytical enquiry into my "mental peculiarities," "idiosyncrasies of thought," "habits of mind," "mental traits," and what not. From his illustrative quotations it appears that some or all of these are manifested wherever there are differences between myself and my critic in the points of view from which we regard the elements of science. Hence they are not properly personal questions at all, but questions specially fitted for discussion here and now. I may, therefore, commence by enquiring what species of "mental peculiarity" my critic himself exhibited when he seriously asked me whether I had proved *by experiment* that a thing might have been what it is not!!

The title of Mr Spencer's pamphlet informs us that it deals with *Criticisms*; and I am the first of the subjects brought up in it for vivisection, albeit I have been guilty (on Mr Spencer's own showing) only of "*tacitly*" expressing an opinion! Surely my vivisector exhibits here also some kind of "mental peculiarity." Does a man become a critic because he quotes, with commendation if you like, a clever piece of analysis or exposition published by another?......

Mr Spencer complains that an American critic (whose estimate is "tacitly" agreed in by Mr Matthew Arnold) says of the "Formula of Evolution":—"This may be all true, but it seems at best rather the blank form for a universe than anything corresponding to the actual world about us." On which I remark, with Mr Kirkman, "Most just, and most merciful!" But mark what Mr Spencer says:

"On which the comment may be that one who had studied celestial mechanics as much as the reviewer has studied the general course of transformations, might similarly have remarked that the formula—'bodies attract one another directly as their masses and inversely as the squares of their distances,' was at best but a blank form for solar systems and sidereal clusters."

We now see why Mr Spencer calls his form of words a *Formula*, and why he is indignant at its being called a *Definition*. He puts his Formula of Evolution alongside of the Law of Gravitation! Yet I think you will very easily see that it is a definition, and nothing more. By the help of the Law of Gravitation (not very accurately quoted by Mr Spencer) astronomers are enabled to predict the positions of known celestial bodies four years beforehand, in the *Nautical Almanac*, with an amount of exactness practically depending merely upon the accuracy of the observations which are constantly being made:—and, with the same limitation, the prediction could be made for 1900 A.D., or 2000 A.D., if necessary. If now Mr Spencer's form of words be a formula, in the sense in which he uses the term as applied to the Law of Gravitation, it ought to enable us to predict, say four years beforehand, the history of Europe, with at least its main political and social changes! For Mr Spencer says that his "formula" expresses "all orders of changes in their general course,—astronomic, geologic, biologic, psychologic, sociologic"; and therefore "could not possibly be framed in any other than words of the highest abstractness."

Of Mr Spencer's further remarks there are but three which are directed specially against myself. (Mr Kirkman is quite able to fight his own battles.) He finds evidence of "idiosyncrasies" and what not, in the fact that, after proclaiming that nothing could be known about the physical world except by observation and experiment, I yet took part in writing the "Unseen Universe"; in which arguments as to the Unseen are based upon supposed analogies with the seen. He says:—"clearly, the relation between the seen and the unseen universes cannot be the subject of any observation or experiment; since, by the definition of it, one term of the relation is absent." I do not know exactly what "mental peculiarity" Mr Spencer exhibits in this statement. But it is a curious one. Am not I, the thinker, a part of the Unseen; no object of sense to myself or to others; and is not that term of relationship between the seen and the Unseen always present? But besides this, Mr Spencer mistakes the object of the book in question. The theory there developed was not put forward as probable, its purpose was attained when it was shown to be conceivable and not inconsistent with any part of our present knowledge.

Mr Spencer's second fault-finding is *à propos* of a Review of *Thomson and Tait's Nat. Phil.* (*Nature*, July 3, 1879) by Clerk Maxwell. Maxwell, knowing of course perfectly well that the authors were literally quoting Newton, and that they had expressly said so, jocularly remarked "Is it a fact that 'matter' has any power, either innate or acquired, of resisting external influences?" Mr Spencer says:—"And to Prof. Clerk Maxwell's question thus put, the answer of one not having a like mental peculiarity with Prof. Tait, must surely be—No." Mr Spencer, not being aware that the passage is Newton's, and not recognising Maxwell's joke, thinks that Maxwell is at variance with the authors of the book!

Finally, Mr Spencer attacks me for inconsistency etc. in my lecture on Force (*Nature*, September 21, 1876). I do not know how often I may have to answer the perfectly groundless charge of having, in that Lecture, given two incompatible definitions of the same term. At any rate, as the subject is much more important than my estimates of Mr Spencer's accuracy or than his estimates of my "mental peculiarities," I may try to give him clear ideas about it, and to show him that there is no inconsistency on the side of the mathematicians, however the idea of force may have been muddled by the metaphysicians. For that purpose I shall avoid all reference to "differentiations" and "integrations"; either as they are known to the mathematicians, or as they occur in Mr Spencer's "Formula." Of course a single line would suffice, if the differential calculus were employed.

Take the very simplest case, a stone of mass M, and weight W, let fall. After it has fallen through a height h, and has thus acquired a velocity v, the Conservation of Energy gives the relation

$$M \frac{v^2}{2} = Wh.$$

Here both sides express *real things*; $M \frac{v^2}{2}$ is the kinetic energy acquired, Wh the work expended in producing it.

But if we choose to divide both sides of the equation by $\frac{v}{2}$ (the average velocity during the fall) we have (by a perfectly legitimate operation)

$$Mv = Wt,$$

where t is the time of falling. This is read:—*the momentum acquired is the product of the force into the time during which it has acted.* Here, although the equation is strictly correct, it is an equation between purely artificial or non-physical quantities, each as unreal as is the product of a quart into an acre. It is often mathematically convenient, but that is all. The introduction of these artificial quantities is, at least largely, due to the strong (but wholly misleading) testimony of the "muscular" sense.

Each of these modes of expressing the same truth, of course gives its own mode of measuring (and therefore of defining) force.

The second form of the equation gives

$$W = \frac{Mv}{t}.$$

Here, therefore, force appears as the time-rate at which momentum changes; or, if we please, as the time-rate at which momentum is produced by the force. In using this latter phrase we adopt the convenient, and perfectly misleading, anthropomorphism of the mathematicians. This is the gist of a part of Newton's second Law.

The first form of the equation gives

$$W = \frac{M\frac{v^2}{2}}{h},$$

so that the same force now appears as the space-rate at which kinetic energy changes; or, if we please, as the space-rate at which energy is produced by the force.

Here are some of Mr Spencer's comments:—"force is that which changes the state of a body; force is a rate, and a rate is a relation (as between time and distance, interest and capital); therefore a relation changes the state of a body."

The contradiction which Mr Spencer detects here, and over which he waxes eloquent and defiant, exists in his own mind only. The anthropomorphism which has misled him is but a convenient and harmless relic of the old erroneous interpretations of the impressions of sense.

In his reply (*Nature*, Dec. 2, 1880) Herbert Spencer reproduced a good deal of his pamphlet, got somewhat indignant over a side issue, and sneered at Tait's mathematics. "If," he remarked, "his mathematics prove that while force is an agent which does work, it is also the rate at which an agent does work, then I say—so much the worse for mathematics." In a brief rejoinder (*Nature*, Dec. 9, 1880) Tait wrote:

Mr Spencer has employed an old remark of Prof. Huxley as to what mathematics can, and cannot, do; but he has not employed it happily, for the question at issue is really this: is it correct to speak, at one time, of force as an agent which changes a body's state of rest or of motion, and again to speak of it as the time-rate at which momentum changes or as the space-rate at which energy is transformed?

I answer that there is not the slightest inconvenience here; except, perhaps, in the eyes of those metaphysicians (if there be any) who fancy they know *what* force *is*. Such phrases as "the wind blows" or "the sun rises" though used by the most accurate even of scientific writers, would otherwise (on account of their anthropomorphism) have to be regarded as absolute nonsense.

Here the controversy ended, for Spencer's later letter of Dec. 16 took no notice of the scientific questions which were supposed to be the subjects of a discussion which he himself had originated. Tait regarded the controversy as a joke; for he knew it was hopeless to convince of any errancy a mind which believed that the greatest physical generalization of modern times could be established as an *à priori* intuition.

DR BALFOUR STEWART, F.R.S.

(From *Nature*, Dec. 29, 1887.)

In the genial Manchester Professor the scientific world has lost not only an excellent teacher of physics but one of its ablest and most original investigators. He was trained according to the best methods of the last generation of experimentalists, in which scrupulous accuracy was constantly associated with genuine scientific honesty. Men such as he was are never numerous; but they are the true leaders of scientific progress: *directly*, by their own contributions; *indirectly*, though (with rare exceptions) even more substantially, by handing on to their students the choicest traditions of a past age, mellowed by time and enriched from the experience of the present. The name of Stewart will long be remembered for more than one striking addition to our knowledge, but his patient and reverent spirit will continue to impress for good the minds and the work of all who have come under its influence.

He was born in Edinburgh, on November 1st, 1828, so that he had entered his sixtieth year. He studied for a short time in each of the Universities of St Andrews and Edinburgh, and began practical life in a mercantile office. In the course of a business voyage to Australia his particular taste for physical science developed itself, and his first published papers: "On the adaptation of the eye to different rays," and "On the influence of gravity on the physical condition of the Moon's surface," appeared in the *Transactions of the Physical Society of Victoria* in 1855. On his return he gave up business for science, and resumed study under Kelland and Forbes, to the latter of whom he soon became Assistant. In this capacity he had much to do with the teaching of Natural Philosophy on occasions when Forbes was temporarily disabled by his broken health. During this period, in 1858, Stewart was led to his well-known extension of Prévost's *Law of Exchanges*, a most remarkable and important contribution to the theory of Radiation. He seems to have been the first even to suggest, from a scientific stand-point, that radiation is not a mere surface phenomenon. With the aid of Forbes' apparatus, then perhaps unequalled in any British University, he fully demonstrated the truth of the conclusions to which he had been led by theory; and the award of the Rumford Medal by the Royal Society, some years later, showed that his work had been estimated at its true value, at least in the scientific world. In fact his proof of the necessary equality between the radiating and the absorbing powers of every substance (when divested of some of the unnecessary excrescences which often mask the real merit of the earlier writings of a young author) remains to this day the simplest, and therefore the most convincing, that has yet been given.

Radiant Heat was, justly, one of Professor Forbes' pet subjects, and was therefore brought very prominently before his Assistant. Another was Meteorology, and to this Stewart devoted himself with such enthusiasm and success that in 1859 he was appointed Director of the Kew Observatory. How, for eleven years, he there maintained and improved upon the memorable labours of Ronalds and Welsh needs

T.

37

only to be mentioned here: it will be found in detail in the *Reports of the British Association*. Every species of inquiry which had to be carried out at Kew: whether it consisted in the testing of Thermometers, Sextants, Pendulums, Aneroids, or Dipping-Needles, the recording of Atmospheric Electricity, the determination of the Freezing-Point of Mercury or the Melting-Point of Paraffin, or the careful study of the peculiarities of the Air-Thermometer: received the benefit of his valuable suggestions and was carried out with his scrupulous accuracy.

About twenty years ago Stewart met with a frightful railway accident, from the effects of which he did not fully recover. He was permanently lamed, and sustained severe injury to his constitution. From the vigorous activity of the prime of life he passed, in a few months, to grey-headed old age. But his characteristic patience was unruffled, and his intellect unimpaired.

His career as Professor of Physics in the Owens College has been, since his appointment in 1870, brilliantly successful. It has led to the production of an excellent treatise on *Practical Physics*, in which every necessary detail is given with masterly precision, and which contains (what is even more valuable, and could only have been secured to the world by such a publication) the matured convictions of a thorough experimenter as to the choice of methods for the attack of each special Problem.

His *Elementary Physics*, and his *Conservation of Energy*, are popular works on physics rather than scientific treatises: but his *Treatise on Heat* is one of the best in any language, a thoroughly scientific work, specially characteristic of the bent of mind of its Author.

Stewart published, in addition to his *Kew Reports*, a very large number of scientific memoirs and short papers. Many of these (notably the article in the *Encyc. Brit.*, 9th ed.) deal with Terrestrial Magnetism, in itself as well as in its relations to the Aurora and to solar disturbances. A valuable series of papers, partly his own, partly written in conjunction with De la Rue and Loewy, deals with Solar Physics. His paper on the "Occurrence of Flint Implements in the Drift" (*Phil. Mag.* 1, 1862) seems to have been ignored by the "advanced" geologists, one of whose pet theories it tends to dethrone; and to have been noticed only by physicists, especially Sir W. Thomson, whose beautiful experiments have done so much to confirm it. His paper on "Internal Radiation in Uniaxal Crystals," to which Stokes alone seems to have paid any attention, shows what Stewart might have done in Mathematical Physics, had he further developed the genuine mathematical power which he exhibited while a student of Kelland's.

I made Stewart's acquaintance in 1861, when he was the first-appointed Additional Examiner in Mathematics in the University of Edinburgh, a post which he filled with great distinction for five years. A number of tentative investigations ultimately based upon our ideas as to possible viscosity of the luminiferous medium, effect of gravitation-potential, on the physical properties of matter, etc., led to the publication of papers on "Rotation of a disc in vacuo, Observations with a rigid spectroscope, Solar spots and planetary configurations," etc. These, as well as our joint work called *The Unseen Universe*, have been very differently estimated by different classes of critics. Of course I cannot myself discuss their value. There is, however, one of these speculations, so closely connected with Stewart's Radiation work as to require particular mention, especially as it seems not yet to have received proper consideration, viz. "Equilibrium of Temperature in an enclosure containing matter in visible motion" (*Nature*, IV. p. 331, 1871). The speculations are all of a somewhat transcendental character, and therefore very hard to reduce to forms in which they can be experimentally tested; but there can be no doubt that Stewart had the full conviction that there is in them all an underlying reality, the discovery of whose exact nature would at once largely increase our knowledge.

Of the man himself I cannot trust myself to speak. What I *could* say will easily be divined by those who knew him intimately; and to those who did not know him I am unwilling to speak in terms which, to them, would certainly appear exaggerated.

PROFESSOR ROBERTSON SMITH.

(From *Nature*, April 12, 1894.)[1]

The death of Prof. Robertson Smith, on March 31, at a comparatively early age, is a profound loss to the whole thinking world.

Unfortunately for Science, and (in too many respects) for himself, his splendid intellectual power was diverted, early in his career, from Physics and Mathematics, in which he had given sure earnest of success. He turned his attention to eastern languages, and acquired a knowledge of Hebrew, Arabic, and other tongues, quite exceptional in the case of a Briton.

Dr Smith was born at Keig, Aberdeenshire, in 1846, and educated at Aberdeen University, the New College, Edinburgh, and the Universities of Bonn and Göttingen. In 1868 he became Assistant to the Professor of Physics in Edinburgh University; in 1870, at the age of twenty-four, he was appointed to the chair of Hebrew in the Free Church College of Aberdeen. A few years later he fell under the suspicion of holding heterodox views concerning Biblical history. Orthodoxy raised her voice against him in the newspapers, in the churches, in the Presbyteries, and finally, in the General Assembly of the Free Church of Scotland, and the clamour culminated in his dismissal from the Professorship at Aberdeen in 1881. This was effected, *not* by a direct condemnation of his published opinions, but by a monstrous (temporary) alliance between ignorant fanaticism and cultivated jesuitry which deplored the "unsettling tendency" of his articles!

He next became successor to Prof. Baynes in the Editorship of the last edition of the *Encyclopaedia Britannica*; and here his business qualities, as well

---

[1] The Article is unsigned; but Tait initialled it as his in his own copy of *Nature*. It seems necessary to mention this in view of the somewhat "hearsay" language of the sixth paragraph. The strong language at the end of the third paragraph embodies Tait's often expressed views of the great "Heresy Hunt."

as his extraordinary range of learning, came prominently before the world. In 1883 Dr Smith was appointed Reader in Arabic at Cambridge, and three years later he succeeded the late Mr Bradshaw as librarian to the University. He was afterwards elected to a Fellowship at Christ's College, and to the Professorship of Arabic.

What Smith might have done in science is shown by his masterly paper "On the Flow of Electricity in Conducting Surfaces" (*Proc. R. S. E.*, 1870), which was rapidly written in the brief intervals of leisure afforded by his dual life as simultaneously a Student in the Free Church College, and Assistant to the Professor of Natural Philosophy in Edinburgh University.

We understand that his engagement as Assistant to Prof. Tait had its origin in the extremely remarkable appearance made by young Smith as a Candidate in the Examination for the Ferguson Scholarships, an examination in which most of the very best men in the four Scottish Universities are annually pitted against one another.

In Edinburgh University he did splendid service in the work of initiating the Physical Laboratory: and there can be no doubt that the *esprit de corps*, and the genuine enthusiasm for scientific investigation, which he was so influential in exciting there, have inaugurated and promoted many a successful career (not in this country alone, but in far regions everywhere), and that, near and far, his death will be heard of with heart-felt sorrow.

A light and playful feature of his too few years of scientific work consisted in his exposures of the hollowness of the pretensions of certain "philosophers," when they ventured to tread on scientific ground. Several of these will be found in the *Proceedings and Transactions* of the Royal Society of Edinburgh (1869–71). Smith treats his antagonist "tenderly" as if he loved him, but the exposure is none the less complete.

A writer in *The Times* thus testifies to Dr Smith's remarkable powers: "In him there has passed away a man who possessed not only one of the most learned but also one of the most brilliant and striking minds in either of the great English Universities, and who was held in the highest regard by the leading orientalists of the continent. His extraordinary range of knowledge, the swiftness and acuteness of his intellect, and his passionate love of truth combined to make an almost unique personality. His talents for mathematics and physical science were scarcely less remarkable than those for linguistic studies, and if he had not preferred the latter, there is no question that he could have reached great eminence in the former."

RELIGION AND SCIENCE.

(From *The Scots Observer*, Dec. 8, 1888.)

"For not by the rays of the sun, nor the glittering shafts of the day,
Must the fear of the gods be dispelled; but by words and their wonderful play."

Thus said Clerk Maxwell, one of the genuine scientific men of the generation just past, in ridicule of the assumptions and pretensions of the hydra-headed pseudo-science. And he said rightly, for true science is not, and cannot be, at variance with religion.

That there is a conflict between Religion and Science, not a mere difference between certain theologians and certain "scientists," is very frequently stated as a matter of fact. It is occasionally spoken of in grandiloquent phrase as the *Incompatibility of Religion and Science!* Of course, if there be even a fragment of truth in this, it is matter of very serious import indeed; but it is prudent in all such cases to ask the simple question, "On what authority is the statement made?" The answer to this will either make careful investigation imperative, or render it altogether unnecessary. To this question, then, and to this alone, the present article is devoted.

There are not very many people who really know what science is, though there are many who think they do, and who even pose successfully, so far as the public is concerned, as scientific men. This may seem a somewhat bold statement, but it is capable of easy proof. Let us take an analogy or two to begin with.

1. *A* is a "medical man." What do you understand by such a phrase? Does it necessarily imply that, if you were seriously ill, you could with confidence submit yourself to *A*'s professional care? Are there no quacks on the Register? If you think there are none, don't read any further. I am not addressing *you*. I look for some common-sense in my readers.

2. *B* is a "military man." Does this imply that he is capable of conducting a campaign, of leading a company, or even of putting an awkward squad through their facings? By no means, and you know it quite well. Yet you persist in calling *B* a "military man," though he may be merely a re-incarnation of Bardolph, Pistol, or Parolles!

3. *C* is a "business man." The term includes the trusty family lawyer, as well as the bankrupt speculator!

4. *D* is a "scientific man." We've got to him at last. Well! the term was applied to Lord Brougham, and to Dionysius Lardner; it is still applied to Isaac Newton.

It was for this reason that, in speaking of Clerk Maxwell, I had to make a slight but important qualification of the term, and to call him a *genuine* scientific man. But in that italicised word there is a whole world of meaning. It opens an abyss of impassable width among the group of so-called "scientific men"—leaving the vast majority hopelessly on the wrong side. There we will leave them for a little, in order to discuss a collateral issue.

What is science? Whatever it is, it is certainly not mere knowledge, even of facts. If I had Babbage's Logarithms by heart, as a school-boy has the multiplication table, I should *know* more than any other human being, besides being the most wonderful calculator that ever existed; but my knowledge would not be science. No more, indeed, than it would be scholarship, if it consisted in my being able to repeat the whole of Liddell and Scott. Every decently-educated Greek knew all the contents of that huge volume (its errors excepted) two thousand years ago, better than did these learned pundits. But it would never occur to the least discerning among us to speak of that Greek's "scholarship." These things are what, now-a-days, we call *cram*. Descriptive botany, natural history, volumes of astronomical observations, etc., are collections of statements, often facts, from which scientific truth may ultimately be extracted, but they are not science. Science begins to dawn, but only to dawn, when a Copernicus, and after him a Kepler or a Galilei, sets to work on these raw materials, and sifts from them their essence. She bursts into full daylight only when a Newton extracts the quintessence. There has been, as yet, but one Newton; there have not been very many Keplers. Thus the great mass of what is commonly called science is totally undeserving of the name. But, the name being given (though in error) to the crude collection of undigested facts, the worker at them gets the name of "scientific man" from an undiscerning public. And it is to be particularly observed that statements as to the so-called incompatibility between religion and science come, all but exclusively, from this class of persons. Not from all—not even from the majority of them; but from their ranks alone. How trustworthy is their judgment, as a general rule, may be seen from some of their recent statements. Nothing like these has been heard since "all, with one voice, about the space of two hours, cried out, Great is Artemis of the Ephesians." Take this one—unique, it is to be hoped, in its absurdity—"When the boldness of (Darwin's) generalisations, and the great school which he has founded, are taken into account, it is perhaps no exaggeration to bracket him with Newton, Kepler, and Tycho Brahe"! So careful and conscientious an observer as Darwin may justly be bracketed with Tycho, who was distinguished for these very qualities; but Tycho himself was no more than the mere hodman of Kepler. Should the hypothesis of Wallace and Darwin turn out to be correct, its authors may perhaps claim the position of Copernicus; otherwise they fall back to that of Ptolemy. The mention of Newton in such a connection is only an ingenuous confession of inability to comprehend the very nature of Newton's work. There cannot be a Newton in natural science until there has been at least one Kepler; and he has not yet appeared, nor do we see much promise of his coming.

Let us revert to Maxwell's lines,—"Words, and their wonderful play." Has any one, in a single phrase, ever more fully, or more fittingly, characterised at once the one talent, and the abundant but temporary success, of the pseudo-scientific? Roar me in King Cambyses' vein, and you have the multitude at your feet. This is their distinctive mark.

I have purposely made the above remarks in a disconnected manner, insidiously working towards the enemy's goal. Hitherto we have had narrative and description only. We will now, gentlemen, have a little logic, if you please, and take the place-kick.

First. The so-called incompatibility of religion and science is proclaimed solely from the ranks of those whose subject has not yet reached the scientific stage, and from the ranks of pseudo-science.

Second. In both of these ranks, "words and their wonderful play" are the chief weapons; and they are employed with a pertinacity truly amazing.

Third. There has not, in all, been any very persistent or even vociferous assertion of the so-called incompatibility.

From these it indubitably follows :—

Fourth. That, even among those whose subject has not yet reached the scientific stage, and among the pseudo-scientific, there can be but few who maintain the so-called incompatibility.

Fifth. As this tenet is held only by a small minority of those who are on the wrong side of the abyss above mentioned, which separates the *genuine* scientific men from the rest of the species, it might be entirely disregarded were it not for its pernicious effect upon those who do not even pretend to be scientific.

Thus the really serious question is, " How can this pernicious effect be neutralised or remedied ? " Not, certainly, by anathematising as infidels those who are rather to be pitied for their ignorance. And, most certainly, not by pulpit denunciations, in which the most transparently absurd dicta of pseudo-science are complacently used as weapons against adversaries armed with the very same—a Kilkenny-cat display without claws or teeth on either side. If the clergy—upon whom, more than on any others, this task ought to fall—would lay much more stress than they have hitherto thought of doing upon the humility which is characteristic of all true knowledge, whether it be religious or scientific, and upon the blatant boasting equally characteristic of ignorance, they would be able easily to convince their flocks that the present outcry is the work of a small minority only, and has no countenance whatever from those who really know science.

CHAPTER VIII

POPULAR SCIENTIFIC ARTICLES

THUNDERSTORMS

Lecture in the City Hall, Glasgow, January 29th, 1880.

WHEN I was asked to give this lecture I was also asked to give a short list of subjects from which your directors might select what they thought most fit. I named three—one being of purely scientific importance, the others being of practical importance as well. Regarded from the scientific point of view, one of them was to be considered as fully understood in principle, and requiring only additional experimental data to make it complete. This was the *Conduction of Heat in Solids*. Another was to a certain extent scientifically understood, but its theory was, and still is, in need of extended mathematical development. This was the popular scientific toy, the *Radiometer*. The third was, and remains, scarcely understood at all. Of course it was at once selected for to-night. I might have foreseen that it would be. I had incautiously forgotten the nursery rule that a slice of bread always falls with the buttered side down, and the stern conviction of sailors that a marlinspike never drops from aloft except point foremost.

You may well ask, then, why I am here to-night. What can I say about a subject which I assert to be scarcely understood at all? To such a question there are many answers, but the most satisfactory are supplied by analogy.

Would interesting and even scientifically attractive matter have failed a lecturer, do you think, if he had chosen *Astronomy* for his subject, in days before Newton, before Kepler, before even Copernicus? Yet he could certainly not have properly described even the arrangement of the planets in the solar system, far less the laws of their motions, and least of all the dynamical basis of these laws. Still, you will grant that he might have given an admirable lecture. But stay: would he now be any better off? After all that Copernicus, Kepler, Newton, and their successors have done, do we yet scientifically understand *why* the planets move as they do? Certainly not. The mechanism of gravitation is still to us, as it was to Newton, an absolute mystery. Only one even *plausible* attempt to explain it has yet been made; and *that*, in spite of Sir W. Thomson's very ingenious attempts to improve it, we cannot yet venture to call *probable*. But, for all that, a lecturer on gravitation has a magnificent field before him. Though he knows nothing of its mechanism he knows its universal law, and he might profitably occupy your attention for many evenings in tracing the consequences of that grand but simple statement of physical fact. Think, for instance, of the many apparently altogether incompatible results to which it leads: planetary perturbations, combined with stability of the solar system; precession and nutation of the earth's axis, yet permanence of the seasons; and constant tidal agitation of the sea, yet no consequent submersion of continents. He might next trace the growth of stars, with their attendant planets, from an original chaotic distribution of matter—calculate even the temperature which each would acquire during its growth—all by the aid of this recognised law, of whose own explanation he yet remains absolutely ignorant.

So it is with the splendid phenomenon of optics. No experimenter has yet, to his certain knowledge, verified by direct proof the existence of that wonderful elastic jelly which we *know* must fill all space without offering perceptible resistance to bodies moving through it—that jelly whose inconceivably rapid quivering is the mechanism by which not only do we see all objects, from the sun to the most minute of stars, but by which comes to us continually from the sun that supply of energy without which vegetable life, and with it that of animals, must at once cease from the earth. Yet a lecturer could never be at a loss on matters connected with light and colour, or with radiation in any of its varied forms. And all the time he is consciously in total ignorance of the nature of this extraordinary luminiferous ether upon which they all entirely depend.

A few years ago no qualified physicist would have ventured an opinion as to the nature of electricity. Magnetism had been (to a certain extent, at least) cleared up by an assumption that it depended on electric currents; and from Örsted and Ampère to Faraday and Thomson, a host of brilliant experimenters and mathematicians had grouped together in mutual interdependence the various branches of electrodynamics. But still the fundamental question remained unsolved, *What is Electricity?* I remember Sir W. Thomson, eighteen years ago, saying to me, "Tell me what electricity is, and I'll tell you everything else." Well, strange as it may appear to you, I may now call upon him to fulfil his promise. And for good reason, as you shall see.

Science and Scotland have lately lost in Clerk Maxwell one of their greatest sons. He was, however, much better known to science than to Scotland. In scientific ranks true merit is almost always certain to be recognised—in popular ranks modest merit scarcely ever is. His was both true and exceptionally modest merit. One grand object which he kept before him through his whole scientific life was to reduce electric and magnetic phenomena to mere stresses and motions of the ethereal jelly. And there can be little doubt that he has securely laid the foundation of an electric theory—like the undulatory theory of light admirably simple in its fundamental assumptions, but, like it, requiring for its full development the utmost resources of mathematical analysis. It cannot but seem strange to the majority of you to be told that we know probably as much about the secret mechanism of electricity as we do about that of light, and that it is more than exceedingly probable that a ray of light is propagated by electric and electromagnetic disturbances. It can be but a small minority of you who have been at college so recently as to have been taught this. Yet, from the purely scientific point of view, it is one of the most remarkable advances made during this century.

But to know what electricity is, in the same sense as we may be said to know what light is, does not necessarily guide us in the least degree to a notion of its source in any particular instance. What, for example, causes the luminosity of fireflies, glowworms, etc., among natural objects, and of phosphorescent watch-dials among artificial ones? The answer is not yet ready, though it may soon come. So we might know quite well *what* is electricity and yet be, as I told you at starting, we *are*, almost entirely uncertain of the exact source of *atmospheric* electricity.

To come to my special subject. I am not going to try to describe a thunderstorm. First, because I am certain that I could not do it without running the risk of overdoing it, and thus becoming sensational instead of scientific; and secondly, because the phenomenon must be quite familiar, except perhaps in some of its more singular details, to every one of you. From the artistic point of view of the poets, who claim the monopoly of the expressions of wonder and awe, you have descriptions without end. Who does not know at least the finest of them, from that of Lucretius of old to that of Byron in modern times?

But science has to deal with magnitudes which are very much larger or smaller than those which such words as huge, enormous, tiny, or minute are capable of expressing. And though an electric spark, even from our most powerful artificial sources, appears to the non-scientific trifling in comparison with a mile-long flash of lightning, the difference (huge, if you like to call it) is as nothing to others with which scientific men are constantly dealing. The nearest star is as much farther from us than is the sun, as the sun is farther from us than is London. The sun's distance is ninety-three millions of miles. If that distance be called enormous, and it certainly is so, what adjective have you for the star's distance? The particles of steam are as much less than a drop of water as an orange is less than the whole earth. How can you fitly characterise their smallness? Ordinary human language, and specially the more poetic forms of it, were devised to fit human feelings and emotions, and not for scientific purposes: for the Imagination, not for the Reason. As there is a limit alike to pleasure and to pain, so there is to wonder and astonishment; and with words expressive of something near these limits ordinary language rests content. A thoroughly scientific account of a thunderstorm, if it were possible to give one, would certainly be at once ridiculed as pedant.

Let us therefore, instead of attempting to discuss the phenomenon as a whole, consider separately some of its more prominent features. And first of all, what are these features when we are *in* the thunderstorm?

By far the most striking, at least if the thunderstorm come on during the day, is the extraordinary darkness. Sometimes at midday in summer the darkness becomes comparable with that at midnight, and in another sense it much resembles midnight darkness, for it is very different in kind as well as intensity from that produced by the densest fog. Objects are distinctly visible through it at distances of many miles, whether when self-luminous or when instantaneously lit up by lightning. The darkness then is simply intense *shadow*, produced by the great thickness and great lateral extension of the cloud-masses overhead. This altogether unusual amount of cloud always forms a prominent feature in a great thunderstorm. We have thus obtained one very important clue to the origin of the phenomenon. In cases where the darkness is not so great, though the storm is visibly raging overhead, it is always observed either that we are not far from the edge of its area or, in rarer cases, that the lower cloud strata lie much higher than usual. These form, therefore, no exception to the general conclusion just given as to the abnormal amount of cloud present. Seen from a distance, the mass of cloud belonging to the storm usually presents a most peculiar appearance, quite unlike any other form of cloud. It seems to boil up, as it were, from below, and to extend through miles of vertical height. The estimated height of its lower surface above the ground varies within very wide limits. Saussure has seen it as much as three miles; and in one case noticed by De l'Isle it may have been as much as five miles. On the other hand, at Pondicherry and Manilla it is scarcely ever more than half a mile. Haidinger gives the full details of an extraordinary case, in which the thunder-loud formed a stratum of only 25 feet thick, raised 30 yards above the ground. Yet two people were killed on this occasion. Other notable instances of a similar extreme character are recorded.

Careful experiment shows us that the air is scarcely ever free from electricity, even in the clearest weather. And even on specially fine days, when large separate cumuli are floating along, each as it comes near produces a marked effect on the electrometer. Andrews obtained by means of a kite, on a fine clear day, a steady decomposition of water by the electricity collected by a fine wire twisted round the string. Thanks to Sir W. Thomson, we can now observe atmospheric electricity in a most satisfactory manner. I will test, to show you the mode of proceeding, the air inside and outside the hall. [The experiment was shown, and the external air gave *negative* indications.]

On several occasions I have found it almost impossible, even by giving extreme directive force to the instrument by means of magnets, to measure the atmospheric potential with such an electrometer, and had recourse to the old electroscope, with specially long and thick gold-leaves. On February 26th, 1874, when the sleet and hail, dashing against the cupola of my class-room, made so much noise as to completely interrupt my lecture, I connected that instrument with the water-dropper, and saw the gold-leaves discharge themselves against the sides every few seconds, sometimes with positive, sometimes, often immediately afterwards, with negative electricity. Such effects would have required for their production a battery of tens of thousands of cells. Yet there was neither lightning nor thunder, and the water was trickling from the can at the rate of only two and a half cubic inches per minute. Probably had there not been such a violent fall of sleet steadily discharging the clouds we should have had a severe thunderstorm. That this is no fancied explanation is evident from the fact that falling rain-drops are often so strongly charged with electricity as to give a spark just before they touch the ground. On such occasions, when the fall occurs at night, the ground is feebly lit up. This "luminous rain," as it has been called, is a phenomenon which has been over and over again seen by competent and trustworthy observers. In the *Comptes Rendus* for November last we read of the curious phenomenon of electrification of the observer's umbrella by a slight fall of snow, to such an extent that he could draw sparks from it with his finger.

In calm clear weather the atmospheric charge is usually positive. This is very commonly attributed to evaporation of water, and I see no reason to doubt that the phenomena are closely connected. I will show you one of the experiments upon which the idea is based. I can take no other form of experiment than a somewhat violent one, as the effects of the more delicate ones could not easily or certainly be made visible to so large an audience. [A few drops of water were sprinkled on a heated crucible, insulated, and connected with the electrometer.]

There can be no doubt that, whatever be the hidden mechanism of this experiment, the steam has carried with it a strong charge of positive electricity, for it has left the rest of the apparatus with a strong negative charge. We might reverse the subject of measurement by connecting the electrometer with the escaping steam, but I omit it, to save time, and because we will now try that form of the experiment in another way. [High-pressure steam escaping from a little boiler was made to play upon an insulated conductor furnished with spikes, and connected with the electrometer, which then showed a strong positive charge.]

There are many substances which produce on evaporation far greater electric developments than water does, some of positive, others of negative, electricity. By far the most remarkable in this respect to which attention has yet been called is an aqueous solution of sulphate of copper. (*Proc. R.S.E.*, 1862.) The smallest drop of this solution thrown on a hot dish gives an intense negative effect—so great, in fact, that it may be occasionally employed to charge a small Leyden jar. But this, like the smaller effect due to water under similar circumstances, is not yet completely explained.

The next striking features are the flashes of lightning which at intervals light up the landscape with an intensity which must in the majority of cases far exceed that produced by the full moon. To the eye, indeed, the flash does not often appear to furnish more than the equivalent of average moonlight, but it must be remembered that it lasts for a period of time almost inconceivably short, and that the full effect of light on the eye is not produced until after the lapse of a considerable fraction of a second. Professor Swan has estimated this interval at about one-tenth of a second; and he has proved that the apparent intensity of illumination for shorter intervals is nearly proportional to the duration. (*Trans. R.S.E.*, 1849.) I can illustrate this in a very simple manner. [Two beams of light were thrown upon the screen by reflection from mirrors, each of which was fixed *nearly* at right angles to an axis. When matters were so adjusted that the brightness of the two illuminated spots was the same, one mirror was made to rotate. The corresponding light spot described a circle about the other, and its brightness became less the larger the circle in which it was made to revolve.] The lightning flash itself on this account, and for the further reason that its whole apparent surface is exceedingly small, must be in some degree comparable with the sun in intrinsic brilliancy— though, of course, it cannot appear so. The fact that its duration is excessively short is easily verified in many ways, but most simply by observing a body in rapid motion. The spokes of the wheels of the most rapidly-moving carriage appear absolutely fixed when illuminated by its light alone. One can read by its light a printed page stuck on a disc revolving at great speed. But the most severe test is that of Sir Charles Wheatstone's revolving mirror. Seen by reflection in such a mirror, however fast it may be rotating, a flash of lightning is not perceptibly broadened, as it certainly would be if its duration were appreciable.

The apparatus which, in our laboratories, enables us to measure the time which light, moving at nearly 200,000 miles per second, takes to pass over a few feet, is *required* to prove to us that lightning is not absolutely instantaneous. Wheatstone has shown that it certainly lasts less than a millionth part of a second. Take this, along with Swan's datum, which I have just given you, and you see that the apparent brightness of the landscape, as lit up by a lightning flash, is *less than one hundred thousandth part* of what it would be were the lightning permanent. We have thus rough materials for instituting a comparison between the intrinsic brightness of lightning and of the sun.

Transient in the extreme as the phenomenon is, we can still, in virtue of the duration of visual impressions, form a tolerably accurate conception of the form of a flash; and in recent times instantaneous processes of photography have given us permanent records of it. These, when compared with photographic records of ordinary electric sparks, bear out to the full the convictions at once forced by appearances on the old electricians, that a flash of lightning is merely a very large electric spark. The peculiar zig-zag form, sometimes apparently almost doubling back on itself, the occasional bifurcations, and various other phenomena of a lightning flash, are all shown by the powerful sparks from an electric machine. [These sparks were exhibited directly; and then photographs, of which one is represented in the woodcut, were exhibited.]

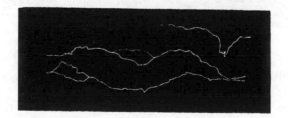

But the spectroscope has recently given us still more convincing evidence of their identity, if any such should be wanted. This is a point of great importance, but I have not now time to discuss it. Thus, though on a very small scale comparatively, we may study the phenomena of lightning at convenience in our laboratories. We thus know by experiment that electricity chooses always the easiest route, the path of least resistance. Hence the danger of taking the otherwise most natural course of standing under a tree during a thunderstorm. The tree, especially when wet, is a much better conductor than the air, and is consequently not unlikely to be chosen by the discharge; but the human body is a conductor much superior to the tree, and therefore is chosen in preference so far as it reaches. The bifurcations of a flash can puzzle no one who is experimentally acquainted with electricity, but the zig-zag form is not quite so easily explained. It is certainly destroyed, in the case of short sparks, by heating the air. [Photographs of sparks in hot and in cold air were exhibited. One of each kind is shown in the woodcut. The smoother is that which passed through the hot air. The other passed through the cold air nearer the camera, and is therefore not quite in focus.]

Now heating in a tube or flame not only gets rid of motes and other combustible materials but it also removes all traces of electrification from air. It is possible, then, that the zig-zag form of a lightning flash may, in certain cases at least, be due to local electrification, which would have the same sort of effect as heat in rarefying the air and making it a better conductor.

A remark is made very commonly in thunderstorms which, if correct, is obviously inconsistent with what I have said as to the extremely short duration of a flash.

Even if we supposed the flash to be caused by a luminous body moving along, like the end of a burning stick whirled around in a dark room, it would pass with such extraordinary rapidity that the eye could not possibly follow its movements. Hence it is clear that when people say they *saw* a flash go upwards to the clouds from the ground, or downwards from the clouds to the ground, they must be mistaken.

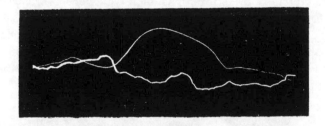

The origin of the mistake seems to be a *subjective* one, viz., that the central parts of the retina are more sensitive, by practice, than the rest, and therefore that the portion of the flash which is seen directly affects the brain sooner than the rest. Hence a spectator looking towards *either* end of a flash very naturally fancies that end to be its starting point.

Before I can go farther with this subject it is necessary that I should give some simple facts and illustrations connected with ordinary machine electricity. These will enable you to follow easily the slightly more difficult steps in this part of our subject which remain to be taken.

Since we are dealing mainly with *motion* of electricity, it is necessary to consider to what that motion is due. You all know that winds, i.e., motions of the air, are due to differences of pressure.

If the pressure were everywhere the same at the same level we should have no winds. Similarly the cause of the motion of heat in a body is difference of temperature. When all parts of a body are at the same temperature there is no change of distribution of heat. Now electricity presents a precisely analogous case. It moves in consequence of difference of *potential*. Potential, in fact, plays, with regard to electricity, a part precisely analogous to the *rôle* of pressure, or of temperature, in the case of motions of fluids and of conducted heat. It would tax your patience too much were I to give an exact definition of potential in a lecture like this; but you may get a sufficiently approximate notion of it in a very simple way —again by analogy. Suppose I wished to specify the power of the pump used to compress air in an air-gun receiver. I should say it can produce a limiting pressure of 40 or 100 atmospheres, as the case may be. If you try to go any farther it leaks. This would be considered quite definite information. But, mark you, nothing is said as to the *capacity* of the receiver. When the pump has done its utmost, the receiver, be it large or small, contains air at the definite pressure of the 40 or 100 atmospheres which measure the power of the pump.

A longer time will be required for a more capacious receiver, but the ultimate pressure is the same in all. And when two receivers contain air at the same pressure you may open a communication between them, but no air will pass, however much they may differ in capacity. Similarly, you may measure the power of a flame or a furnace by the highest temperature it can produce. It will take a longer time to effect this the greater is the thermal capacity of the body to be heated; but when two bodies are at the same temperature no heat passes from one to the other. Similarly, the power of an electrical machine may be measured by the utmost potential it can give to a conductor. The greater the capacity of the conductor the longer time will be required for the machine to charge it; but no electricity passes between two conductors charged to the same potential. Hence the power of a machine is to be measured by using the simplest form of a conductor, a sphere, and finding the utmost potential the machine can give it. It is easily shown that the potential of a solitary sphere is directly as the quantity of electricity, and inversely as the radius. Hence electricity is in equilibrium on two spheres connected by a long thin wire when the quantities of electricity on them are proportional—not to their surfaces, nor to their volumes, as you might imagine—to their radii. In other words, the capacity is proportional to the radius. This, however, is only true when there are no other conductors within a finite distance. When a sphere is surrounded by another concentric sphere, which is kept in metallic connection with the ground, its capacity is notably increased, and when the radii of the spheres are nearly equal the capacity of the inner one is directly as its surface, and inversely as the distance between the two spheres. Thus the capacity is increased in the ratio of the radius of one sphere to the difference of the radii of the two, and this ratio may easily be made very large. This is the principle upon which the Leyden jar depends.

We may usefully carry the analogy of the pump a good deal further. Supposing the piston to be fully pressed home every stroke, the amount of work spent, even if the whole be kept cool, on each stroke continually increases, so that more than double the amount of work is required to charge the receiver to 40 atmospheres instead of 20. The same holds with electricity. Each successive unit of the charge requires more work to force it in than did the preceding one, because the repulsion of all already in has to be overcome. It is found, in fact, that the work required to put in a charge is proportional to the square of the charge. Of course less work is required for a given charge the greater is the capacity of the receiver. Conversely, the damage which can be done by the discharge, being equal to the work required to produce the charge, is proportional to the square of the charge, and inversely to the capacity of the receiver. Or, what comes to the same thing, it is proportional to the square of the potential and to the capacity of the conductor directly. Thus a given quantity of electricity gives a greater shock the smaller the capacity of the conductor which contains it. And two conductors, charged to the same potential, give shocks proportional to their capacities. But in every case, a doubling of the charge, or a doubling of the potential, in any conductor, produces a fourfold shock.

The only other point I need notice is the nature of the distribution of electricity on a conductor. I say *on* a conductor, because it is entirely confined to the surface. The law is—it is always so arranged that its attractions or repulsions in various directions exactly balance one another at every point in the *substance* of the conductor. It is a most remarkable fact that this is always possible, and in every case in one way only. When the conductor is a single sphere the distribution is uniform. When it is elongated the quantity of electricity per square inch of its surface is greater at the ends than in the middle; and this disproportion is greater the greater is the ratio of the length to the transverse diameter. Hence on a very elongated body, terminating in a point, for instance, the electric density—that is, the quantity per square inch of surface—may be exceedingly great at the point while small everywhere else. Now in proportion to the square of the electric density is the outward pressure of the electricity tending to escape by forcing a passage through the surrounding air. It appears from experiments on the small scale which we can make with an electrical machine, that the electric density requisite to force a passage through the air increases under given circumstances, at first approximately as the square root of the distance which has to be traversed, but afterwards much more slowly, so that it is probable that the potential required to give a mile-long flash of lightning may not be of an order very much higher than that producible in our laboratories.

But from what I have said you will see at once that under similar circumstances an elongated body must have a great advantage over a rounded one in effecting a discharge of electricity. This is easily proved by trial. [The electric machine being in vigorous action, and giving a rapid series of sparks, a pointed rod connected with the ground was brought into the neighbourhood, and the sparks ceased at once.] In this simple experiment you see the whole theory and practical importance of a lightning conductor. But, as a warning, and by no means an unnecessary one, I shall vary the conditions a little and try again. [The pointed rod was now insulated, and produced no observable effect.] Thus you see the difference between a proper lightning-rod and one which is worse than useless, positively dangerous. There is another simple way in which I can destroy its usefulness—namely, by putting a little glass cap on the most important part of it, its point, and thus rendering impossible all the benefits it was originally calculated to bestow. [The pointed rod was again connected with the ground, but furnished with a little glass cap. It produced no effect till it was brought within four or five inches of one of the conductors of the machine, and then *sparks* passed to it.] You must be strangely well acquainted with the phases of human perversity if you can anticipate what I am now going to tell you, namely, that this massive glass cap, or *repeller*, as it was fondly called, was only a year or two ago taken off from the top of the lightning-rod employed to protect an important public building. [The repeller was exhibited. It resembled a very large soda-water bottle with a neck much wider than the usual form.] From the experiments you have just seen it must be evident to you that the two main requisites of an effective lightning-rod are that it should have a sharp point (or, better, a number of such points, lest one should be injured), and that it should be in excellent communication with the ground. When it possesses these, it does not require to be made of exceptionally great section ; for its proper function is *not*, as is too commonly supposed, to parry a dangerous flash of lightning: it ought rather, by silent but continuous draining, to prevent any serious accumulation of electricity in a cloud near it. That it may effectually do this it must be thoroughly connected with the ground, or (if on a ship or lighthouse) with the sea. In towns this is easily done by connecting it with the water mains, at sea by using the copper sheathing of the ship, or a metal plate of large surface fully immersed. Not long ago a protected tower was struck by lightning. No damage was done in the interior, but some cottages near its base were seriously injured. From a report on the subject of this accident it appears that the lower end of the lightning-rod was "jumped" several feet into the solid rock! Thus we see, in the words of Arago, how "False science is no less dangerous than complete ignorance, and that it *infallibly* leads to consequences which there is nothing to justify."

That the lightning-rod acts as a constant drain upon the charge of neighbouring clouds is at once proved when there is, accidentally or purposely, a slight gap in its continuity. This sometimes happens in ships, where the rod consists of separate strips of metal inlaid in each portion of the mast. If they are not accurately fitted together, a perfect torrent of sparks, almost resembling a continuous arc of light is seen to pass between them whenever a thunderstorm is in the neighbourhood.

I cannot pass from this subject without a remark upon the public as well as private duty of having lightning-rods in far greater abundance than we anywhere see them in this country. When of proper conducting power, properly pointed, properly connected with the ground and with every large mass of metal in a building they afford absolute protection against ordinary lightning—every single case of apparent failure I have met with having been immediately traceable to the absence of one or other of these conditions. How great is their beneficial effect you may gather at once from what is recorded of Pietermaritzburg, viz., that till lightning-rods became common in that town it was constantly visited by thunderstorms at certain seasons. They still come as frequently as ever, but they cease to give lightning-flashes whenever they reach the town, and they begin again to do so as soon as they have passed over it.

A knight of the olden time in full armour was probably as safe from the effects of a thunderstorm as if he had a lightning-rod continually beside him ; and one of the Roman emperors devised a perfectly secure retreat in a thunderstorm in the form of a subterraneous vault of iron. He was probably led to this by

T.

thinking of a mode of keeping out missiles, having no notion that a thin shell of soft copper would have been quite as effective as massive iron.   But those emperors who, as Suetonius tells us, wore laurel crowns or sealskin robes, or descended into underground caves or cellars on the appearance of a thunderstorm, were not protected at all.   Even in France, where special attention is paid to the protection of buildings from lightning, dangerous accidents have occurred where all proper precautions seemed to have been taken.   But on more careful examination it was usually found that some one essential element was wanting.   The most common danger seems to lie in fancying that a lightning-rod is necessarily properly connected with the earth if it dips into a mass of water.   Far from it.   A well-constructed reservoir full of water is *not* a good " earth " for a lightning-rod.   The better the stone-work and cement the less are they fitted for this special purpose, and great mischief has been done by forgetting this.

A few years ago the internal fittings of the lighthouse at Skerryvore were considerably damaged by lightning, although an excellent lightning-rod extended along the whole height of the tower.   But a long copper stove-pipe, rising through the whole interior of the tower, and the massive metallic ladder rising from the ground to the lowest chamber, though with a considerable gap between them, offered less resistance than the rod, for the lower end of the ladder was nearer to the sea than was the pool on the reef into which the lightning-rod plunged.   Hence the main disruptive discharge took place from the stove-pipe to the ladder, blowing the intervening door to pieces.   The real difficulty in these situations, exposed to tremendous waves, lies in effecting a permanent communication between the lightning-rod and the sea.   But when this is done the sea makes far the best of " earths."

When a lightning-rod discharges its function imperfectly, either from insufficient conducting power or because of some abnormally rapid production of electricity, a luminous brush or glow is seen near its point.   This is what the sailors call St Elmo's Fire, or Castor and Pollux.   In the records of mountain climbing there are many instances of such discharges to the ends of the alpenstocks or other prominent pointed objects.   One very remarkable case was observed a few months ago in Switzerland, where at dusk, during a thunderstorm, a whole forest was seen to become luminous just *before* each flash of lightning, and to become dark again at the instant of the discharge.

Perhaps the most striking of such narratives is one which I will read to you from the memoirs of the Physical and Literary Society, from which sprang the *Royal Society of Edinburgh.*   These Essays are rare and curious, and the names of Maclaurin, James and Matthew Stewart, Whytt, and Monro, appear among their authors.

The following observations on Thunder and Electricity by Ebenezer McFait, M.D., show how, in the search for truth, men may unwittingly put themselves in the gravest danger:—

" The experiment proposed by Mr Franklin, to prove that lightning and the electrical fire are the same, has often been repeated with success both in England and abroad, so that the most noted electrical experiments have been performed by fire drawn from the clouds.

" Mr Franklin also first discovered that sharp points attract and discharge the electrical matter most copiously; and from thence supposes, that a very sharp-pointed rod, fixed to the extremity of the top-mast of a ship, with a wire conducted down from the foot of the rod round one of the shrouds, and over the ship's side into the sea, would silently lead off the electrical fire, and save the ship from thunder in hot countries; and that, by a similar method, buildings might be preserved.

"So useful a proposal deserves to be examined: variety of experiments may give hints for new improvements. For this reason the following observations are communicated, though not so complete as might be wished, being the result of one trial only.

"It seldom thunders in this northern clime. In June, 1752, there seemed to be some thunder at a distance from Edinburgh; but from the beginning of July to the beginning of October we had nothing almost but continual rains. The last summer was uncommonly warm and dry; and yet we had only a few claps of thunder at Edinburgh, one evening, and my attempts for making any of those experiments were entirely unsuccessful until Saturday night, September 15, when we had a very great storm.

"I used a round iron rod, two-tenths of an inch diameter, about eleven feet long, sharpened at one end; the other end was inserted in a glass tube, and that tube stood in a common glass bottle, which I held in my hand.

"I used also another rod about three feet long, sharpened in like manner at one end, which stood with the other end in a glass tube, which was stuck in the ground. I began upon the Calton Hill. The lightning and fire in the air abounded greatly, and yet it was some time before anything else appeared. At last some rain began to fall, and the air turned moister; then fire appeared upon the extremities of each of the rods in a small pretty blaze, very like the fire which is discharged from the point of a sword in the dark, when the person that holds it is electrified, and stands upon glass or resin; or like that which appears upon any sharp point, when presented to an electrified gun barrel, but in greater quantity. I touched the long rod with my finger, but had no sparks from it. The short rod was accidentally taken out of its tube, and yet continued to burn and blaze as formerly. In like manner the flame continued upon the end of the long rod, though I took hold of it anywhere at pleasure above the glasses, until I moved my hand or finger along, within a few inches of the flame; then it was attracted by my hand, and vanished.

"I went from the Calton Hill to the Castle Hill, at the other end of the town; and in passing through the streets no fire appeared upon either of the rods; but almost immediately when I got clear of the houses, upon the open hill, the point of the longer iron rod took fire. In the dark I had lost the tube belonging to the shorter rod, and the point of it did not catch fire when the longer one was kindled. Perhaps I did not wait long enough for a proper trial, for I soon touched the flame upon the long rod with the sharp point of the short one, and then it also took flame, and continued burning, as before, without any further dependence upon the longer one.

"I held the shorter rod by the sharp end, and approached the blunt end of it to the flame upon the point of the longer rod; then this blunt end caught the fire, and the flame upon the points of the two rods continued rather stronger than on the single one before, so long as I kept them in contact, and the fires within three or four inches of one another; but when I drew them farther asunder the flame upon the extremity of the blunt rod vanished. This happened as often as I tried it, and it is evident that in like manner I could have got the fire to fix upon the points of a great many rods, and so have had them all flaming together. Once or twice a flash of lightning seemed to dart directly against the point of the rod; then the fire, as I thought, expanded itself and united with the lightning, but it immediately began to shine again when the lightning was past.

"Though it rained much in time of these observations, yet the fire upon the ends of the rods did not go out until it became so heavy as if it were pouring down out of funnels.

"After this I went home for some time, resolving to come abroad again when the storm was more tolerable; but it continued to rain all night, so violently, that I was obliged, with regret, to leave several experiments to the chance of some future opportunity. For example, I suspected that the glass tubes had not been of great use on this occasion, and wanted to have tried whether I should have had the same appearances by using the rods alone, without any other apparatus. This is very probable, as also that the glasses, by being wet, allowed the electrical fire to flow off as it was attracted.

"I beg leave to add a few remarks relative to this subject. It would seem that experiments of this kind may be made without danger when the thunder is at a moderate distance. The lightning expands itself, as it flies, and by expansion loses its vigour. Perhaps there is one simple and easy way of protecting masts and spires from thunder, viz., to fix horizontally upon the highest parts of them a flat round piece of wood, of a foot diameter or more, in order to prevent those blazing fires from fixing upon them, and accumulating.

"This storm passed directly over Edinburgh, and came on from the south by west, as nearly as could be estimated. There was a great deal of lightning that night, above sixty miles to the westward, but no thunder heard. At Glasgow there was very much lightning, and a few distant faint claps of thunder. On the road from Belford or Berwick it lightened incessantly, but two claps of thunder only were heard, and those very faint, so that there is reason to think that the fire of this storm spread over the breadth of 130 miles at least. I wish I could also give some account where this thunder began, and how far it ran before it was extinguished.

"On September 3rd there was a great deal of streamers, which rose nearly from the same point that the thunder afterwards came from, and gradually worked north till they descended below the horizon. The air had a thunder-like appearance for several days before this storm; and for some nights after it the streamery vapour appeared equally diffused, muddy, inert, and languid, and not vibrating any variety of colours, as if the more volatile parts had been consumed. It is highly probable that lightning and the *aurora borealis* are of the same materials. In hot countries streamers are not seen, or but rarely, because they are kindled into thunder and flashes of lightning. In cold countries streamers abound, and it seldom thunders. The streamers have served to predict thunder to follow next day, in summer, and they have been also seen to break out into flashes of lightning. Thunder disturbs the motion of the magnetic needle, and it has been lately found in Sweden that streamers do the same. Thus thunder, electricity, magnetism, and the *aurora borealis*, appear all wonderfully related; and many things remain undiscovered in this vast field, which is but just newly opened.

"As it is probable that the height which some philosophers have assigned for the streamers in the atmosphere is by several hundreds of miles too much, it were to be wished that people in various latitudes would carefully observe their altitude at different times of the night, that by comparing simultaneous observations this matter may be determined with more certainty."

At first reading one is inclined to regard this as a joke, but there is nothing whatever to justify the notion. That Dr McFait was not killed on this occasion was certainly not due to any want of precautions on his part, well calculated to make such an event all but certain. He wanted only a knob on the blunt end of his short rod. We are reminded of the remark made by one of the seconds in a well-known duel about his principal, "To come on horseback to a fight with pistols! and in a white waistcoat, too! Couldn't he have got a bull's eye painted on it, just

over his heart? It would have expedited matters, and made them still more simple."
Richmann, of St Petersburg, had just before been killed while apparently in far less
danger than Dr McFait, and other incautious experimenters have since similarly
suffered.

The destructive effects of lightning are familiar to all of you, so that I need
not spend time in illustrating them on a puny scale by the help of Leyden jars.

All the more ordinary effects can thus easily be reproduced on a small scale.
How small you may easily conceive, when I tell you that a three-foot spark is
considered a long one, even from our most powerful machines, while it is quite
certain that lightning flashes often exceed a mile in length, and sometimes extend
to four and five miles. One recorded observation, by a trustworthy observer, seems
to imply a discharge over a total length of nearly ten miles.

When a tree is struck by a violent discharge it is usually split up laterally into
mere fibres. A more moderate discharge may rupture the channels through which
the sap flows, and thus the tree may be killed without suffering any apparent external
damage. These results are usually assigned to the sudden vaporisation of moisture,
and the idea is probably accurate, for it is easy to burst a very strong glass tube,
if we fill it with water and discharge a jar by means of two wires whose extremities
are placed in the water at a short distance from one another. The tube bursts
even if one end be left open, thus showing that the extreme suddenness of the
explosion makes it act in all directions, and not solely in that of least resistance.
When we think of the danger of leaving even a few drops of water in a mould
into which melted iron is to be poured, we shall find no difficulty in thus accounting
for the violent disruptive effects produced by lightning.

Heated air is found to conduct better than cold air, probably on account of
the diminution of density only. Hence we can easily see how it is that animals are
often killed in great numbers by a single discharge, as they crowd together in a
storm, and a column of warm air rises from the group.

Inside a thundercloud the danger seems to be much less than outside. There
are several instances on record of travellers having passed *through* clouds from
which, both before and after their passage, fierce flashes were seen to escape. Many
remarkable instances are to be found in Alpine travel, and specially in the reports
of the officers engaged in the survey of the Pyrenees. Several times it is recorded
that such violent thunderstorms were seen to form round the mountain on which
they were encamped that the neighbouring inhabitants were surprised to see them
return alive.

Before the use of lightning-rods on ships became general great damage was
often done to them by lightning. The number of British ships of war thus wholly
destroyed or much injured during the long wars towards the end of the last and
the beginning of the present century is quite comparable with that of those lost
or injured by gales, or even in battle. In some of these cases, however, the damage
was only indirectly due to lightning, as the powder magazines were blown up. In
the powder magazine of Brescia, in 1769, lightning set fire to over two million
pounds of gunpowder, producing one of the most disastrous explosions on record.

A powerful discharge of lightning can fuse not only bell-wires, but even stout rods of iron. It often permanently magnetises steel, and in this way has been the cause of the loss of many a good ship; for the magnetism of the compass-needles has been sometimes destroyed, sometimes reversed, sometimes so altered that the compass pointed east and west. And by the magnetisation of their steel parts the chronometers have had their rates seriously altered. Thus two of the sailor's most important aids to navigation have been simultaneously rendered useless or, what is worse, misleading; and this, too, at a time when, because of clouds, astronomica observations were generally impossible. All these dangers are now, however, easily and all but completely avoidable.

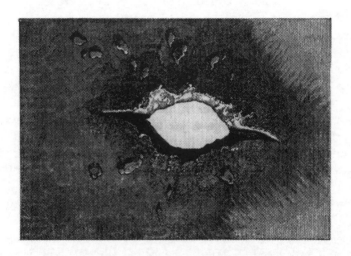

Sheet-lead punctured by lightning.

A very singular effect of lightning sometimes observed is the piercing of a hole in a conducting-plate of metal, such as the lead-covering of a roof. In such a case it is invariably found that a good conductor well connected with the ground approaches near to the metal sheet at the part perforated. We can easily repeat the experiment on a small scale with tinfoil. [A thick piece of sheet-lead from the lower buildings of Nelson's monument, Edinburgh, punctured by lightning, was exhibited. It is figured in the woodcut, reprinted (by permission) from *Proc. R.S.E.*, 1863.]

The name *thunderbolt*, which is still in use, even by good writers, seems to have been introduced in consequence of the singular effects produced when lightning strikes a sandhill or sandy soil. It bores a hole often many feet in length, which is found lined throughout with vitrified sand. The old notion was that an intensely hot, solid mass, whose path was the flash of lightning, had buried itself out of sight, melting the sand as it went down. It is quite possible that this notion may have been strengthened by the occasional observation of the fall of aerolites, which are sometimes found, in the holes they have made, still exceedingly hot. And at least many of the cases in which lightning is said to have been seen in a perfectly clear sky are to be explained in the same way. Everyone knows Horace's lines:—

"Diespiter
Igni corusco nubila dividens
Plerumque, per purum tonantes
Egit equos volucremque currum."

But Virgil's remark is not so commonly known. He is speaking of prodigies of various kinds, and goes on :—

"Non alias coelo ceciderunt plura sereno
Fulgura; nec diri toties arsere cometae."

It is very singular that he should thus have associated comets and meteorites which quite recent astronomical discovery has shown to have a common origin.

Another remarkable peculiarity, long ago observed, is the characteristic smell produced when lightning strikes a building or a ship. In old times it was supposed to be sulphurous; now-a-days we know it to be mainly due to ozone. In fact, all the ready modes of forming ozone which are as yet at the disposal of the chemist depend upon applications of electricity. But besides ozone, which is formed from the oxygen of the air, there are often produced nitric acid, ammonia, and other compounds derived from the constituents of air and of aqueous vapour. All these results can be produced on a small scale in the laboratory.

Hitherto I have been speaking of lightning discharges similar in kind to the ordinary electric spark, what is commonly called *forked* or *zig-zag* lightning. Our nomenclature is very defective in this matter, and the same may be said of the chief modern European languages. For, as Arago remarks, by far the most common form of lightning flash observed in thunderstorms is what we have to particularise, for want of a better term, as *sheet-lightning*. He asserts that it occurs thousand-fold as often as forked lightning; and that many people have never observed the latter form at all! It is not at all easy to conceive what can be the nature of sheet-lightning, if it be not merely the lighting up of the clouds by a flash of forked lightning not directly visible to the spectator. That this is, at least in many cases, its origin is evident from the fact that its place of maximum brightness often takes the form of the *edge* of a cloud, and that the *same* cloud-edge is occasionally lit up several times in quick succession. You will remember that we are at present dealing with the appearances observed *in* a thunderstorm, so that I do not refer to that form of sheet-lightning which commonly goes by the name of *summer-lightning*, and which is not, audibly, at least, followed by thunder.

The next remarkable feature of the storm is the thunder, corresponding, of course, on the large scale to the snap of an electric spark. Here we are on comparatively sure ground, for sound is very much more thoroughly understood than is electricity. We speak habitually and without exaggeration of the *crash* of thunder, the *rolling* of thunder, and of a *peal* of thunder; and various other terms will suggest themselves to you as being aptly employed in different cases. All of these are easily explained by known properties of sound. The origin of the sound is in all cases to be looked for in the instantaneous and violent dilatation of the air along the track of the lightning flash; partly, no doubt, due to the disruptive effects of electricity of which I have already spoken, but mainly due to the excessive rise of temperature which renders the air for a moment so brilliantly incandescent. There is thus an extremely sudden compression of the air all round the track of the spark; and a less sudden, but still rapid, rush of the air into the partial vacuum which it produces. Thus the sound wave produced must at first be of the nature of a bore or breaker. But as such a state of motion is unstable, after proceeding a moderate distance the sound becomes analogous to other loud but less violent sounds, such as those of the discharge of guns. Were there few clouds, were the air of nearly uniform density, and the flash a short one, this would completely describe the phenomenon, and we should have a thunder *crash* or thunder *clap* according to the greater or less proximity of the seat of discharge. But, as has long been well known, not merely clouds but surfaces of separation of masses of air of different density, such as constantly occur in thunderstorms, *reflect* vibrations in the air; and thus we may have many successive echoes, prolonging the original sound. But there is another cause, often more efficient than these. When the flash is a long one, all its parts being nearly equidistant from the observer, he hears the sound from all these parts simultaneously; but if its parts be at very different distances from him, he hears *successively* the sounds from portions farther and farther distant from him. If the flash be much zig-zagged, long portions of its course may run at one and the same distance from him, and the sounds from these arrive simultaneously at his ear. Thus we have no difficulty in accounting for the *rolling* and *pealing* of thunder. It is, in fact, a mere consequence, sometimes of the reflection of sound, sometimes of the finite velocity with which it is propagated. The usual rough estimate of five seconds to a mile is near enough to the truth for all ordinary calculations of the distance of a flash from the observer.

The extreme distance at which thunder is heard is not great, when we consider the frequent great intensity of the sound. No trustworthy observation gives in general more than about nine or ten miles, though there are cases in which it is possible that it may have been heard 14 miles off. But the discharge of a single cannon is often heard at 50 miles, and the noise of a siege or naval engagement has certainly been heard at a distance of much more than 100 miles. There are two reasons for this: the first depends upon the extreme suddenness of the production of thunder; the second, and perhaps the more effective, on the excessive variations of density in the atmosphere, which are invariably associated with a thunderstorm. In certain cases thunder has been propagated, for moderate distances from its apparent source, with a velocity far exceeding that of ordinary sounds. This used to be attributed to the extreme suddenness of its production; but it is not easy, if we adopt this hypothesis, to see why it should not occur in all cases. Sir W. Thomson has supplied a very different explanation, which requires no unusual velocity of sound, because it asserts the production of the sound *simultaneously* at all parts of the air between the ground and the cloud from which the lightning is discharged.

We now come to an exceedingly strange and somewhat rare phenomenon, to which the name of *fire-ball* or *globe-lightning* has been given. As we are as yet unable to produce anything of this kind by means of our electrical machines, some philosophers have tried to cut the Gordian knot of the difficulty by denying that any such thing can exist. But, as Arago says, "*Où en serions nous, si nous nous mettions à nier tout ce qu'on ne sait pas expliquer?*" The amount of trustworthy and independent evidence which we possess as to the occurrence of this phenomenon is such as must convince every reasonable man who chooses to pay due attention to the subject. No doubt there is a great deal of exaggeration, as well as much imperfect and even erroneous observation, in almost all of these records. But the existence of the main feature (the fire-*ball*) seems to be proved beyond all doubt.

The most marked peculiarities of this species of lightning-discharge are its comparatively long duration and its comparatively slow motion. While a spark, or lightning flash, does not last longer than about a millionth part of a second, if so long, globe-lightning lasts from one to ten seconds, sometimes even longer, so that a sufficiently self-possessed spectator has time carefully to watch its behaviour. The general appearance is that of a luminous ball, which must be approximately spherical, because it always appears circular in outline, slowly and steadily descending from a thundercloud to the ground. It bursts with a loud explosion, sometimes before reaching the ground, sometimes as it impinges, and sometimes after actually rebounding. Its size varies from that of a child's head to a sphere of little less than a yard in diameter. On some occasions veritable flashes of lightning were seen to proceed from large fire-balls as they burst. It is difficult to imagine what these balls can be if they be not a species of natural Leyden jar very highly charged. If it be so, no ordinary lightning-rod can possibly prevent danger from it; and we may thus be able to explain the very few cases in which damage has been done by lightning to thoroughly protected buildings. To guard against this form nothing short of a pretty close net-work of stout copper wires would suffice. Meanwhile I give a brief sketch of *two* out of the long series of descriptions of such phenomena which Arago has patiently collected. The first is given on the authority of Babinet, who was deputed by the Academy of Sciences to make inquiries into the case.

Shortly, but not immediately, after a loud peal of thunder, a tailor who was sitting at his dinner saw the paper ornament which covered his fire-place blown down as if by a gentle breeze, and a globe of fire, about the size of a child's head, came gently out and moved slowly about at a slight elevation above the floor. It appeared bright rather than hot, and he felt no sensation of warmth. It approached him like a little kitten which desired to rub itself in play against his legs; but he drew his feet away, and by slow and cautious movements avoided contact with it. It remained several seconds near his feet, while he leaned forward, and carefully examined it. At last it rose vertically to about the level of his head, so he threw himself back in his chair and continued to watch it. It then became slightly elongated, and moved obliquely towards a hole pierced to the chimney about a yard above the mantel piece. This hole had been made for the chimney of a stove which was used in winter. "But," as the tailor said, "the globe could not *see* the hole, for paper had been pasted over it." The globe went straight for the hole, tore off the paper, and went up the chimney. After the lapse of time which at the rate at which he had seen it moving, it would have required to get to the top of the chimney, a terrific explosion was heard, and a great deal of damage was done to the chimney and the roofs around it.

The next is even more striking: In June, 1849, in the evening of one of the days when cholera was raging most formidably in Paris, the heat was suffocating, the sky appeared calm, but summer lightning was visible on all sides. Madame Espert saw from her window something like a large red globe, exactly resembling the moon, when it is seen through mist. It was descending slowly towards a tree. She at first thought it was a balloon, but its colour undeceived her; and while she was trying to make out what it was, she saw the lower part of it take fire ("*Je vis le feu prendre au bas de ce globe*"), while it was still some yards above the tree. The flames were like those of paper burning slowly, with sparks and jets of fire. When the opening became twice or thrice the size of one's hand, a sudden and terrific explosion took place. The infernal machine was torn to pieces, and a dozen flashes of zig-zag lightning escaped from it in all directions. The *débris* of the globe burned with a brilliant white light, and revolved like a catherine-wheel. The whole affair lasted for at least a minute. A hole was bored in the wall of a house, three men were knocked down in the street, and a governess was wounded in a neighbouring school, besides a good deal of other damage.

As another instance, here is a description (taken from Dove) of one which fell at Barbadoes, in 1831, during a terrific hurricane: At three o'clock in the morning the lightning ceased for a few moments, and the darkness which enveloped the town was indescribably terrible. Fiery meteors now fell from the clouds; one in particular, of spherical form, and of a deep red colour, fell perpendicularly from a considerable height. This fire-ball fell quite obviously by its own weight, and not under the influence of any other external force. As with accelerated velocity it approached the earth it became dazzlingly white, and of elongated form. When it touched the ground it splashed all about like melted metal, and instantly disappeared. In form and size it resembled a lamp globe; and the splashing about at impact gave it the appearance of a drop of mercury of the same size.

I have never seen one myself, but I have received accounts of more than one of them from competent and thoroughly credible eye-witnesses. In particular on a stormy afternoon in November, 1868, when the sky was densely clouded over, and the air in a highly electrical state, there was heard in Edinburgh one solitary short, but very loud, clap of thunder. There can be no doubt whatever that this was due to the explosion of a fire-ball, which was seen by many spectators in different parts of the town, to descend towards the Calton Hill, and to burst whilst still about a hundred feet or so above the ground. The various accounts tallied in most particulars, and especially in the very close agreement of the positions assigned to the ball by spectators viewing it from different sides, and in the intervals which were observed to elapse between the explosion and the arrival of the sound.

The remaining phenomena of a thunderstorm are chiefly the copious fall of rain and of hail, and the almost invariable lowering of the barometer. These are closely connected with one another, as we shall presently see.

Almost all the facts to which I have now adverted point to water-substance, in some of its many forms, as at least one of the chief agents in thunderstorms. And when we think of other tremendous phenomena which are undoubtedly due to water, we shall have the less difficulty in believing it to be capable of producing thunderstorms also.

First of all let us think of some of the more obvious physical consequences of a fall of a mere tenth of an inch of rain. Suppose it to fall from the lowest mile of the atmosphere. An inch of rain is 5 lb. of water per square foot, and gives out on being condensed from vapour approximately 3,000 units of heat on the centigrade scale. The mass of the mile-high column of air a square foot in section is about 360 lb., and its specific heat about a quarter. Thus its temperature throughout would be raised by about 33° C., or 60° F. For one-tenth inch of rain, therefore, we should have a rise of temperature of the lowest mile of the atmosphere amounting to 3·3° C., quite enough to produce a very powerful ascending current. As the air ascends and expands it cools, and more vapour is precipitated, so that the ascending current is further accelerated. The heat developed over one square foot of the earth's surface under these conditions is equivalent to work at the rate of a horse-power for twelve minutes. Over a square mile this would be ten million horse-power for half an hour. A fall of one-tenth of an inch of rain over the whole of Britain gives heat equivalent to the work of a million millions of horses for half an hour! Numbers like these are altogether beyond the limits of our understanding. They enable us, however, to see the full explanation of the energy of the most violent hurricanes in the simplest physical concomitants of the mere condensation of aqueous vapour.

I have already told you that the source of atmospheric electricity is as yet very uncertain. Yet it is so common and so prominent a phenomenon in many of its manifestations that there can be little doubt that innumerable attempts have been made to account for it. But when we consult the best treatises on meteorology we find it either evaded altogether or passed over with exceedingly scant references to evaporation or to vegetation. Not finding anything satisfactory in books, I have consulted able physicists, and some of the ablest of meteorologists, in all cases but one with the same negative result. I had, in fact, the feeling which every one must experience who attempts to lecture on a somewhat unfamiliar subject, that there *might* be much known about it which I had not been fortunate enough to meet with. Some years ago I was experimentally led to infer that mere *contact* of the particles of aqueous vapour with those of air, as they fly about and impinge according to the modern kinetic theory of gases, produced a separation of the two electricities, just as when zinc and copper are brought into contact the zinc becomes positively electrified and the copper negatively. Thus the electrification was supposed to be the result of chemical affinity. Let us suppose, then, that a particle of vapour, after impact on a particle of air, becomes electrified positively (I shall presently mention experiments in support of this supposition), and see what further consequences will ensue when the vapour condenses. We do not know the mechanism of the precipitation of vapour as cloud, and we know only partially that of the agglomeration of cloud-particles into rain-drops; but of this we can be sure that, if the vapour-particles were originally electrified to any finite potential the cloud-particles would be each at a potential enormously higher, and the rain-drops considerably higher still. For, as I have already told you, the potential of a free charged sphere is proportional directly to the quantity of electricity on it and inversely to its radius; so when eight equal and equally charged spheres unite into one sphere of double the radius, its potential is four times that of each of the separate spheres. The

40—2

potential in a large sphere, so built up, is in fact directly proportional to its surface as compared with that of any one of the smaller equal spheres of which it is built.

Now, the number of particles of vapour which go to the formation of a single average rain-drop is expressed in billions of billions; so that the potential of the drop would be many thousands of billion times as great as that of a particle of vapour. On the very lowest estimate this would be incomparably greater than any potential we can hope to produce by means of electrical machines.

But this attempt at explanation of atmospheric electricity presents two formidable difficulties at the very outset.

1. How should the smaller cloud-particles ever unite if they be charged to such high potentials, which of course must produce intense repulsions between them?

2. Granting that, in spite of this, they do so unite, how are they separated from the mass of negatively electrified air in which they took their origin?

I think it is probable that the second objection is more imaginary than real, since there is no doubt that the diffusion of gases would speedily lead to a great spreading about of the negatively electrified particles of air from among the precipitated cloud-particles into the less highly electrified air surrounding the cloud. And if the surrounding air were equally electrified with that mixed with the cloud, there would be no electric force preventing gravity from doing its usual work. This objection, in fact, holds only for the *final* separation of the whole moisture from the oppositely electrified air; and gravity may be trusted to accomplish this. That gravity is an efficient agent in this separation is the opinion of Prof. Stokes. It must be observed that as soon as the charge on each of the drops in a cloud rises sufficiently, the electricity will pass by discharge to those which form the bounding layer of the cloud.

The first objection is at least partially met by the remark that in a cloud-mass when just formed, if it be at all uniform, the electric attractions and repulsions would approximately balance one another at every point, so that the mutual repulsion of any two water-drops would be almost compensated except when they came very close to one another.

But there is nothing in this explanation inconsistent with the possibility that the particles of water may be caused to fly about repeatedly from cloud to cloud, or from cloud to an electrified mass of air; and in many of these regions the air, already in great part deprived of its moisture, may have become much cooled by expansion as it ascends, so that the usual explanation of the production of hail is not, at least to any great extent, interfered with.

I may here refer to some phenomena which seem to offer, if closely investigated, the opportunity for the large scale investigations which, as I shall presently show, will probably be required to settle the source or sources of atmospheric electricity.

First, the important fact, well known nearly 2,000 years ago, that the column of smoke and vapour discharged by an active volcano gives out flashes of veritable lightning. In more modern times this has been repeatedly observed in the eruptions of Vesuvius and other volcanos—such as, for instance, the island of Sabrina. It is recorded by Sir W. Hamilton (British ambassador at Naples at the end of last century) that in the eruption of 1794 these flashes were accompanied by violent peals of thunder. They destroyed houses in the neighbourhood of the mountain, and are said to have done considerable damage even at places 250 miles off, to which the cloud of volcanic dust and vapour was carried by the wind. I will read a couple of extracts from Hamilton's paper. It is in the *Philosophical Transactions* for 1795 :—

"The electric fire in the year 1779, that played constantly within the enormous black cloud over the crater of Vesuvius, and seldom quitted it, was exactly similar to that which is produced on a very small scale by the conductor of an electrical machine communicating with an insulated plate of glass, thinly spread over with metallic filings, etc., when the electric matter continues to play over it in zig-zag lines without quitting it. I was not sensible of any noise attending that operation in 1779; whereas the discharge of the electrical matter from the volcanic clouds during this eruption, and particularly the second and third days, caused explosions like those of the loudest thunder; and, indeed, the storms raised evidently by the sole power of the volcano resembled in every respect all other thunder-storms, the lightning falling and destroying everything in its course. The house of the Marquis of Berio at S. Jorio, situated at the foot of Vesuvius, during one of these volcanic storms, was struck with lightning, which, having shattered many doors and windows, and damaged the furniture, left for some time a strong smell of sulphur in the rooms it passed through. Out of these gigantic and volcanic clouds, besides the lightning, both during this eruption and that of 1779, I have, with many others, seen balls of fire issue, and some of a considerable magnitude, which, bursting in the air, produced nearly the same effect as that from the air-balloons in fire-works, the electric fire that came out having the appearance of the serpents with which those fire-work balloons are often filled. The day on which Naples was in the greatest danger from the volcanic clouds, two small balls of fire, joined together by a small link like a chain-shot, fell close to my *casino* at Posilipo. They separated, and one fell in the vineyard above the house, and the other in the sea, so close to it that I heard a splash in the water; but, as I was writing, I lost the sight of this phenomenon, which was seen by some of the company with me, and related to me as above."

"The Archbishop of Taranto, in a letter to Naples, and dated from that city the 18th of June, said, 'We are involved in a thick cloud of minute volcanic ashes, and we imagine that there must be a great eruption either of Mount Etna or of Stromboli.' The bishop did not dream of their having proceeded from Vesuvius, which is about 250 miles from Taranto. We have had accounts also of the fall of the ashes during the late eruption at the very extremity of the province of Leece, which is still farther off; and we have been assured likewise that those clouds were replete with electrical matter. At Martino, near Taranto, a house was struck and much damaged by the lightning from one of these clouds. In the accounts of the great eruption at Vesuvius in 1631 mention is made of the extensive progress of the ashes from Vesuvius, and of the damage done by the *ferilli*, or volcanic lightning, which attended them in their course."

Sabine, while at anchor near Skye, remarked that the cloud-cap on one of the higher hills was permanently luminous at night, and occasionally gave out flashes resembling those of the aurora. I have not been able to obtain further information as to this very important fact; but I have recently received a description of a very similar one from another easily accessible locality.

My correspondent writes from Galway, to the following effect, on the 2nd of the present month :—

"At the commencement of the present unprecedently long and severe storm the wind blew from south-west and was very warm. After blowing for about two days it became, *without change of direction*, exceedingly bitter and cold; and the rain was, from time to time, mixed with sleet and hail, and lightning was occasional. This special weather is common for weeks together in March or early April. The air is (like what an east wind brings in Edinburgh) cold, raw, dry, and in every way uncomfortable, especially to people accustomed to the moist Atlantic winds. During these weeks a series of small clouds, whose shadows would only cover a field of a few acres, seem to start at regular intervals from the peaks of hills in Connemara and Mayo. They are all more or less charged with electricity. From high ground, behind the city, I have at one time seen such a cloud break into lightning over the spire of the Jesuits' church. At another, I have seen such a cloud pour down in a thin line of fire, and fall into the bay in the shape of a small incandescent ball. On one occasion I was walking with a friend, when I remarked, 'Let us turn and make a run for it. We have walked unwittingly right underneath a little thundercloud.' I had scarcely spoken when a something *flashed on the stony ground at our very feet*, a tremendous crash pealed over our heads, and the smell of sulphur was unmistakable. I fancy that I have been struck with these phenomena more than others, from the circumstance that they have always interfered with my daily habits. My walks often extended to considerable distances and to very lonely districts. Now these small local spurts of thunderstorms would hardly excite attention in the middle of a town, all the less as the intervening weather is bright, though raw—these spurts coming on every three or four quarters of an hour. Neither would they excite much attention in the country, as, while such a little storm was going on in one's immediate neighbourhood, you would see at no great distance every sign of fine weather. In fact they always seem to me like the small change of a big storm."

My correspondent, though a good observer and eloquent in description, is not a scientific man[1]. But it is quite clear from what he says that a residence of a few weeks in Galway, at the proper season, would enable a trained physicist to obtain, with little trouble, the means of solving this extremely interesting question. He would require to be furnished with an electrometer, a hygrometer, and a few other simple pieces of apparatus, as well as with a light suit of plate armour, not of steel but of the best conducting copper, to insure his personal safety. Thus armed he might fearlessly invade the very nest or hatching-place of the phenomenon, on the top of one of the Connemara hills. It is to be hoped that some of the rising generation of physicists may speedily make the attempt, in the *spirit* of the ancient chivalry, but with the offensive and defensive weapons of *modern* science.

Another possible source of the electricity of thunderstorms has been pointed out by Sir W. Thomson. It is based on the experimental fact that the lower air is usually charged with negative electricity. If ascending currents carry up this lower air the electricity formerly spread in a thin stratum over a large surface may, by convection, be brought into a very much less diffused state, and thus be raised to a potential sufficient to enable it to give a spark.

[1] The correspondent was probably the late Professor D'Arcy Thompson, of Galway, already referred to on p. 238.

However the electrification of the precipitated vapour may ultimately be accounted for, there can be no doubt of the fact that at least as soon as *cloud* is formed the particles are electrified ; and what I have said as to the immense rise of potential as the drops gradually increase in size remains unaffected. I have tried various forms of experiment, with the view of discovering the electric state of vapour mixed with air. For instance, I have tested the vapour which is suddenly condensed when a receiver is partially exhausted; the electrification of cooled bodies exposed to moist air from a gas-holder ; and the deposition of hoar-frost from a current of moist air upon two polished metal plates placed parallel to one another, artificially cooled, and connected with the outer and inner coatings of a charged jar. All have given results, but as yet too minute and uncertain to settle such a question. These experiments are still in progress. It appears probable, so far, that the problem will not be finally solved until experiments are made on a scale much larger than is usual in laboratories.

A great thunderstorm in summer is in the majority of cases preceded by very calm sultry weather. The atmosphere is in a state of unstable equilibrium, the lower strata are at an abnormally high temperature, and highly charged with aqueous vapour. It is not easy, in a popular lecture like this, to give a full account of what constitutes a state of stable equilibrium, or of unstable, especially when the effects of precipitation of vapour are to be largely taken into account. It is sufficient for my present purpose to say that in all cases of thoroughly stable equilibrium, a slight displacement *tends to right itself*; while, in general, in unstable equilibrium, a slight displacement tends to increase. Now, if two cubic feet of air at different levels could be suddenly made to change places, without at first any other alteration, and if, on being left to themselves, each would, under the change of pressure which it would suddenly experience, and the consequent heating or cooling, with its associated evaporation or precipitation of moisture, tend to regain its former level, the equilibrium would be stable. This is not the case when the lower strata are very hot, and fully charged with vapour. Any portion accidentally raised to a higher level tends to rise higher, thus allowing others to descend. These, in consequence of their descent, tend still farther to descend, and thus to force new portions up. Thus, when the trigger is once pulled, as it were, we soon have powerful ascending currents of hot moist air, precipitating their moisture as cloud as they ascend, cooling by expansion, but warmed by the latent heat of the vapour condensed. This phenomenon of ascending currents is strongly marked in almost every great thunderstorm, and is precisely analogous to that observed in the centre of a West Indian tornado and of a Chinese typhoon.

When any portion of the atmosphere is ascending it must be because a denser portion is descending, and whenever such motions occur *with acceleration* the pressure must necessarily be diminished, since the lower strata are not then supporting the whole weight of the superincumbent strata. If their whole weight were supported they would not descend. Thus even a smart shower of rain must directly tend to lower the barometer. [A long glass tube, filled with water, was suspended in a vertical position by a light spiral spring, reaching to the roof of the hall. A number of bullets hung at the top of the water column, attached to the tube by a thread. When the thread was burned, by applying a lamp, the bullets descended in the water, and *during their descent* the spring contracted so as to raise the whole tube several inches.]

In what I have said to-night I have confined myself mainly to *great* thunder-storms, and to what is seen and heard by those who are within their sphere of operation. I have said nothing of what is commonly called *summer-lightning*, which is probably, at least in a great many cases, merely the faint effect of a distant thunderstorm, but which has also been observed when the sky appeared tolerably clear, and when it was certain that no thunderstorm of the ordinary kind had occurred within a hundred miles. In such cases it is probable that we see the lightning of a storm which is taking place in the upper strata of the atmosphere, at such a height that the thunder is inaudible, partly on account of the distance, partly on account of the fact that it takes its origin in air of small density.

Nor have I spoken of the aurora, which is obviously connected with atmospheric electricity, but in what precise way remains to be discovered. Various theories have been suggested, but decisive data are wanting. Dr Balfour Stewart inclines to the belief that great auroras, visible over nearly a whole terrestrial hemisphere, are due to inductive effects of changes in the earth's magnetism. This is not necessarily inconsistent with the opinion that, as ordinary auroras generally occur at times when a considerable change of temperature takes place, they are phenomena due to the condensation of aqueous vapour in far less quantity, but through far greater spaces, than the quantities and spaces involved in ordinary thunderstorms.

In taking leave of you and of my subject I have two remarks to make. First, to call your attention to the fact that the most obscure branches of physics often present matter of interesting reflection for all, and, in consequence, ought not to be left wholly in the hands of professedly scientific men. Secondly, that if the pre-cautions which science points out as, at least in general, sufficient, were recognised by the public as *necessary*, the element of danger, which in old days encouraged the most debasing of superstitions, would be all but removed from a thunderstorm. Thus the most timid would be able to join their more robust fellow-creatures in watching fearlessly, but still of course with wonder and admiration, one of the most exquisite of the magnificent spectacles which Nature from time to time so lavishly provides.

STATE OF THE ATMOSPHERE WHICH PRODUCES THE FORMS OF MIRAGE OBSERVED BY VINCE AND BY SCORESBY.

(From *Nature*, Vol. XXVIII, May 24, 1883.)

IN 1881, when I wrote the article *Light* for the *Encyc. Brit.*, I had not been able to meet with any detailed calculations as to the probable state of the atmosphere when multiple images are seen of objects situated near the horizon. I had consulted many papers containing what are called "general" explanations of the phenomena, but had found no proof that the requisite conditions could exist in nature: except perhaps in the case of the ordinary mirage of the desert, where it is obvious that very considerable temperature-differences may occur in the air within a few feet of the ground. But this form of mirage is essentially unsteady, for it involves an unstable state of equilibrium of the air. In many of Scoresby's observations, especially that of the solitary inverted image of his father's ship (then thirty miles distant, and of course far below the horizon), the *details* of the image could be clearly seen with a telescope, showing that the air must have been in equilibrium. The problem seemed to be one well fitted for treatment as a simple example of the application of Hamilton's *General Method in Optics*, and as such I discussed it. The details of my investigation were communicated in the end of that year to the Royal Society of Edinburgh, and will, I hope, soon be published. The paper itself is too technical for the general reader, so that I shall here attempt to give a sketch of its contents in a more popular form. But a curious little historical statement must be premised.

It was not until my calculations were finished that I found a chance reference to a great paper by Wollaston (*Phil. Trans.* 1800). I had till then known only of Wollaston's well-known experiment with layers of different liquids in a small vessel. But these, I saw, could not reproduce the proper mirage phenomena, as the rays necessarily enter and emerge from the transition strata by their *ends* and not by their lower *sides*. This experiment is by no means one of the best things in Wollaston's paper, so far at least as the immediate object of the paper is concerned. That so much has been written on the subject of mirage during the present century, with only a casual reference or two to this paper, is most surprising. It may perhaps be accounted for by the fact that Wollaston does not appear to have had sufficient confidence in his own results to refrain from attempting, towards the end of his paper, a totally different (and untenable) hypothesis, based on the effects of aqueous vapour. Be the cause what it may, there can be no doubt that the following words of Gilbert were amply justified when they were written, early in the present century: "In der That ist Wollaston der Erste und Einzige, der die *Spieglung* aufwärts mit Glück zu erklären unternommen hat." For his methods are, in principle, perfectly correct and sufficiently comprehensive; while some of his experiments imitate closely the state of the air requisite for the production of Vince's phenomena. Had Wollaston only felt the necessary confidence in his own theory, he could hardly have failed to recognise that what he produced by the extreme rates of change of temperature in the small airspace close to a red-hot bar of metal, could be produced by natural rates of change in some ten or twenty miles of the atmosphere: and he would have deserved the credit of having completely solved the problem.

Six months after my paper was read, another happy chance led me to seek for a voluminous paper by Biot, of which I had seen no mention whatever in any of the books I had previously consulted. The probable reason for the oblivion into which this treatise seems to have fallen is a curious one. It forms a considerable part of the volume for 1809 of the *Mém. de l'Institut*. But in the three first great libraries which I consulted, I found this volume to be devoid of plates. In all respects but this, each of the sets of this valuable series appeared to be complete. Without the figures, which amount to no less than sixty-three, it is practically impossible to understand the details of Biot's paper. The paper was, however, issued as a separate volume, *Recherches sur les Réfractions extraordinaires qui ont lieu près de l'horizon* (Paris, 1810), which contains the plates, and which I obtained at last from the Cambridge University Library. I have since been able to procure a copy for the Edinburgh University Library. Biot's work is an almost exhaustive one, and I found in it a great number of the results which follow almost intuitively from my methods: such as the possible occurrence of *four* images, under the conditions usually assumed for the explanation of the ordinary mirage; the effects of (unusual) refraction on the apparent form of the setting sun, etc. But it seems to me that Biot's long-continued observations of the phenomena as produced over extensive surfaces of level sand at Dunkirk have led him to take a somewhat one-sided view of the general question. And, in particular, I think that his attempted explanation of Vince's observations (so far as I am able to understand it; for it is very long, and in parts extremely obscure and difficult, besides containing some singular physical errors) is not satisfactory. His general treatment of the whole question is based to a great extent upon the properties of caustics, though he mentions (as the *courbe des minima*) the "locus of vertices" which I had employed in my investigations, and which I think greatly preferable. There can be no doubt, however, that Biot's paper comes at least next in importance to that of Wollaston: though in his opinion Wollaston's work was complete only on the physical side of the problem. "Sous le rapport de la physique son travail ne laisse rien à désirer."

But, if the chief theoretical papers on the subject have thus strangely been allowed to drop out of notice, the case is quite different with several of those which deal with the observed phenomena. Scoresby's *Greenland*, his *Arctic Regions*, and his *Voyage to the Northern Whale Fishery*, are still standard works; and in them, as well as in Vols. IX and XI of the *Trans. R.S.E.*, he has given numerous careful drawings of these most singular appearances. The explanatory text is also peculiarly full and clear, giving all that a careful observer could be expected to record. It is otherwise with the descriptions and illustrations in Vince's paper (*Phil. Trans.* 1799). In fact the latter are obviously not meant as *drawings* of what was seen; but as *diagrams* which exhibit merely the general features, such as the relative position and magnitude of the images: the details being filled in at the option of the engraver. That such was the view taken by Brewster, is obvious from the illustrations in his *Optics* (Library of Useful Knowledge), for while one of Scoresby's drawings is there *copied*, one of Vince's is treated in a highly imaginative style by the reproducer.

Scoresby's sketches are composite, as he takes care to tell the reader, so that in the reproduction below (Fig. 1), I have simply selected a few of the more remark-

Fig. 1.

able portions which bear on the questions to be discussed. It is to be remarked that the angular dimensions of these phenomena are always of *telescopic* magnitude: the utmost elevation of an image rarely exceeding a quarter or a third of a degree.

Because the rays concerned are all so nearly horizontal, and (on the whole) *concave* towards the earth; and because they must also have on the whole considerably greater curvature than the corresponding part of the earth's surface, especially if they happen to have points of contrary flexure; it is clear that, for a preliminary investigation, we may treat the problem as if the earth were a plane. This simplifies matters very considerably, so that definite numerical results are easily obtained; and there is no difficulty in afterwards introducing the (comparatively slight) corrrections due to the earth's curvature. But these will not be further alluded to here.

Of course I began, as almost every other person who has thought of the production of the ordinary mirage of the desert must naturally have begun, by considering the well-known problem of the paths of projectiles discharged from the same gun, with the same speed but at different elevations of the piece. This corresponds, in the optical problem, to the motion of light in a medium the square of whose refractive index is proportional to the distance from a given horizontal plane. Instead, however, of thinking chiefly of the different elevations corresponding to a given range, I sought for a simple criterion which should enable me to decide (in the optical application) whether the image formed would, in any particular case, be a direct or an inverted one. And this, I saw at once, could be obtained, along with the number and positions of the images, by a study of the form of the locus on which lie the *vertices* of all the rays issuing from a given point. Thus, in the ballistic problem, the locus of the vertices of all the paths from a given point, with different elevations but in the same vertical plane, is an ellipse.

Its minor axis is vertical, the lower end being at the gun; and the major axis (which is twice as long) is in the plane of projection. Now, while the inclination of the piece to the horizon is less than 45°, the vertex of the path is in the *lower* half of this ellipse, where the tangent leans forward from the gun; and in this case a small increase of elevation *lengthens* the range, so that the two paths do not intersect again above the horizon. In the optical problem this corresponds to an *erect* image. But, when the elevation of the piece is greater than 45°, the vertex of the path lies in the *upper* half of the ellipse, where the tangent leans back over the gun; and a small increase of elevation *shortens* the range. Two contiguous paths, therefore, intersect one another again above the horizon. And, in the optical problem, this corresponds to an *inverted* image. In symbols, if the eye be taken as origin and the axis of x horizontal, there will be a direct image for a ray at whose vertex dy/dx and x (in the curve of vertices) have the *same* sign, an inverted image when the signs are different.

Hence, whatever be the law of refractive index of the air, provided only it be the same at the same distance from the earth's surface (i.e. the surfaces of equal density parallel planes, and therefore the rays each symmetrical about a vertical axis), all we have to do, in order to find the various possible images of an object at the same level as the eye, is to *draw the curve of vertices for all rays passing through the eye, in the vertical plane containing the eye and the object, and find its intersections with the vertical line midway between the eye and the object.* As soon as this simple idea occurred to me, I saw that it was the very kernel of the matter, and that all the rest would be mere detail of calculation from particular hypotheses. Each of the intersections in question is the vertex of a ray by which the object can be seen, and the corresponding image will be erect or inverted, according as the curve of vertices leans from or towards the eye at the intersection. Thus, in Fig. 2,

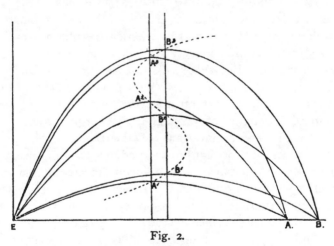

Fig. 2.

let E be the eye, and the dotted line the curve of vertices for all rays in the plane of the paper, and passing through E. Let A be an object at the level of the eye, $A^1A^2A^3$ the vertical line midway between E and A. Then A^1, A^2, A^3 are the vertices of the various rays by which A can be seen. If we make the same construction for a point B, near to A, we find that whereas the contiguous rays through A^1, B^1 and through A^3, B^3 do not intersect, those through A^2, B^2 do intersect. At A^1 and A^3 the curve of vertices leans *from* the eye, and we have erect images; at A^2 it leans back *towards* the eye, and we have an inverted image. And thus, if this curve be continuous, the images will be alternately erect and inverted. The sketch above is essentially the same as one given by Vince, only that he does not

employ the curve of vertices. If the object and eye be not at the same level, the
construction is not *quite* so simple. We must now draw a curve of vertices for rays
passing through the eye, and another for rays passing through the object. Their
intersections give all the possible vertices. (This construction of course gives the
same result as the former, when object and eye are at the same level.) But the
images are now by no means necessarily alternately erect and inverted, even though
the curve of vertices be continuous. However, I merely note this extension of the
rule, as we shall not require it in what follows.

I then investigated the form of the curve of vertices in a medium in which the
square of the refractive index increases by a quantity proportional to the square of
the distance from a plane in which it is a minimum, and found that (under special
circumstances, not however possible in air) three images could be produced in
such a medium. But the study of this case (which I could not easily explain
here without the aid of mathematics) led me on as follows.

As the curvature of a ray is given by the ratio of the rate of change of index
per unit of length perpendicular to the ray, to the index itself (a result which I find
was at least virtually enunciated by Wollaston); and as all the rays producing the
phenomena in question are very nearly horizontal: i.e. perpendicular to the direction
in which the refractive index changes most rapidly: their curvatures are all
practically the same at the same level. Hence if the rate of diminution of the
refractive index, per foot of ascent, were nearly constant, through the part of the
atmosphere in which the rays travel, the rays we need consider would all be
approximately arcs of equal circles; and the curve of vertices would (so far as these
rays are concerned) lean wholly *from* the eye; being, in fact, the inferior part of
another equal circle which has its lowest point at the eye. Hence but one image,
an erect one, would be formed; but it would be seen elevated above the true
direction of the object. This is practically the ordinary horizontal refraction, so far
as *terrestrial* objects on the horizon are concerned. The paths of the various rays

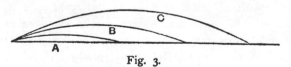

Fig. 3.

would be of the form in Fig. 3 (the drawing is, of course, immensely exaggerated)
and the locus of vertices, *A, B, C,* obviously leans *from* the eye. But now suppose
that, below a stratum of this kind, there were one of constant density, in which of

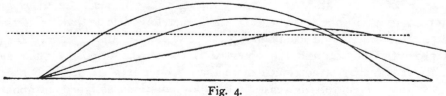

Fig. 4.

course the rays would be straight lines. Then our sketch takes the form Fig. 4
(again exaggerated); each of the portions of the ray in the upper medium being

congruent to the corresponding one in the former figure (when the two figures are drawn to the same scale), but *pushed farther to the right as its extremities are less inclined to the horizon*. In its new form the curve of vertices *ABC* leans back *towards* the eye, and we have an inverted image. The lower medium need not be uniform as, for simplicity, we assumed above. All that is required is that the rate of diminution of density upwards shall be less in it than in the upper medium.

Those who have followed me so far will at once see that, as a more rapid increase of density, commencing at a certain elevation, makes the curve of vertices lean back, so a less rapid decrease (tending to a "stationary state") at a still higher elevation will make the curve of vertices again lean forward from the eye. I need not enlarge upon this.

Thus to repeat: the conditions requisite for the production of Vince's phenomenon, at least in the way conjectured by him, are, a stratum in which the refractive index diminishes upwards to a nearly stationary state, and below it a stratum in which the upward diminution is either less or vanishes altogether. The former condition secures the upper erect image, the latter the inverted image and the lower direct image.

In my paper read to the Royal Society of Edinburgh I have given the mathematical details following from the above statement; and have made full calculations for the effect of a transition stratum, such as must occur between two uniform strata of air of which the upper has the higher temperature. From Scoresby's remarks it appears almost certain that something like this was the state of affairs when the majority (at least) of his observations were made. When two masses of the same fluid, at different temperatures, rest in contact; or when two fluids of different refractive index, as brine and pure water, diffuse into one another; the intervening layer must have a practically "stationary" refractive index at each of its bounding surfaces, and a stratum of greatest rate of change of index about midway between them. The exact law of change in the stratum is a matter of comparatively little consequence. I have assumed (after several trials) a simple harmonic law for the change of the square of the refractive index within the stratum. This satisfies all the above conditions, and thus cannot in any case be very far from the truth. But its special merit, and for my purpose this was invaluable, is that it leads to results which involve expressions easily calculated numerically by means of Legendre's Tables of Elliptic Integrals. This numerical work can be done once for all, and then we can introduce at leisure the most probable hypotheses as to the thickness of the transition stratum, the height of its lower surface above the ground, and the whole change of temperature in passing through it. I need not now give the details for more than one case, and I shall therefore select that of a transition stratum 50 feet thick, and commencing 50 feet above the ground. From the physical properties of air, and the observed fact that the utmost angular elevation of the observed images is not much more than a quarter of a degree, we find that the upper uniform layer of air must under the conditions assigned be about 70° C. warmer than the lower. Hence by the assumed law in the stratum, the maximum rise of temperature per foot of ascent (about the middle of the transition stratum) must be about 0°·2 C. per foot. Such changes have actually been observed by Glaisher in his balloon ascents, so that thus far the hypothesis is justified. But we have an independent means of testing it. The form of the curve of vertices is now somewhat like the full lines in the following cut (Fig. 5), where O is the eye, and the lines RS, TU represent the boundaries of the transition stratum. It is clear, that, if PM be the vertical tangent, there can be but one image of an object unless its distance from O is at least twice OM. This will therefore be called the "*Critical distance.*" If the distance be greater than this there are three images:—one erect, seen directly through the lower uniform stratum—then an inverted one, due to the diminution of refractive index above the lower boundary of the transition stratum— and finally an erect image, due to the approximation to a stationary state towards the upper boundary of that stratum. Now calculation from our assumed data gives OM about six miles, so that the nearest objects affected should be about twelve miles off. Scoresby says that the usual distance was from ten to fifteen miles. Thus the hypothesis passes, with credit, this independent and severe test. A slight reduction of the assumed thickness of the transition stratum, or of its height above the ground, would make the agreement exact.

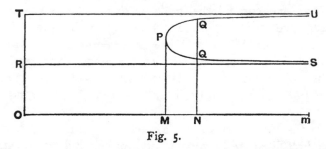

Fig. 5.

All the phenomena described in Vince's paper of 1799, as well as a great many of those figured in Scoresby's works, can easily be explained by the above assumptions. Scoresby's remarkable observation of a single inverted image of his father's ship (when thirty miles off, and of course far below the horizon) requires merely a more rapid diminution of density at a definite height above the sea. His figure is the second in Fig. 1 above. But Scoresby figures, as above shown, several cases in which two or more inverted images, without corresponding erect ones, were seen above the ordinary direct image. The natural explanation is, of course, a series of horizontal layers of upward diminishing density and without a "stationary state" towards their upper bounding planes. I find that, by roughly stirring (for a very short time) a trough in which weak brine below is diffusing into pure water above, we can reproduce this phenomenon with great ease. In fact, when temporary equilibrium sets in, the fluids are arranged in a number of successive parallel strata with somewhat abrupt changes of density.

But the mathematical investigation, already spoken of, shows that it is quite possible that there may be layers tending to a stationary state without any corresponding *visible* images.

This depends on the fact that, while the inverted image (due to the lower part of a stratum) is *always* taller than the object seen directly (though not much taller unless the object is about the critical distance), the numerical calculation shows that the erect image is in general extremely small, and can come into notice only when the object is not far beyond the critical distance. Thus there may have been, in *all* of Scoresby's observations (though he has only occasionally noticed and depicted them) an erect image above each inverted one, but so much reduced in vertical height as to have been invisible in his telescope, or at least to have formed a mere horizontal line so narrow that it did not attract his attention. The greatly superior number of inverted images, compared with that of the direct ones, figured by Scoresby, may thus be looked upon as another independent confirmation of the approximate correctness of the hypothetical arrangement we have been considering.

To obtain an experimental repetition of the phenomena in the manner indicated by the above hypothesis, a tank, with parallel glass ends, and about 4 feet long, was half-filled with weak brine (carefully filtered). Pure water was then cautiously introduced above it, till the tank was nearly filled. After a few hours the whole had settled down into a state of slow and steady diffusion, and Vince's phenomenon was beautifully shown. The object was a metal plate with a small hole in it, and a lamp with a porcelain globe was placed behind it. The hole was triangular, with one side horizontal (to allow of distinction between direct and inverted images) and was placed near one end of the tank, a little below the surface-level of the unaltered brine, the eye being in a corresponding position at the other end. A little vertical adjustment of object and eye was required from time to time as the diffusion progressed. The theoretical results that the upper erect image is usually much less than the object, and that it is seen by slowly convergent rays, while the inverted image is larger than the object and is seen by diverging rays, were easily verified.

To contrast Wollaston's best-known experiment with this, a narrow tank with parallel sides was half-filled with very strong brine, and then cautiously filled up with pure water. (The strong brine was employed to make up, as far as possible, for the shortened path of the rays in the transition stratum.) Phenomena somewhat resembling the former were now seen, when object and eye were nearly at the same distance apart as before, and the tank about half-way between them. But in this case the disparity of size between the images was not so marked—the upper erect image was always seen by diverging rays, the inverted image by rays diverging or converging according as the eye was withdrawn from, or made to approach, the tank. In this case, the curvature of each of the rays in the vessel is practically constant, but is greatest for the rays which pass most nearly through the stratum of most rapid change of refractive index. Hence, when a parallel beam of light fell horizontally on the tank and was received on a sufficiently distant screen, the lower boundary of the illuminated space was blue—and the progress of the diffusion could be watched with great precision by the gradual displacement of this blue band. I propose to employ this arrangement for the measurement of the rate of diffusion, but for particulars I must refer to my forthcoming paper.

Wollaston's experiment with the red-hot poker was probably, his experiment with the red-hot bar of iron almost certainly, similar to that above described with the long tank, and the weak brine; and *not* to that with the short tank, though the latter is usually cited as Wollaston's contribution to the explanation of the Vince phenomenon. We have seen how essentially different they are, and that the latter does not correspond to the conditions presented in nature.

LONG DRIVING

(From the *Badminton Magazine*[1], March, 1896.)

Error ubique patet; falsa est doctrina periti:
Sola fides numeris intemerata manet.

In the great drama, familiarly known as a *Round of Golf*, there are many Acts, each commonly but erroneously called a Stroke. Besides Acts of Driving, to which this article is devoted, there are Acts of Approaching, Acts in (not *always* out of) Hazards, and Acts of Holing-out. There is another class of Acts, inevitable as human beings are constituted—Acts of Negligence, Timidity, or Temerity. Of these we cannot complain, and they give much of its interest to the game. A philosophic professional, after missing an easy putt, put this aspect of the game in words which could scarcely be improved : " If we cud a' aye dae what we wantit, there wud be nae fun in't." Besides these, there are, too frequently, other Acts wholly superfluous and in general injurious, to the game: Acts of Gambling, Fraud, and Profanity. These, however, belong to the domain of the moral, rather than of the natural, philosopher.

Each Act of golf proper has several Scenes. An Act of Driving essentially contains *four* ; besides the mere preliminary work of the caddie, such as teeing the ball (when that is permissible), handing his master the proper club, and clearing loafers and nursery maids out of the way. These are :—

SCENE I.—*Stance, Waggle, and Swing.*

Here the only *dramatis personae* directly engaged are the player and his club. This scene is rarely a brief one, even with the best of players, and is often absurdly protracted.

SCENE II.—*The Stroke Proper.*

The club and the ball practically share this scene between them; but the player's right hand, and the resistance of the air, take *some* little part in it. It is a very brief one, lasting for an instant only, in the sense of something like one ten-thousandth of a second. Yet in that short period most important events take place. [Sometimes, it is true, this scene does not come off at all, the club passing, instead of meeting, the ball. It is called a stroke for all that, and is sedulously noted on the scorer's card.]

SCENE III.—*The Carry.*

Here the action is confined to the ball, gravity, and the atmosphere. The scene may last for a second or two only, if the ball be topped, or if a poor player is at work; but with good drivers it usually takes six seconds at the very least.

---

[1] The parts in square brackets were on a first proof preserved by Tait, but were deleted in the final proof.

SCENE IV.—*The Run.*

Here pure chance is the main actor. The scene has no measurable duration when the ball lands in mud or soft sand. It may continue for two, three, or more seconds if the ball be topped or get a running fall, and the links be hard and keen. The ball's progress may be by mere rolling or by a series of leaps. This is usually (at least on a "sporting" course) a most critical scene, and the player feels himself breathing more freely when it is safely concluded.

Our chief concern is with the second and third of these scenes; the fourth, from its very nature, being of such a capriciously varying character that it would be vain to attempt a discussion of it, and the first being of interest to our present purpose only in so far as it is a necessary preliminary to the second. Its result, as far as we are concerned, is merely to bring the club face into more or less orderly and rapid contact with the ball. We say advisedly the club *face* (not merely the club *head*), for operations such as hacking and sclaffing, however interesting in themselves, form no part of golf, properly so-called.

All the resources of the pen, the pencil, and the photographic camera have been profusely employed on behalf of the public to convey to it some idea of the humours of Scene I., so that we may omit the discussion of it also. [Yet, after all, sketching and word painting, even when the most amazing vagaries of the most fertile imagination have been freely taken advantage of, are sometimes far less effective in this matter than is the simple verity as recorded by the imperturbably truthful camera. For there are many things, sad or laughable, unimportant or of the most immense consequence to the discovery of peculiarities of style, which escape the keenest vision, and yet are seized upon and preserved for leisurely after-study in a single wink, as it were, of that terrible photographic Eye.]

Brief as is the duration of the second scene, the analysis of even its main features requires considerable detail if it is to be made fully intelligible. I will attempt to give this in as popular a form as possible. But before I do so, it may be well to show its importance by a passing reference to some of its consequences, as we shall thus have a general notion of what has to be explained. When the ball parts company with the club this scene ends, and the third scene begins. Now, at that instant, having by its elasticity just recovered from the flattening which it suffered from the blow, the ball must be moving as a free rigid solid. It has a definite speed, in a definite direction, and it *may* have also a definite amount of rotation about some definite axis. The existence of rotation is manifested at once by the strange effects it produces on the curvature of the path, so that the ball may skew to right or left, soar upwards as if in defiance of gravity, or plunge headlong downwards, instead of slowly and reluctantly yielding to that steady and persistent pull. The most cursory observation shows that a ball is hardly ever sent on its course without *some* spin, so that we may take the fact for granted, even if we cannot fully explain the mode of its production. And the main object of this article is to show that LONG CARRY ESSENTIALLY INVOLVES UNDERSPIN.

Now, if golf balls and the faces of clubs were both perfectly hard (i.e. not deformable) the details of the effects of the blow would be a matter of simple dynamical calculation, and Scene II. would be absolutely instantaneous. If, in addition, the ball were perfectly spherical, smooth, and of homogeneous material, no blow could possibly set it in rotation; if it were defective in any of these particulars, we could easily calculate the direction of the axis of rotation and the amount of spin produced in it by any assigned blow. But, unfortunately, neither balls nor clubs can make an approach to perfect hardness. For there is never, even in a gentle stroke, a mere *point* of contact between ball and club. In good drives the surface of contact may often, as we see by an occasional trace from undried paint or by the pattern impressed by the first drive on a new leather face, be as large as a shilling. The exact mathematical treatment of so large a distortion is an exceedingly complex and difficult matter. But fortunately we are not called upon to attack it, for it is obvious, from the facts of common observation already cited, that the final effect on the ball is *of the same general character* as if it had been perfectly hard, though the speed of projection, and notably that of spin, will be materially less. And it is with the character rather than with the amount of the effect that we are mainly concerned, as will be seen farther on. Thus, in a great part of what follows, we will argue as if both club and ball were hard, since we seek to explain the *character* of the results and are not for the time concerned with their *magnitudes*.

When we reflect on the very brief duration of the impact, during which the average force exerted is about three tons' weight (while the player is, for an instant, working at two or three horse-power at least), we see at once that we may practically ignore the effects of gravity, of the continued *pushing* forward of the club head, and even of the resistance of the air (though amounting to, say, five-fold the weight of the ball) during that short period; so that we are concerned only with the velocity and the orientation of the club face at the moment of impact.

The simplest case which we have to consider is that in which the club face, at the instant of meeting the ball, is moving perpendicularly to itself. If the ball be spherical and homogeneous, there can be no spin; and thus we are concerned only with the steps of the process by which the ball ultimately leaves the club in the common direction of motion. The first effect is the impulsive pushing forward of the part of the ball which is struck, the rest, by its inertia, being a little later in starting. Thus the ball and club face are both distorted until they, for an instant, form, as it were, one body, which has the whole momentum which the club head originally possessed. As the club head is usually about five times more massive than the ball, the common speed is five-sixths only of the original speed of the head. [In the case of a more massive club head a correspondingly less fraction of its speed would be lost, but a proportionately greater effort would be required to give it a definite speed to begin with. A sort of compromise must thus be made, and experience has led to the proportion cited above—so long, at least, as we are dealing with balls of about ten or eleven to the pound.] But the ball and club both tend to recover from their distortion, and experiment shows that they exert on one another, during this recovery, an additional impulsive pressure which is a definite fraction of that already exerted between them. This fraction is technically called the "co-efficient of restitution," and it is upon its magnitude that the higher or lower quality of a ball, and of a club face, mainly depend. Its value, when good materials are employed, is usually about 0·6. Thus the club and ball at last separate with a *relative* speed six-tenths of that with which the club approached the ball. The ball, therefore, finally acquires a speed about one-third greater than that which the club head originally had. Thus the head must have a pace of about 180 feet per second in order that it may drive the ball at the rate of 240 feet per second. And, for various values of the co-efficient of restitution, the ultimate pace of the ball can never be less than five-sixths, nor as much as five-thirds, of the initial speed of the club head. We thus get at present about four-fifths of what is (theoretically) attainable; and we may perhaps, by means of greatly improved materials, some day succeed in utilising a considerable part of this wasted fifth.

Where, as is almost invariably the case, the face of the club is not moving perpendicularly to itself at impact there is always one perfectly definite plane which passes through the centre of the ball and the point of first contact, and is parallel to the direction of motion of the head. It is in this plane, or parallel to it, that the motions of all parts of the ball and the club head (except, of course, some of the small relative motions due to distortion) take place. Hence if the ball acquire rotation it must be about an axis perpendicular to this plane. The whole circumstances of the motion can therefore be, in every case, represented diagrammatically by the section of the ball and club face made by this plane. The diagram may take one or other of the two forms below, either of which may be derived from the other by perversion and inversion.

[Thus, if the page be turned upside-down and held before a mirror, the result will be simply to make the first figure into the second, and the second into the first —merely, in fact, altering their order. Holding the page *erect*, before a mirror, we get diagrams specially suited for a left-handed player.]

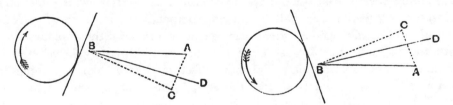

In each of the figures the velocity of the club head at impact is represented by the line AB, and the dotted lines AC and CB represent its components parallel and perpendicular to the club face respectively. By properly tilting the figures, AB may be made to take any direction we please, i.e.—the club head may be represented as moving in any direction whatever—but it is quite sufficient for our purpose to treat it as moving horizontally. It is the existence of the component velocity AC, in a direction parallel to the club face, which (alone) makes the difference between this case and the simple one which we have just treated. And if the ball were perfectly smooth this component would lead to no consequences. But because of friction this component produces a tangential force whose effect is partly to give the ball as a whole a motion parallel to AC, partly to give it rotation in the direction indicated by the curved arrow. The direction of motion of the ball when free lies somewhere *between* the directions of *AB* and *CB*, say in the line *DB*. [The reader must take this statement for granted, if it be not pretty obvious to him; for its proof, even in the simple case in which the ball is regarded as perfectly hard, involves the consideration of moment of inertia, which I must not introduce.]

As already remarked, one or other of the diagrams above applies to any possible case. But there are two special cases which are of paramount importance, and if these be fully understood by him the reader can easily make for himself the application to any other.

In the first of these special cases the plane of the diagram is to be regarded as *horizontal*, and the club face (perpendicular to it by the conditions of the diagram) consequently vertical, while the rotation given to the ball is about a vertical axis. The spectator is, therefore, supposed to be looking down upon the club and ball from a station high *above* them. The interpretation of the indicated result thus depends upon the direction of the line joining the player's feet. If that line be (as it ought to be) perpendicular to the face of the club, it is parallel to *BC*; so that the club (when it reaches the ball) is being *pulled in* (first figure), or *pushed out* (second figure), in addition to sweeping past in front of the player parallel to the line joining his feet. The first of these is the very common fault called "slicing." The second is not by any means so common, and I am not aware that it has ever been dignified by a special name. If, on the other hand, the line of the feet be parallel to *AB*, the sweep of the club head is in the correct line, but the face is turned outwards (first figure), or inwards (second figure), and we have what is called "heeling" or "toeing." These terms must not be taken literally, for heeling may be produced by the toe of the club and toeing by the heel. Slicing and heeling have thus precisely the same effect, so far as the rotation (and consequent "skewing") of the ball is concerned; but the position of the line *DB* shows that, other things being correct, a sliced ball starts a little to the left of the intended direction, while the heeled ball commences its disastrous career from the outset by starting a little to the right. It is most important to the player that he should be able to distinguish between these common faults, because, though their (ultimate) results are identical, the modes of cure are entirely different. This, of course, is obvious from what we have said above as to the intrinsic nature of each. Toeing, and the innominate fault mentioned above, both give the opposite rotation to that produced by heeling, and therefore the opposite skew. If slicing and toeing occur together, each tends to mitigate the evil effects of the other; so with heeling and the innominate. But slicing and heeling together will produce aggravation of each other's effects.

In the second special case the plane of the diagram is regarded as *vertical*, and the spectator's line of sight passes horizontally between the player's feet from a point behind him. The first diagram, therefore, corresponds to under-cutting, and the second to topping, if *AB* be horizontal; or to jerking, and bringing the club upwards behind the ball[1], respectively, if the face be vertical. The first diagram

---

[1] This suggests a very favourite diagram employed by many professed instructors in the game. It is usually embellished with a full circle, intended to show the proper path of the head, and the ball is placed (on a high tee) a good way in front of the lowest point of the circle. It will be seen from the text above that this virtual pulling in of the club head produces, in a vertical plane, the same sort of result as does slicing in a horizontal plane.

also represents the natural action of a spoon, or a "grassed" play club, *AB* being horizontal. In all these cases the spin is about a horizontal axis, and therefore the skewing is upwards or downwards. Thus, we have traced out generally, and also specially for the most important cases, the processes of the second scene, which usher the ball into the third with a definite speed and a definite rotation.

In the discussion of the third scene, in which the ball is left to its own resources, to struggle as best it can against the persistent downward pull of gravity and the ever-varying resistance of the air, we will treat fully of really good drives —i.e., those in which the spin, if there be any, is about a horizontal axis perpendicular to the plane of flight, and is such as to cause the ball to "soar," not to "dook." Incidentally, however, we will notice (though with much less detail) the causes which produce departures of various kinds from a high standard of driving.

We will treat, *first*, of the path as affected by gravity alone; *second*, of the path under gravity and resistance alone (the ball having no rotation); *third*, of the path as it would be if the ball were spinning, but not affected by gravity; *fourth*, as it is when all these agents are simultaneously at work; and, *finally*, the effects of wind. The first and third of these, in each of which one of the most important agents is left wholly out of account, though of less consequence than the others, are necessary to the proper development of the subject, inasmuch as their preliminary treatment will enable us to avoid complications which might embarrass the reader.

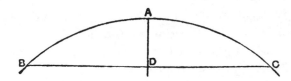

1. If there were no resistance, the path of a golf ball would be part of a parabola, *BAC*, whose axis, *AD*, is vertical. The vertex, *A*, of the path would be always *midway* along the range, *BC*; and the ball would reach the ground with the speed given it from the tee. A golf ball would therefore be an exceedingly dangerous missile. For fairish but high driving would easily make the range *BC* something like a quarter of a mile! And at that distance the ball would fall with precisely the same speed as that with which it left the tee. The range for any definite "elevation" (i.e. angle at which the path was inclined to horizon at starting) would be proportional to the square of the initial speed, so that double speed would give quadruple range; and for any given speed, it would increase with the elevation up to 45°, and thence diminish with greater elevation. Any one can test this last result by means of the jet from a garden engine. The speed of such a jet is so small that the resistance is inconsiderable.

For comparison with some of the numerical results to be given below, we will here give a few simple particulars.

Suppose the ball to have an initial speed of 200 feet per second; we have

Elevation	Range	Height	Time
75°	621	579	12
45	1,242	310·5	8·8
15	621	41·6	3·2
7·5	321	10·6	1·6

The lengths are in feet and the times in seconds. Notice that for elevation 15° we get a range of 207 yards, with a maximum height of 14 yards. These, so far, are not *very* unlike what may sometimes occur in an actual drive. But we must look to *all* the facts; and this closer comparison shows the resemblance to be only superficial. For, *first*, the vertex is midway along the path; *second*, the ball comes down, as it rose, at 200 feet per second. These are utterly contrary to experience. But, *third*, this long journey is effected in little more than three seconds. A golfer finds that it requires nearly seven seconds. The unresisted projectile theory is thus completely at fault, so far as application to golf is concerned.

2. Let us next consider the effect of atmospheric resistance, the ball having no spin. [This was, and unfortunately must continue to be, a matter of grave concern to myself. For when I began to learn golf, my instructor (an elderly man, but a very fair player for all that) urged me to bear constantly in mind that "*all spin is detrimental.*" This was, he told me, the definite result of his long experience. It cost me much thought, and long practice, to carry out his recommendation, and it is possible that I have more personal experience of the behaviour of balls almost free from spin than has any other player. The more nearly I approached this ideal the greater was the proportion of run to carry in my driving. I understand it now—too late by thirty-five years at least.]

It has already been said that want of homogeneity in a spherical ball almost certainly leads to its getting spin from the very tee. But, even should it be projected without rotation, it will soon acquire some as it moves through the air. The spin so acquired will be of an uncertain and variable nature, and the flight of the ball will be unsteady and erratic. [I have elsewhere explained how to test balls for this defect, by merely making them oscillate while floating on mercury. Any which oscillate quickly are absolutely useless.]

It seems to be pretty well established that, for the range of speed common in golf, the resistance is as the square of the speed. In fact, the faster a ball moves the more air does it displace in a given time and also the faster does it make that air move. The most convenient mode of expressing the amount of resistance is to assign the "terminal velocity" of the ball, i.e. its speed when the resistance is just equal to its weight. If a sack full of golf balls were emptied at a height of three or four miles, the balls would reach the ground with their terminal velocity.

Robins, long ago, gave data from which we can assign 114 feet per second as the terminal velocity of a golf ball. His experiments were all made at moderate speeds. The comparatively recent experiments of Bashforth, though mainly directed to very high speeds, give for speeds under 800 feet per second the means of assigning 95 feet per second as the terminal velocity of a golf ball. When I began to make calculations on this subject, I naturally took the more recent determination as the correct one; and was thus forced to assume at least 300 feet per second as the initial speed of a golf ball in order to account for some of the simplest facts. I have since found that this estimate is very considerably in excess of the truth, and therefore that the terminal velocity, as assigned by Bashforth's data, is considerably too small. I still, however, think that Robins' estimate is somewhat too low, so that in what follows I shall assume 108 feet as the terminal velocity for an ordinary golf ball. Trifling as may appear the differences among these numbers (114, 95, and 108), experiments on initial speed seem to show that, if the first were correct, we ought to drive somewhat further than we do; while, if the second be correct, it is quite certain that, on the open links, considerably greater initial speeds are given than any which I have been fortunate enough to measure in my laboratory, though I have had the kind assistance of some of the most slashing drivers of the day[1].

Assuming, then, 108 as the terminal velocity, it is easy to calculate by the help of Bashforth's tables (which can be adapted to any amount of resistance) the following sufficiently approximate results for different elevations, the initial speed being assumed to be 240 feet per second.

Elevation	Range	Height	Speed at Vertex	Time	Speed of Descent
50°	440	206	59	6·95	90
43	467	175	68	6·5	88
37·5	474	144	76	6·16	84
34	475	119	83	5·43	82
29	467	99	90	4·9	82
23	445	70·6	102	4·1	82
18·5	412	50	113	3·47	84
15·3	389	38	122·7	3·03	86
13·3	341	28·5	124·3	2·71	87·4
10·2	324	20·5	140·4	2·24	100

One feature, at least, in this table is much more consonant with experience than that corresponding to the parabolic path. The ball reaches the ground in all these courses with speed far inferior to that with which it started.

[1] This paragraph and the paragraph and Table following were considerably abridged in the article as published in the *Badminton Magazine*.

I purposely assumed 240 feet per second for the initial speed, though it is greater by ten per cent. than any I have yet actually measured[1], in order to give this form of the theory as fair play as could be equitably conceded. Yet the utmost attainable range, as shown by the table, is short of 160 yards, while the players whose pace I measured had habitually carried about 180 yards the previous day, or were to do it the next. But I measured their habitual elevation as well as their pace. It was always small, rarely more than 1 in 6, i.e. less than 10°. The average was about 1 in 7, or little over 8°. Look again to the table, and we see that the *maximum* range for a speed of 240 (though only about 160 yards at best) involves an elevation of some 34°, altogether unheard of in long driving, while the elevation of 10° gives a range of 108 yards only! Thus *this* form of the theory also breaks down completely. But, before we altogether dismiss it, let us test it in two other and perfectly different ways.

It might be objected to the reasoning above that I have taken initial data from experiments on a limited number of exceptionally good players; and that I have somewhat arbitrarily assumed a value for the resistance of the air greater than that given by one recognised authority and considerably less than that given by another. One good, though only partial, answer is to change my authority so far as elevation is concerned; and, keeping to the same co-efficient of resistance, find the characteristics of drives with various initial speeds.

Some six years ago, Mr Hodge kindly measured for me, by means of a clinometer, the average elevation of drives made by a great number of good players from the first tee at St Andrews. He estimated it at about 13·5°. Bashforth's tables usually involve data for whole degrees only, so I shall assume 14° as the standard elevation. Here are the results for various initial speeds, some wholly unattainable:

Initial speed ...	66	115	142	174	215	275	382	739
Final speed ...	56·5	80	84	88	90	91·3	91·2	90·7
Range	58	145	197	257	329	421	554	833
Height	3·7	10·35	14·7	20·3	27·6	44·6	56	101
Time	0·96	1·6	1·9	2·2	2·6	3	3·6	4·7
Position of vertex	0·51	0·53	0·55	0·56	0·57	0·6	0·63	0·67

Note, *first*, that in all these paths the time is much too short; *second*, that the initial speed required for a carry of 184 yards is 382 feet per second, which must

[1] In my laboratory experiments the players could not be expected to do *full* justice to their powers. They had to try to strike, as nearly as possible in the centre, a ten-inch disc of clay, the ball being teed about six feet in front of it. Besides this preoccupation, there was always more or less concern about the possible consequences of rebound, should the small target be altogether missed.

be regarded as totally unattainable; also that the position of the vertex in that case is little more than three-fifths of the range from the tee.

Next, let us simply take 10° for elevation and 240 for initial speed, and find the results for various amounts of resistance or terminal velocities. Here they are:

Terminal velocity	431	245	176	133	114	98	82
Range	566	506	440	363	328	287	240
Height	26	24·4	22·5	20·6	18·9	17·2	15·3
Time	2·5	2·46	2·36	2·23	2·14	2·05	2·01

The only approach to the practical range of good driving is in the first column. But the corresponding resistance is about one-sixteenth only of our estimate, and the time is absurdly too small. Thus the ordinary resistance theory also fails to explain the facts. When compared with them it breaks down almost as completely as did the parabolic theory.

In what precedes I have endeavoured to make it perfectly clear that *something else* besides mere speed and elevation is required, if *all* the ordinary facts of long driving are to be simultaneously accounted for. Great initial speed is required if the resistance is great, and the larger these are the further is the vertex from midway, *but* then the time taken for a range of 180 yards will be much too short. In order that 180 yards may be covered in six seconds the average horizontal speed must be only 90 feet per second, and gravity would cut short the ball's flight long before it had reached the goal; unless, by way of preventing this, we gave it an extravagant elevation at starting. And, in all cases, the path will be concave downwards throughout its whole extent. In many fine drives it is concave *upwards* for nearly half of the range. The sole additional consideration to which we can have recourse to help us in reconciling these apparently inconsistent facts is *rotation of the ball*—to which we are thus *compelled* to have recourse!

I have been very, perhaps even unnecessarily, cautious in leading up to this conclusion; and I have consequently been careful to fortify my position from time to time by an appeal to the recognised maxim, *Mundum regunt numeri*. I have a vivid recollection of the "warm" reception which my heresies met with, some years ago, from almost all of the good players to whom I mentioned them. The general feeling seemed to be one in which incredulity was altogether overpowered by disgust. To find that his magnificent drive was due merely to what is virtually a toeing operation—performed, no doubt, in a vertical and not in a horizontal plane—is too much for the self-exalting golfer!

The fact, however, is indisputable. When we fasten one end of a long untwisted tape to the ball and the other to the ground, and then induce a good player to drive the ball (perpendicularly to the tape) into a stiff-clay face a yard or two off, we find that the tape is *always* twisted; no doubt to different amounts by different players, but in such wise as to show that the ball makes usually from about 1 to 3 turns in six feet—say from 40 to 120 turns or so per second. This is clearly a circumstance not to be overlooked.

3. Some 230 years ago, Newton employed the analogy of the curved path of a tennis ball "struck with an oblique racket" to aid him in explaining the separation of the various constituents of white light by a prism. And he says, in words which apply aptly to the behaviour of a golf ball, "a circular as well as a progressive motion being communicated to it by that stroke, its parts, on that side where the motions conspire, must press and beat the contiguous air more violently than on the other, and there excite a reluctancy and reaction of the air proportionally greater." In other words, the pressure of the air is greater on the advancing than on the retreating side of the ball, so that it is deflected from its course in the same direction as that of the motion of its front part, due to the rotation. This explanation has not since been improved upon, though the fact itself has been repeatedly verified by many experimenters, including Robins and Magnus.

That the deflecting force thus called into play by the rotation of the ball may be of considerable magnitude is obvious from the fact of the frequently observed *upward* concavity of the earlier part of the path. For this shows that, at first, *the new force is greater than the weight of the ball.* It is thus greater than one-fifth of the direct resistance when the latter has its maximum value. Its magnitude depends upon the rate of spin, and also upon the speed of the ball, and may be regarded as directly proportional to their product. And we know, from the way in which the ball behaves after falling, that the spin does not diminish very rapidly, for a good deal of it remains at the end of the carry. It is probable that the spin contributes to the direct resistance also; and this was one of my reasons for assuming a terminal velocity somewhat less than that deduced from the datum of Robins. Two important effects of hammering, or otherwise roughening, the ball are now obvious: it enables the club to "grip" the ball firmly, so as to secure as much spin as possible, and it enables the ball, when free, to utilise its spin to the utmost.

Some of the effects due to resistance and spin alone are very curious. Thus a top or "pearie" spinning with its axis vertical on a smooth horizontal plane is practically free from the effect of gravity. If it receive a blow which tends to give it horizontal motion only, it moves in an endless spiral, coming back, as it were, to receive a second blow. The sense of the spiral motion is the same as that of the rotation. If its spin were to fall off at the same rate as does its speed of translation, the spiral path would, as it were, uncoil itself into a circular one.

Closely related to this is Robins' experiment with a pendulum whose bob is supported by two strings twisted together, so that they set it in rotation as they untwist. The plane of the pendulum's vibration constantly turns round in the same sense as does the bob.

If the bob be supported by a fine wire to whose upper end torsion can be applied, it may be made to move as a conical pendulum. Then its path will shrink, or open out, as the bob is made to rotate in, or against, the sense of the revolution.

When a narrow, rectangular, slip of paper is let fall, with its greater sides horizontal, it usually begins to spin about its longer axis, and at a rate which is generally greater the narrower it is. *Then* it falls almost uniformly in a nearly straight path, considerably inclined to the vertical. The deflection is always *towards* the side to which the edge of the strip which is at any instant the lower, and therefore the foremost, is being carried by the rotation. If the longer edges be not quite horizontal, the path is usually a nearly perfect helix, the successive positions of the upper surface being arranged very much like the steps of a spiral staircase. This is an exceedingly simple, as well as a beautiful and instructive experiment; and, besides, it has an intimate relation to our subject.

Finally, we need only refer to Robins' musket, which virtually solved the problem of shooting round a corner. The barrel was slightly curved to the left near the muzzle; and the bullet (made purposely to fit loosely) rolled on the concave (right-hand) side of the bore, and thus behaved precisely like a sliced golf ball, starting a little to the left, and then skewing away to the right.

4. As the transverse force due to the spin is always in a direction perpendicular to that of the ball's motion, it has no direct influence on the speed of the ball. Its only effect is on the curvature of the path. Thus, so long as we are dealing only with paths confined to one vertical plane, the axis of rotation must be perpendicular to that plane, and the effect of the transverse force is merely, as it were, an *unbending* of the path which would have been pursued had there been no rotation. From this (very inadequate) point of view we see at once why, other things being the same, even a moderate underspin greatly lengthens the carry, especially in the case of a low trajectory. But such analogies give us no hint as to the actual amount of the lengthening in any particular case. They lead us, however, to suspect that too great a spin may, in its turn, tend to shorten the carry; and that, if of sufficiently great amount, it might even bend the path over backwards and thus lead to the formation of a kink. Nothing but direct calculation, however, can give us definite information on these questions. And, unfortunately, we must trust implicitly in the accuracy of the computer, for we have no independent means of checking, from stage to stage of his work, the results of his calculations. In dealing with the case of an unresisted projectile, such numerical work can be checked at any stage by a simple and exact geometrical process. Even the more complex conditions of a path in which the resistance is as the square of the speed admit of exact analytical expression; but when the transverse force due to rotation is taken account of, the equations do not admit of integration in finite terms, so that the computer has to work out the approximate details of the path by successive little stages, say 6 feet at a time (or somewhere about 90 in all), and any errors of approximation he may make *at any stage* will not only themselves be faithfully represented in the final results, but will necessarily introduce other errors by furnishing incorrect data for each succeeding stage of the calculation. As all of his work was carried out to *four* places of figures at least, such errors are unlikely to have any serious consequence. I have endeavoured to obtain a rough estimate of the probable amount of error thus inevitably introduced by testing my computer (as well as the formula which I gave him) upon examples in which I had the means of independently calculating the exact result at each stage. His work bore this severe test very well. Unfortunately for my present application, it was based through-

out upon too high an estimate of the resistance, and the initial data also were chosen with reference to this. Thus, when reduced to 108 feet as the terminal velocity, the paths have been virtually worked out for intervals of about 9 feet instead of 6, and the initial data appear of a rather haphazard character.

We see from what precedes that the full investigation of the path of a golf ball, even when it is restricted to a vertical plane, would require voluminous tables of at least *triple* entry; for the form and dimensions of the path are now seen to depend quite as essentially upon the amount of spin as upon the initial speed and elevation. There are now no longer two, but innumerable, paths, which involve a definite carry even when they are confined to a vertical plane, and the initial speed is given. Of course no tables for their computation are in existence, and it would prove a somewhat laborious, and therefore costly, work to prepare such tables, even for a few judiciously selected values of elevation, speed, and initial diminution of weight. In their absence it is impossible to make any statements more definite than such very general ones as those above. But (as soon as trustworthy determinations of the resistance of air and of the rate at which the spin of a golf ball falls off are obtained) a couple of good computers, working in duplicate for a month or two, would supply sufficient material for at least a rough approximation to any path affected under the ordinary limits of the initial data of fairly good driving. Let us hope that some wealthy club, or some enthusiastic patron of the game, may be induced to further such an undertaking, at least so far as to enable us to give a fairly approximate answer to such a question as, "Other things being the same, what values of elevation and of initial diminution of weight will together secure the maximum carry?" When we are in a position to give an answer, the clearing up of the whole subject may be regarded as at least fairly commenced.

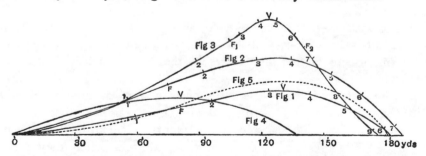

Meanwhile, as a specimen of what may be done in this direction, I give in the annexed plate an approximate sketch of the path of a golf ball under the following initial conditions, the spin being regarded as unaltered during the flight:

Initial speed ...	...	...	...	240 feet-seconds
Initial relief of weight		...	...	2 fold
Elevation	...	...	...	5·2°

This is fig. 1 on the plate, and it will be at once recognised as having at least considerable resemblance to that class of really good, raking drives in which the ball's path is concave upwards for more than a third of the range. Its one obvious defect, the too great obliquity of the descent, is due to the fact that, not knowing the law according to which the spin falls off, I have assumed it to continue unchanged throughout the path. The dotted curve, fig. 5, which gives a very close approximation to the observed path, was obtained by rough calculations (little more than estimates) from the same initial speed as fig. 1, but with no elevation to start with. The spin is initially about 50 per cent. greater than in fig. 1, but it has been assumed to fall off in geometric progression with the lapse of time. From the mode in which this curve was obtained, I cannot insert on it, as I have done on figs. 1, 2 and 3, the points reached by the ball in each second of its flight; but they will probably coincide pretty closely with those on fig. 1. In the last-mentioned figure, F is the point of contrary flexure and V the vertex. We have, farther,

Range	186 yards
Time	6·2
Greatest height	60 feet
Position of vertex	0·71 of range

In fig. 2 the initial speed and rotation are the same as in fig. 1, but the elevation is increased to 12°. It will be seen that little additional carry is gained in consequence. [Had there been no spin, the increase of elevation from 5° to 12° would have made a *very* large increase in the range.] In fig. 3 the elevation is 9·6° only, but the initial diminution of weight is treble of the weight. In this figure we see well shown the effect of supposing the spin to be constant throughout, for it has *two* points of contrary flexure, F_1 and F_2, and only *between* these is it concave downwards.

For contrast with these I have inserted, as fig. 4, a path with the same initial speed, but without spin. Though it has the advantage of 15° of elevation, it is obviously *far* inferior to any of the others in the transcendently important matter of range.

By comparing figs. 2 and 3, we see the effect of further increase of initial spin, especially in the *two* points of contrary flexure in 3. Still further increasing the spin, these points of flexure close in upon the vertex of the path, and, when they meet it, the vertex becomes a cusp as in the second of the cuts shown. The tangent at the cusp is vertical, and the ball has no speed at that point. This is a specially interesting case, *the path of a gravitating projectile nowhere concave downwards.* With still further spin, the path has a kink, as in the first of these figures.

I have not yet been able to realise the kink (though I have reached the cusp stage) with an ordinary golf ball. It would not be very difficult if we could get an exceedingly *hard* ball, made hollow if necessary, and if we were to tee on a steepish slope, and use a well-baffed cleek with a roughened face. I have, however, obtained good kinks with other projectiles; the first was one of the little French humming tops, made of very thin metal, the most recent being in the majestic flight of a large balloon of very thin indiarubber. This is a very striking experiment; eminently safe, and thoroughly demonstrative.

As to the genesis of exceptionally long carries (in the absence of wind), it will be seen from what precedes that I am not in a position to pronounce any very definite opinion. How much may be due to an accidentally happy combination of elevation and spin, how much to extravagant initial speed, can only be decided after long and laborious calculation. That extra speed has a great deal to do with the matter (*always provided there be spin enough*) is obvious from the numerical data given above. Whatever the initial speed, it is cut down by resistance to half in 83 yards, so that to increase the carry by 83 yards requires doubled speed. More particularly to lengthen the carry by thirty yards in fig. 1, the ball must have an initial speed of 300 instead of 240. But such exceptional drives never occur in really careful play:

> Vis consili expers mole ruit sua,
> Vim temperatam di quoque provehunt
> In majus, etc.

would almost seem to have been written for golf. The *vis temperata* is the only passport to a medal or a championship. Its congener, but also its opposite, usually comes into play when two good drivers, playing for amusement and ready for a "lark," find two other swipers ahead of them. Then the temptation is almost irresistible to that "harmless pastime, sport fraternal," which consists in "tickling up" the party in front as soon as they have "played their second." The law which permits this furnishes in itself the strongest possible incentive to outrageously long driving; and thus, in one sense at least, tends to lower the standard of the game.

However this may be, a long drive is not essentially a long carry. In fact, with luck and a hard, keen green, the veriest topper or skittles may *occasionally* pass the best driver, provided he hits hard enough. But it is not golf, as rightly understood, recklessly to defy hazards on the mere chance of being lucky enough to escape them.

5. On the effects of wind little can be learned from calculation until we have full data. For it is almost invariably the case that the speed of the wind varies within very wide limits with the height above the ground. Even when the players themselves feel none, there may be a powerful current sixty or a hundred feet above them.

We will, therefore, simply in order to combat some current prejudices, treat only of the case in which the wind is in the plane of the drive (i.e. a head wind

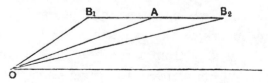

or a following wind) and is of the same speed at all levels within the usual rise of the ball. Then the matter is easy enough. For the air, so far as we are concerned with it, is then *moving as a whole*, and *in it* the path of a golf ball depends only upon the *relative* speed and elevation with which it was started. Find, then, with these data, the path of the ball relatively to the air, and then compound with the results the actual motion of the air, and we have the path of the ball as it appears to a spectator. If, then, the ball be struck from O with velocity represented by OA, and the *reversed* velocity of the air be represented by AB, the velocity of the ball *relatively to the air* is given by OB_1 or OB_2 according as the wind is with the ball or against it. Trace the successive positions of the ball in the moving air for each of these, say at intervals of a second, and then displace these horizontally, forward or backward, to the amount by which the air itself has advanced during the time elapsed. The result is of course merely to compress or to lengthen each portion of the path in proportion to the time which the ball took in traversing it. *There is no effect on the height of any part of the path, nor on the time of passing through it.* It is clear that the path, whose initial circumstances are shown by OB_1 in the figure, will rise higher than that corresponding to OB_2. Hence a ball which has no spin rises higher when driven with a following wind than against an equally strong head wind. This is in the teeth of the general belief, which is probably based on the fact that the vertex of the path against a head wind is brought closer to the spectator at the tee, and therefore its *angular* elevation is increased. When the ball has spin, the conditions of this question become very complex and no general statement can be made; though a calculation can, of course, be carried out for the data of each particular case.

I conclude, as I began, with the much-needed warning:

> False views abound, the "cracks" are all mistaken;
> In figures, only, rests our trust unshaken.

It seems appropriate to note, with regard to Tait's article on long driving, how completely his theory has stood the test of later investigations. In an interesting lecture on the dynamics of a golf ball delivered at the Royal Institution on March 18, 1910, Sir J. J. Thomson[1] went over much of the ground covered by Tait in the preceding article and in his papers on the path of a rotating spherical projectile, reproducing also by way of illustration some of the experiments due to Robins and to Magnus. The most novel feature of Sir J. J. Thomson's lecture was the practical realisation of the possible paths calculated by Tait and shown on page 341. The cusp and kink figured on page 342 were also demonstrated by Sir J. J. Thomson by means of the same ingenious experiment. A stream of negatively charged particles from a red-hot piece of platinum with a spot of barium oxide upon it was caused to travel along the vacuum tube in which the platinum strip was contained. This stream of particles was

[1] See *Nature* of December 22, 1910, Vol. LXXXV, pp. 251–257.

luminous and visible. The stream passed between two plates which could be brought at will to different electrical potentials. When the electric field was established with the lines of force passing upwards, each negatively charged particle was subject to a vertical force analogous to the force of gravity in the case of the golf ball. The luminous stream showed the path described by each particle. This corresponded to the golf ball path when there was no underspin, as in fig. 4 on p. 341. A force analogous to the upward force which acts on the properly spinning golf ball was then applied to the moving electrified particles by introducing a magnetic field, with lines of magnetic force passing horizontally across the stream. The moving particles were driven at right angles to their own motion and to the line of magnetic force. By suitable adjustment of direction and strength of magnetic field, the luminous stream could be made to assume forms identical with those figured by Tait in the curves on page 341. By increasing the strength of the field Sir J. J. Thomson obtained not only the kink figured on page 342, which Tait had demonstrated with the light rubber balloon, but he also obtained a succession of loops or kinks in the luminous stream.

The following paragraphs contain additional notes on three aspects of Tait's life and work. The first note discusses some recent developments of thermoelectric theory, the second touches on the social side of Tait's character, and the third supplies further information regarding the production of one of Tait's most characteristic books.

THERMOELECTRICITY.

Mr J. D. Hamilton Dickson, M.A., of Peterhouse, Cambridge, has recently discussed with remarkable skill and ingenuity the data supplied by Sir James Dewar and Professor Fleming's measurements[1] of the thermo-electric properties of various metals from $-200°$ C. to $+100°$ C. His conclusions, which are of great interest and form an important extension of Tait's early results, were read before the Royal Society of Edinburgh on Nov. 7, 1910, and are being published in full in the *Transactions* of that Society.

As noted on page 79 above, one of Tait's main results was that through a considerable range of temperature the electromotive force between a given pair of metals was in general a parabolic function of the difference of

[1] *Phil. Mag.*, July 1895, Vol. XL, pp. 95-119.

temperature of the junctions. Was this law fulfilled through the range of low temperatures which were the feature of Dewar and Fleming's experiments? To settle this question Dickson first reduced the platinum thermometer temperature readings to absolute scale, and then plotted with great care on section paper the electromotive force of each metal lead couple against the temperature of the junction whose temperature was varied. The curves were parabolas with axis perpendicular to the temperature axis for gold, silver, zinc and the alloy German silver, which metals therefore follow Tait's law strictly.

It was obvious at a glance that in the other cases the electromotive-force curve was not a parabola with axis perpendicular to the temperature axis. By graphical construction of the loci of the points of bisection of sets of parallel chords, Dickson proved that these curves must be curves of the second degree; for the loci mentioned were very approximately straight lines. In the case of the antimony-lead curve the loci of different sets of parallel chords passed approximately through one point, indicating that the curve was a hyperbola. In every other case the loci of mid-points of different sets of parallel chords were parallel straight lines, indicating that the curve was a parabola with axis parallel to these loci but not necessarily perpendicular to the temperature axis. If following Dickson we designate by the "Tait-line" the graph which gives the relation between the temperature and the rate of change of the electromotive force per unit change of temperature, the results obtained by him may be thus summarised :—The lead line being laid down horizontal, the Tait-lines for gold, silver, zinc and German silver are straight, in accordance with Tait's theory; for platinum (two kinds, both of very great purity), copper, cadmium, nickel, magnesium, palladium, and aluminium, the Tait-lines are cubical hyperbolas; and for antimony the Tait-line is a quartic curve. The two last statements may be easily verified as properties of the parabola and hyperbola with axes inclined to the temperature axis.

In one of his early notes on thermoelectricity, embodied in his " First Approximation to the Thermoelectric Diagram" (see *Sci. Pap.*, Vol. I, p. 220), Tait remarked that

When the temperatures were very high, the parabola was slightly steeper on the hotter than on the colder side. This, however, was a deviation of very small amount, and quite within the limits of error introduced by the altered resistance of the circuit at the hotter parts, the deviations of the mercury thermometers from absolute temperature, and the non-correction of the indication of the thermometers for the long column of mercury not immersed in the hot oil round the junction.

As mentioned on p. 77 above, Tait's "First Approximation to the Thermoelectric Diagram" was based on the experiments made by C. E. Greig and myself according to a method which Tait's earlier experiments had shown to be convenient and sufficiently accurate. The thermoelectric curves when drawn were all very approximately parabolic; but generally when a well-marked vertex was obtained a slight lack of perfect symmetry was observable. This lack of symmetry Tait regarded as due to the mercurial thermometers not giving a scale of temperature accurately in harmony with the absolute scale; and, since the deviation from symmetry in any particular case was always very slight, the matter was passed over as of comparatively small moment in obtaining what Tait called the First Approximation. Dickson's results, however, show that this asymmetry may be an essential feature of the curve itself, and not due to any errors involved in the measurement either of the temperature or of the electromotive force, for he gives similar examples of inclined axis from Tait's work, as well as from the work of other experimenters.

THE EVENING CLUB.

As already noted (see above, pp. 33 and 49) Tait mingled very little in general society during the last twenty years of his life. To those who knew him only in these later years he appeared to be more or less of a recluse, and (except during his holiday months in St Andrews) could not in any sense be regarded as a Club-man. Yet he was splendid company when occasion offered; and it is well in this connection to recall that he was one of the founders of the Evening Club, which was organised in 1869 on the model of the "Cosmopolitan" and "Century" Clubs in London. The original prospectus was signed by fourteen well known citizens who formed themselves into a provisional committee. Their names in order of signature were: Dr John Muir, Professor P. G. Tait, Professor David Masson, James Drummond, R.S.A.[1], J. Matthews Duncan, M.D., Robert Wallace, D.D.[2], Robert Cox, W.S., James Donaldson, LL.D.[3], A. Findlater,

---

[1] Curator of the National Gallery.
[2] Afterwards M.P. for Edinburgh.
[3] At present Principal of St Andrews University.

LL.D.[1], D. Douglas, F.R.S.E., J. F. Maclennan[2], Alex. Nicolson[2], Aeneas J. G. Mackay[2], John Maitland[2]. These promoters of the Club, by personal appeals among their friends, quickly gathered together more than a hundred names of those willing to become members. Tait received the following characteristic response from Macquorn Rankine, whose propensity for forming Greek derived words was irrepressible:

59 St Vincent Street,
Glasgow,
29th October, 1869.

My dear Tait

I shall be very happy to join your Capnopneustic Club (as it may appropriately be termed). I beg pardon for not having answered you sooner; but I have been greatly engrossed by business, both engineering and academic.

Believe me,

very truly yours,

W. J. Macquorn Rankine.

In 1870 the membership was 140, and in 1874, 150, including many of the more prominent Edinburgh lawyers, artists, physicians, clergymen, teachers both in college and school, bankers, commercial men, publishers, engineers, etc. There was also a selection of members not resident in Edinburgh, such as Professor Lewis Campbell, St Andrews, Sir M. E. Grant Duff, Professor T. M. Lindsay, Glasgow, Sir William Stirling Maxwell, Professor Nichol, Glasgow, Professor G. G. Ramsay, Glasgow, Sir George Reid, R.S.A., then living in Aberdeen, Professor W. Robertson Smith, Aberdeen, Sir William Thomson, Glasgow, Principal Tulloch, St Andrews.

The Club met every Saturday and Tuesday evening and on Monday evenings immediately after the statutory fortnightly meetings of the Royal Society, for purely social intercourse, cards and serious subjects of debate being taboo. During the first twelve years of its existence Tait frequently attended the gatherings. The roll of guests introduced by

---

[1] Editor of *Chambers' Encyclopaedia.*

[2] Advocate. Maclennan was the author of *Primitive Marriage.* Mackay was afterwards Professor of History.

members has been preserved, and we find Tait personally responsible for introducing some seventy guests between 1870 and 1884. In August 1871, when the British Association met in Edinburgh, the Club held several meetings; and among the guests introduced by Tait were Cayley, Clerk Maxwell, Huxley, Bierens de Haan, Colding, Sylvester and Clifford.

After a vigorous existence for about twenty-five years, the Evening Club began to lose vitality. The members attended irregularly and fitfully, mainly on account of the growing lateness of the dinner hour; and after an effort to hold it together in a less formal manner the Club was disbanded in 1897.

It formed an important episode in the life of Tait, bringing him into close social touch with many men whom otherwise he would never have met. Those who attended the meetings during the seventies recall Tait as one of the great personalities, taking his full share in the talk, and enjoying the relaxation from the hard thinking in which he usually passed his evenings.

But the Club also had a direct bearing on scientific activity; for it was probably in the free and easy conversation of this Evening Club that there germinated in the minds of George Barclay, the first treasurer of the Club, and Thomas Stevenson, the well known engineer, the idea which finally came to fruition in Tait's Lectures on some Recent Advances in Physical Science.

RECENT ADVANCES IN PHYSICAL SCIENCE.

The recent death of Mr George Barclay at the advanced age of 91 has led to the discovery among his papers of further information regarding the course of lectures just referred to (see also above, p. 327). On February 14, 1874, Thomas Stevenson issued the circular to the subscribers announcing that the lectures would be delivered in St George's Hall, Randolph Place, on Tuesdays and Thursdays, at a quarter past four o'clock, during the months of February, March, and April, the first lecture to be given on Thursday, the 19th February. This hall, although convenient as regards situation, was found to be so unsuitable in other respects that after a vain endeavour to find a better place of meeting in the New Town, Thomas Stevenson and George Barclay, who acted respectively as secretary and treasurer, arranged for the delivery of the lectures in the Natural Philosophy Class Room of the University. This was announced in a circular issued by George Barclay on February 24th immediately after the second lecture; so that it appears that all save two of the lectures were given in the University.

The lectures were not published in book form till nearly two years after they were delivered. The reporter had lost his short-hand note of one of them; and Tait had to redeliver this lecture in his retiring room to an audience of one. The book on Recent Advances is the only book of Tait's which contains a dedication. It is in these words: "With this work I desire to associate the names of George Barclay and Thomas Stevenson, Fellows of the Royal Society of Edinburgh, by whose efforts these Lectures were organised, and at whose wish they are published as delivered. P. G. T."

BIBLIOGRAPHY

BEFORE his last illness Tait had begun to prepare material for the third volume of his Scientific Papers. A special feature was to have been a complete bibliography of his contributions to scientific journals, with short explanatory remarks after the titles of the shorter papers which were not included either in the first or second volume of the collected papers. In filling in these remarks on a prepared sheet Tait had got as far as the year 1870. I have endeavoured to continue the work thus begun by Tait himself, giving in approximately chronological order all his published papers and articles. Since by far the greater number of these appear in the Proceedings of the Royal Society of Edinburgh, I have used the letters *R.S.E.* to mean this publication. When a paper has appeared in the Transactions of that Society the reference is to *Trans. R.S.E.* Any other journals are indicated by sufficiently intelligible contractions. The two volumes of the republished Scientific Papers are referred to simply as *S.P.* The first Sixty (LX) appear in Volume I and the rest from LXI to CXXXIII in Volume II. When any reference is made to Tait's books, the characteristic name of the book is given in Italics.

1. Note on the density of Ozone. In conjunction with Thomas Andrews, 1857. *Proc. Roy. Soc.*, VII. Reprinted in Andrews' *Scientific Papers*.

2. Second Note on Ozone. In conjunction with Thomas Andrews, 1857–59. *Proc. Roy. Soc.*, IX. Reprinted in Andrews' *Scientific Papers*.

3. On the volumetric relations of Ozone and the action of the electric discharge on oxygen and other gases. In conjunction with Thomas Andrews, 1859. *Proc. Roy. Soc.*, X. *Phil. Trans.*, 1860. Reprinted in Andrews' *Scientific Papers*.

4. Quaternion investigations connected with Fresnel's wave-surface. 1859. *Quart. Journ. Math.* S.P., I. Tait's *Quaternions*, §§ 432–452.

5. Note on the Cartesian equation of the wave-surface. 1859. *Quart. Journ. Math.*, III. *S.P.*, II.

6. Quaternion investigations connected with electrodynamics and magnetism. 1860. *Quart. Journ. Math.*, III. *S.P.*, III. Tait's *Quaternions*, §§ 458–472.

7. Mathematical Notes: A spherical nebula consists of concentric shells of uniform density; find the law of the latter that to a spectator at a great distance the nebula may appear uniformly bright. The density in shell of radius r is inversely as $\sqrt{(a^2 - r^2)}$. 1860. *Quart. Journ. Math.*, IV.

8. Quaternion investigation of the potential of a closed circuit. 1861. *Quart. Journ. Math.*, IV. *S.P.*, IV. Tait's *Quaternions*, § 471.

9. Note on a modification of the apparatus employed for one of Ampère's fundamental experiments in electrodynamics. 1861. *R.S.E.*, IV. *S.P.*, V.

10. Note on molecular arrangement in crystals. 1862. *R.S.E.*, IV. Crystalline properties illustrated by use of piles of marbles of equal size. A horizontal layer may be arranged in square or triangular order, and successive layers may be simply superposed or inserted into interstices in the preceding ones. With interstitial arrangement the square and triangular order give same density.

11. Formulae connected with continuous displacements of the particles of a medium. 1862. *R.S.E.*, IV. *S.P.*, VI.

12. Note on electricity developed during evaporation and during effervescence from chemical action. In conjunction with J. A. Wanklyn. 1862. *R.S.E.*, IV. Large charges were produced by evaporation of a drop of bromine, etc., from a hot platinum dish.

13. On Determinants. 1862. *Mess. of Math.*, I.

14. Quaternions. (A series of three papers.) 1862. *Mess. of Math.*, I.

15. Energy. In conjunction with W. Thomson (Lord Kelvin). 1862. *Good Words*.

16. Note on central forces. 1862. *Mess. of Math.*, I.

17. Note on a quaternion transformation. 1862. *R.S.E.*, V. *S.P.*, VII. Tait's *Quaternions*, §§ 145, 146, 474.

18. Reply to Prof. Tyndall's remarks on a paper on "Energy" in *Good Words*. 1863. *Phil. Mag.*, XXV.

19. On the Conservation of Energy. 1863. *Phil. Mag.*, XXVI. (Two letters, June and August.)

20. Note on the Hodograph for Newton's law of force. 1863. *Mess. of Math.*, II.

21. Elementary physical applications of quaternions. 1863. *Quart. Journ. Math.*, VI. Tait's *Quaternions*, Chap. XII.

22. On the Conservation of Energy. 1863. Address before Royal Society of Edinburgh. Second interpretation of Newton's Third Law given for first time in modern form. *R.S.E.*, IV.

23. On the history of thermodynamics. 1864. *Phil. Mag.*, XXVIII.

24. The dynamical theory of heat. 1864. *North British Review*.

25. Energy. 1864. *North British Review*.

26. Note on Fermat's Theorem. 1864. *R.S.E.*, V.

27. Note on the history of Energy. 1864. *Phil. Mag.*, XXIX (1865).

28. Note on action. 1865. *R.S.E.*, V. Area about empty focus of planet's orbit represents action.

29. On the law of frequency of error. 1865. *Trans. R.S.E.*, XXIV. *S.P.*, VIII.

30. On the application of Hamilton's Characteristic Function to special cases of constraint. 1865. *R.S.E.*, V. *Trans. R.S.E.*, XXIV. *S.P.*, IX.

31. Note on the behaviour of iron filings strewn on a vibrating plate, and exposed to the action of a magnetic pole. 1865. *R.S.E.*, V. The filings are kept on the ventral segments by a pole placed above the plate, but instantly dispersed when it is below.

32. Preliminary note on the heating of a disk by rapid rotation *in vacuo*. In conjunction with Balfour Stewart. 1865. *Proc. Roy. Soc.*, XIV. *Phil. Mag.*, XXIX.

33. On the heating of a disk by rapid rotation *in vacuo*. In conjunction with Balfour Stewart. 1865. *Proc. Roy. Soc.*, XIV. *Phil. Mag.*, XXX.

34. Note on the compression of air in an air bubble under water. 1865. *R.S.E.*, V. A propos of solution of air in water, and of the vesicular vapour of Clausius. Because of surface tension, an air bubble of radius 0·00001 of an inch contains air at pressure of 11 atmospheres.

35. On some geometrical constructions connected with the elliptic motion of unresisted projectiles. 1866. *R.S.E.*, V. *Tait and Steele*, § 121.

36. On the heating of a disk by rapid rotation *in vacuo*. In conjunction with Balfour Stewart. 1866. *Proc. Roy. Soc.*, XV. *Phil. Mag.*, XXXIII.

37. On some capillary phenomena. 1866. *R.S.E.*, V. Study of process of dividing soap bubble into two, and of causing two to unite. Exhibition of motions in film by using its posterior surface as a concave mirror reflecting selected portions of sunlight.

38. On the Value of the Edinburgh Degree of M.A. An Address delivered to the Graduates in Arts, April 24, 1866. Published by Maclachlan and Stewart, Edinburgh.

39. Sir William Rowan Hamilton. 1866. Biographical article in *North British Review*. Chief source of article in *Encyclopaedia Britannica*, 1880. *S.P.*, cxxviii.

40. Note on formulae representing the fecundity and fertility of women. 1867. *Trans. R.S.E.*, xxiv. Fecundity is found to depend linearly and fertility parabolically on age. Reprinted with additions in Duncan's *Fecundity and Fertility* (Black, 1871). The formulae are known to Statisticians as Tait's Laws.

41. Note on determinants of the third order. 1866. *R.S.E.*, vi. Various quaternion transformations leading to theorems in determinants.

42. Translation of Helmholtz's paper on "Vortex Motion." *Phil. Mag.*, 1867.

43. Note on the reality of the roots of the symbolical cubic which expresses the properties of a self-conjugate linear and vector function. 1867. *R.S.E.*, vi. *S.P.*, x.

44. Note on a celebrated geometrical problem proposed by Fermat. 1867. *R.S.E.*, vi. *S.P.*, xi.

45. Note on the radiant spectrum. 1867. *R.S.E.*, vi.

46. Note on the hodograph. 1867. *R.S.E.*, vi. *S.P.*, xii.

47. Note on an inequality. 1868. *R.S.E.*, vi. The conditions for the coalescence of two soap bubbles lead to a mathematical inequality.

48. Mode of demonstrating equality of radiation and absorption. 1868. *R.S.E.*, vi. Ink letters on platinum strip.

49. Physical proof that the geometrical mean of any number of quantities is less than the arithmetic mean. 1868. *R.S.E.*, vi. *S.P.*, xiii.

50. On the Dissipation of Energy. 1868. *R.S.E.*, vi. *S.P.*, xiv.

51. Notice of Sir David Brewster. *The Scotsman*, 11 Feb. 1868.

52. On the rotation of a rigid body about a fixed point. 1868. Abstract in *Proc. R.S.E.*, vi. Full Paper in *Trans. R.S.E.*, xxv. *S.P.*, xv.

53. On the motion of a pendulum affected by the rotation of the Earth and other disturbing causes. 1869. *R.S.E.*, vi. Tait's *Quaternions*, §§ 427–430.

54. On the heating of a disk by rapid rotation *in vacuo*. In conjunction with Balfour Stewart. Reply to O. E. Meyer. 1869. *Phil. Mag.*, xxxvii. *Ann. d. Phys. u. Ch.*, 136, pp. 165–6.

55. Notice of Principal J. D. Forbes. *The Scotsman*, 11 Jan. 1869.

56. On comets. 1869. *R.S.E.*, vi. *British Assoc. Reports*, xxxix. The sea bird Analogy.

57. Note on electrolytic polarisation. 1869. *R.S.E.*, vi. *S.P.*, xvi. *Phil. Mag.*, xxxviii.

58. Provisional Report of a Committee consisting of Professor Tait, Professor Tyndall, and Dr Balfour Stewart, appointed for the purpose of repeating Principal J. D. Forbes's Experiments on the Thermal Conductivity of Iron and of extending them to other Metals. *Brit. Ass. Rep.*, xxxix.

59. The Progress of Natural Philosophy. *Nature*, 1869.

60. On the Steady Motion of an incompressible perfect fluid in two dimensions. 1870. *R.S.E.*, vii. *S.P.*, xvii. In a letter to Cayley (6 Oct. 1877) Tait refers to this paper as containing a "very curious hydrokinetic theorem which I remember greatly surprised not only Rankine and Stokes but even Thomson."

61. On the most general motion of an incompressible perfect fluid. 1870. *R.S.E.*, vii. *S.P.*, xviii.

62. Mathematical Notes. 1870. *R.S.E.*, vii. Every cube is difference of two squares, etc.

63. On Green's and other allied Theorems. 1870. *Trans. R.S.E.*, xxvi. *S.P.*, xix. Abstract in *Proc. R.S.E.*, vii.

T.

64. Note on linear partial differential equations. 1870. *R.S.E.*, VII. *S.P.*, XX.

65. Notes from the Physical Laboratory of the University. 1870. *R.S.E.*, VII. Account of J. P. Nichol's experiments on radiation and convection of heat from blackened and bright surfaces at various pressures of surrounding gas. At 10 mm. pressure the effect mainly due to radiation. The effect of lowering the pressure in lessening the rate of cooling was more marked in the case of the bright than of the blackened surface.
A. Brebner continued experiments of No. 57 above.
P. W. Meik and John Murray experimented on change of electric resistance of two kinds of copper with increase of load. The change was proportional to the load and was different in the two cases.

66. Address forming part of the Report of the Proceedings at the 32nd Annual General Meeting of the Scottish Provident Institution on 30th March 1870 (Professor Tait, Convener).

67. Laboratory Notes. On thermoelectricity. 1870. *R.S.E.*, VII. Incorporated in *Transactions* paper, No. 108 below.

68. Note on linear differential equations in Quaternions. 1870. *R.S.E.*, VII. *S.P.*, XXI.

69. Energy and Professor Bain's "Logic." 1870. *Nature*, III.

70. On some Quaternion Integrals. 1870. *R.S.E.*, VII. *S.P.*, XXII.

71. Laboratory Notes. 1871. (1) On thermoelectricity. (2) On Phyllotaxis. Develops formula for the fundamental divergence of the spiral. *R.S.E.*, VII.

72. Review of Balfour Stewart's "Elementary Physics." 1870. *Nature*, III.

73. Laboratory Notes. 1871. (1) On anomalous spectra, and on a simple Direct Vision spectroscope. Suggests a combination of two liquids. (2) On a method of illustrating to a large audience the composition of Simple Harmonic Motions under various conditions. Instead of Lissajoux's apparatus suggests mirrors rotating about axes not quite perpendicular to the planes of the mirrors. (3) On a simple mode of explaining the optical effects of mirrors and lenses. Use of term divergence instead of reciprocal of distance. *R.S.E.*, VII.

74. Review of Tyndall's "Use and Limits of the Imagination in Science." 1871. *Nature*, III.

75. Review of Highton's "On the Relations between Chemical Change Heat and Fire." 1871. *Nature*, III.

76. Sensation and Science. 1871. *Nature*, IV.

77. Mathematical Notes. (1) On a quaternion integration. Given in Tait's *Quaternions*, § 431. (2) On the Ovals of Descartes. Results partly got by quaternions—according to Cayley mostly to be found in Chasles' *Aperçu Historique*. 1871. *R.S.E.*, VII.

78. On Spectrum Analysis. 1871. Abstract of an Address, important because of its historic statements. 1871. *R.S.E.*, VII.

79. Mathematical Notes. 1871. (1) On a property of self-conjugate linear and vector functions. Tait's *Quaternions*, Chap. V, Question 18 at end. (2) Relation between corresponding ordinates of two parabolas. Suggested by Thermoelectric Diagram. (3) On some quaternion transformations. Tait's *Quaternions*, §§ 514–517. *R.S.E.*, VII.

80. On an expression for the potential of a surface distribution and on the operator $T\nabla$. Tait's *Quaternions*, §§ 516, 517. 1871. *R.S.E.*, VII.

81. Reply to Dr Ingleby. 1871. *Nature*, V.

82. Note on Spherical Harmonics. Suggested by a Quaternion mode of attack—referred to by Maxwell as "ravishing." 1871. *R.S.E.*, VII.

83. Review of "Physikalisches Repetitorium." 1871. *Nature*, V.

84. Address to the Mathematical and Physical Section of the British Association at Edinburgh. 1871. *Brit. Ass. Reports*, XLI. *S.P.*, XXIII.

85. On thermoelectricity. 1871. *Brit. Ass. Reports*, XLI.

86. Reply to Professor Clausius. 1872. *Phil. Mag.*, XLIII.

87. On the history of the Second Law of thermodynamics in reply to Professor Clausius. 1872. *Phil. Mag.*, XLIV.

88. Reply to Professor Clausius. 1872. *Phil. Mag.*, XLIV. German translation in *Annalen d. Physik u. Chemie*, CXLV.

89. Laboratory Notes. On thermoelectricity. 1872. *R.S.E.*, VII. Incorporated in *Transactions* paper, No. 108 below.

90. Note on a singular property of the retina. 1872. *R.S.E.*, VII. *S.P.*, XXIV.

91. On the operator $\phi(\nabla)$. 1872. *R.S.E.*, VII. Tait's *Quaternions*, § 508.

92. Note on pendulum motion. 1872. *R.S.E.*, VII. Tait's *Dynamics*, § 143.

93. Note on the Strain Function. 1872. *R.S.E.*, VII. *S.P.*, XXVI.

94. On orthogonal isothermal surfaces. 1866–1872. *Trans. R.S.E.*, XXVII. *S.P.*, XXV.

95. Laboratory Notes. 1872. *R.S.E.*, VII. (1) On thermoelectricity; circuits with more than one neutral point. Included in *Transactions* paper, No. 108 below. (2) On a method of exhibiting the sympathy of pendulums. Magnets suspended horizontally end on to one another in the same straight line. "Interesting illustration of the linear propagation of disturbances in a medium consisting of discrete massive particles when only contiguous ones act on each other... It might be curious to enquire whether a certain form of the equation of motion might not give some hints as to the formation of a dynamical hypothesis of the action of transparent solids on the luminiferous ether."

96. On some quaternion integrals, Part II. 1872. *R.S.E.*, VII. *S.P.*, XXII.

97. Scintillation (signed G. H.). 1872. *Nature*, VI, p. 181.

98. On a question of arrangement and probabilities. 1872. *R.S.E.*, VIII. *S.P.*, XXVII.

99. Artificial Selection. 1872. *Macmillan's Magazine*, XXV.

100. On the stiffness of wires. 1872. *R.S.E.*, VIII.

101. Comets' tails (signed G. H.). 1872. *Nature*, VII, p. 105.

102. Sensation and Science. 1872. *Nature*, VI.

103. Laboratory Notes. 1872. (1) On the relation between thermal and electric conductivity. (2) On electric conductivity at a Red Heat. Contains probably first mention of remarkable fall off of conductivity (or increase of resistance) of iron at a red heat as compared with platinum. (3) On the thermoelectric relations of pure iron; contained in later *Transactions* paper, No. 108 below.

104. Notice of Professor Macquorn Rankine. *The Glasgow Herald*, Dec. 28, 1872.

105. Note on Ångström's Method for the conductivity of bars. 1873. Gives the harmonic analysis necessary with numerical details for certain bars deduced from experimental data obtained by C. E. Greig and A. L. MacLeish. *R.S.E.*, VIII.

106. Additional Note on the Strain Function. 1873. *R.S.E.*, VIII. *S.P.*, XXVI.

107. On the thermoelectric properties of pure nickel. 1873. First time the nickel line was obtained—showed same bend as iron but at lower temperature. *R.S.E.*, VIII.

108. First approximation to a Thermoelectric Diagram. 1873. *Trans. R.S.E.*, XXVII. *S.P.*, XXIX.

109. Thermoelectricity. The Rede Lecture delivered at Cambridge in 1873. Abstract in *Nature*, Vol. VIII. *S.P.*, XXVIII.

110. Laboratory Note. On the flow of water through fine tubes. 1873. Experiments by

C. G. Knott and C. M. Smith with tubes of elliptic and circular section. Elliptic tube of axes 4 to 1 let through about two-thirds of the amount which passes through circular bored tube of same section. *R.S.E.*, VIII (see above, p. 114).

111. Review of Maxwell's "Electricity and Magnetism." 1873. *Nature*, VII.

112. Review of De Morgan's "Budget of Paradoxes." 1873. *Nature*, VII.

113. Tyndall and Forbes. 1873. *Nature*, VIII.

114. Note on the transformation of double and triple integrals. 1873. *R.S.E.*, VIII. *S.P.*, XXX.

115. Forbes and Tyndall. 1873. An Article intended originally for the *Contemporary Review*, in reply to Tyndall's earlier Article. It appeared in Rendu's *Glaciers of Savoy*, translated by A. Wills, Q.C., and edited by George Forbes, B.A. Macmillan and Co., London. 1874.

116. On the heating of a disk by rapid rotation *in vacuo*. In conjunction with Balfour Stewart. 1873. *Proc. Roy. Soc.*, XXII.

117. Note on the various possible expressions for the force exerted by an element of one linear conductor on an element of another. 1873. *R.S.E.*, VIII. *S.P.*, XXXI.

118. Note on Mr Sang's communication of 7th April 1873, on a singular property possessed by the fluid in crystal cavities in Iceland Spar. 1873. *R.S.E.*, VIII. "Carbonic acid, partly liquid, partly gaseous, fills the cavity. Distillation, when one end is heated ever so slightly above the other, the circumstances being of almost unexampled favourability for such an effect. Hence the apparent motion of the bubble. *It is not the same bubble as it moves.*"

119. Preliminary note on a new method of obtaining very perfect vacua. In conjunction with Sir James Dewar. 1874. *R.S.E.*, VIII.

120. Todhunter's Experimental Illustrations. 1874. *Nature*, IX.

121. Laboratory Notes. (1) On Atmospheric electricity: observed alternations of positive and negative electrifications during hail storm. (2) On the thermoelectric position of sodium. 1874. *R.S.E.*, VIII.

122. Herbert Spencer v. Thomson and Tait. 1874. *Nature*, IX.

123. "Cram." 1874. *Nature*, IX.

124. Double Rainbow. 1874. *Nature*, X.

125. Mathematical Notes. (1) On a singular theorem given by Abel. *S.P.*, XXXII. (2) On the equipotential surfaces for a straight line: simplified proof of well-known theorem. (3) On a fundamental principle in Statics. 1874. *R.S.E.*, VIII. *S.P.*, XXXIII.

126. On the thermoelectric positions of sodium and potassium. 1874. *R.S.E.*, VIII.

127. Review of Sedley Taylor's "Sound and Music." 1874. *Nature*, X.

128. Laboratory Notes. (1) Photographic records of sparks from the Holtz machine. (2) Determination of surface tension of liquids by ripples produced by tuning fork. (3) Capillary phenomena at the surface of separation of two liquids. 1875. *R.S.E.*, VIII.

129. Further Researches in very perfect vacua. In conjunction with (Sir) James Dewar. 1875. *R.S.E.*, VIII. Report in *Nature*, XII, p. 217, under title "On Charcoal Vacua." Contained complete explanation of radiometer (see above, p. 81).

130. Photographs of electric sparks in hot and cold air. 1875. *Trans. R.S.E.*, XXVII. Zigzagness removed when air was heated. The photographs were taken by A. Matheson.

131. Cosmical Astronomy. A series of Lectures published in *Good Words* of 1875.

132. Laboratory Notes. (*a*) On the application of Sir W. Thomson's dead-beat arrangement to chemical balance. (*b*) Photographs of electric sparks taken in cold and in heated

air. (*c*) On the electric resistance of iron at high temperatures. 1875. *R.S.E.*, VIII. (*a*) in *S.P.*, XXXIV.

133. Laboratory Notes. (*a*) On the origin of atmospheric electricity. (*b*) Preliminary experiments on the thermal conductivity of some dielectrics (by C. M. Smith and C. G. Knott).

134. On the linear differential equation of the second order. 1876. *R.S.E.*, IX. *S.P.*, XXXV.

135. Notice of Professor G. G. Stokes. 1876. *Nature*, XII.

136. Review of Miss Buckley's "Natural Science." 1876. *Nature*, XIII.

137. Laboratory Notes. 1876. (*a*) On a possible influence of magnetism on the absorption of light and some correlated subjects. *S.P.*, XXXVI. (*b*) On a mechanism for integrating the general linear differential equation of the second order. A combination of two equal modifications of Amsler's planimeter. (*c*) The electric conductivity of nickel. Experiments by C. Michie Smith and J. Gordon MacGregor. *R.S.E.*, IX.

138. Note on the origin of thunderstorms. 1876. Discusses production of vortex columns coming down from above. *R.S.E.*, IX.

139. Translation of F. Mohr's "Views of the Nature of Heat." *Phil. Mag.* (The paper appeared originally in Liebig's *Annalen*, Vol. XXIV, 1837.)

140. Force. Evening Lecture at British Association (Glasgow Meeting). 1876. *Nature*, XIV. *S.P.*, XXXVII. Printed also as Appendix to Second Edition of *Recent Advances*.

141. Some elementary properties of plane closed curves. *Brit. Ass. Reports*, 1876. *Messenger of Math.*, VI. *S.P.*, XXXVIII.

142. Applications of the theorem that two closed plane curves intersect an even number of times. 1876. *R.S.E.*, IX. This marks the beginning of what became a series of papers on Knots.

143. Definition and Accuracy. 1876. *Nature*, XV.

144. Notice on some recent atmospheric phenomena. 1876. *R.S.E.*, IX. Description of an aurora of unusual form.

145. Article "Heat" in Handbook to the Loan Collection of Scientific Apparatus (at South Kensington). 1876. Basis of the book on *Heat* (1884).

146. Note on the measure of beknottedness. 1877. *R.S.E.*, IX.

147. Notice of James Lindsay. *The Scotsman*, Jan. 5, 1877.

148. Note on the effect of heat on infusible impalpable powders. 1877. *R.S.E.*, IX. *S.P.*, XLII.

149. On knots. With remarks by Listing. 1877. *R.S.E.*, IX.

150. On links. 1877. *R.S.E.*, IX.

151. On the relative percentages of the atmosphere and of the ocean which would flow into a given rent on the Earth's surface. 1877. *R.S.E.*, IX. Answer to question, are we drying up? Problem suggested as to equilibrium arrangement of water poured into a shaft already full of air so deep as to have its density exceeding that of water.

152. Sevenfold knottiness. 1877. *R.S.E.*, IX.

153. On amphicheiral forms and their relations. 1877. *R.S.E.*, IX.

154. Preliminary Note on a new method of investigating the properties of knots. 1877. *R.S.E.*, IX.

155. On knots. 1877. *Trans. R.S.E.*, XXVIII. *S.P.*, XXXIX.

156. Note on an identity. 1877. *R.S.E.*, IX. *S.P.*, XLIII.

157. Review of Maxwell's "Matter and Motion." 1877. *Nature*, XVI.

158. Review of Wormell's "Thermodynamics." 1877. *Nature*, XVII.

159. Review of Zöllner's "Scientific Papers." 1877. *Nature*, XVII.

160. Note on vector conditions of integrability. 1877. *R.S.E.*, IX. *S.P.*, XLIV.

161. Review of Scottish Universities Commission Report. 1877. *Nature*, XVII.

162. Note on a geometrical theorem as to the properties of three concentric circles, which have the same common difference of radii, and which intersect one another. 1878. *R.S.E.*, IX. *S.P.*, XLV.

163. Exhibited a double mouthpiece by means of which it is easy for two players to produce chords from a French horn. 1878. *R.S.E.*, IX.

164. Note on the surface of a body in terms of a volume integral. 1878. *R.S.E.*, IX. *S.P.*, XLVI.

165. On the teaching of Natural Philosophy. *The Contemporary Review*, 1878.

166. Explanation of white rainbow observed by Sir Robert Christison. 1878. *R.S.E.*, IX. Due to the greatly increased effective surface from which the light came—in this case bright clouds near the sun.

167. Review of Clifford's "Dynamic." 1878. *Nature*, XVIII.

168. On the strength of currents required to work a telephone. 1878. *R.S.E.*, IX. *S.P.*, XLVII.

169. Review of Freeman's "Fourier's Theory of Heat." 1878. *Nature*, XVIII.

170. Thermal and electric conductivity. 1878. *R.S.E.*, IX. *Trans. R.S.E.*, XXVIII. *S.P.*, XLVIII.

171. Review of Proctor's "Pleasant Ways in Science." 1878. *Nature*, XIX.

172. On some definite integrals. 1878. *R.S.E.*, IX. Suggested by electric image problems leading to the solution of a definite integral.

173. Note on electrolytic conduction. 1878. *R.S.E.*, IX. *S.P.*, XLIX.

174. Letter from Hermann Stoffkraft. 1878. *Nature*, XIX.

175. Review of Bain's "Education." 1878. *Nature*, XIX.

176. On certain effects of periodic variation of intensity of a musical note. In conjunction with A. Crum Brown. 1878. An organ note was made discontinuous by being sounded through a partition and a revolving disk cut into separate sectors. The different notes indicated by theory were picked out by means of resonators. *R.S.E.*, IX.

177. Note on a mode of producing sounds of very great intensity. 1878. *R.S.E.*, IX. *S.P.*, L.

178. Does Humanity require a new Revelation? *The International Review*, 1878.

179. Quaternion Proof of Minding's Theorem. 1879. *London Math. Soc. Proc.*, X.

180. On the Dissipation of Energy. (Letter to Sir W. Thomson.) *Phil. Mag.*, VII, 1879.

181. On the Measurement of Beknottedness. 1879. *R.S.E.*, X.

182. Review of Beckett's "On the Origin of the Laws of Nature." 1879. *Nature*, XX.

183. Laboratory Notes. 1879. (1) Measurements of the electromotive force of the Gramme Machine at different speeds. (2) Elasticity of india-rubber at different temperatures. *R.S.E.*, X.

184. Quaternion investigations connected with Minding's Theorem. 1879. *R.S.E.*, X.

185. On methods in definite integrals. 1879. *R.S.E.*, X.

186. Application of certain formulae to sum infinite series. 1879. *R.S.E.*, X.

187. Obituary notice of Professor Kelland. In conjunction with Professor Chrystal. 1879. *R.S.E.*, X.

188. Obituary notice of Clerk Maxwell. 1879. *R.S.E.*, X.

189. Appreciation of T. Andrews (Scientific Part). 1879. *Nature*, XX.

190. On comets. 1879. Discussion of the forms of paths described by particles ejected from the comet. *R.S.E.*, X.

191. Maxwell's Scientific Work. 1880. *Nature*, XXI.

192. Mathematical Notes. 1880. (1) On a problem in arrangements. (2) On a graphical solution of the equation $V\rho\phi\rho = 0$. *R.S.E.*, X. *S.P.*, LII.

193. Note on the velocity of gaseous particles at the negative pole of a vacuum tube. 1880. *R.S.E.*, X. The idea was to measure the velocity of the particles by viewing their spectra in directions perpendicular to and parallel to the lines of motion of the incandescent particles. Method failed through lack of light.

194. Scientific jokes (signed G. H.). 1880. *Nature*, XXI, pp. 349, 368, 396.

195. On some applications of rotatory polarisation. 1880. *R.S.E.*, X. See below, No. 204.

196. On the colouring of maps. 1880. Various proofs that four colours suffice.

197. On the accurate measurement of high pressures. 1880. *R.S.E.*, X. Given in *Challenger Report*, see below, No. 213. *S.P.*, LX.

198. Note on the theory of the 15 Puzzle. 1880. *R.S.E.*, X. *S.P.*, LIII.

199. Further remarks on the colouring of maps. 1880. *R.S.E.*, X.

200. Note on a theorem in geometry of position. 1880. *Trans. R.S.E.*, XXIX. *S.P.*, LIV.

201. On Minding's Theorem. 1880. *Trans. R.S.E.*, XXIX. *S.P.*, LV.

202. The Tay Bridge (signed G. H.). *Nature*, XXII, p. 265.

203. Thunderstorms. A Lecture. 1880. *Nature*, XXII. Reprinted in this volume.

204. A rotatory polarisation spectroscope of great dispersion. 1880. *Nature*, XXII. *S.P.*, LVI.

205. On the Formula of Evolution. 1880. *Nature*, XXIII.

206. Sir W. R. Hamilton. 1880. Biography in *Encyclopaedia Britannica*. Vol. XI. *S.P.*, CXXVIII.

207. Note on the temperature changes due to compression. 1881. *R.S.E.*, XI.

208. On some effects of rotation in liquid jets. 1881. *R.S.E.*, XI. A jet of mercury from a funnel falling horizontally and nearly tangentially on a slightly inclined glass plate rolls up the plate. The effect was produced with water escaping from a rapidly escaping tube. *R.S.E.*, XI.

209. Note on thermal conductivity, and on the effects of temperature-change of specific heat and conductivity on the propagation of plane heat waves. 1881. *R.S.E.*, XI. *Phil. Mag.*, XII.

210. Note on a simple method of showing the diminution of surface tension in water by heat. 1881. A hot bar of iron was brought near the surface of a thin sheet of water covered with Lycopodium seed. *R.S.E.*, XI.

211. Note on a singular problem in Kinetics. 1881. *R.S.E.*, XI. *S.P.*, LVII.

212. On some space loci. 1881. *R.S.E.*, XI. Reprinted as Appendix to No. 195. *S.P.*, LV.

213. On the crushing of glass by pressure. 1881. Facts described and theory given. Referred to in *Challenger Report*. *R.S.E.*, XI. See *S.P.*, LX.

214. The pressure errors of the Challenger thermometers. 1881. *Nature*, XXV. *Challenger Reports*. *S.P.*, LX.

215. On mirage. 1881. *R.S.E.*, XI. *Trans. R.S.E.*, XXX. *S.P.*, LVIII.

216. Solar chemistry. 1881. A letter to *Nature* signed G. H. Vol. XXIV. *S.P.*, LIX.

217. Memoir of J. Macquorn Rankine. Prefixed to *Scientific Papers*. 1881.

218. Graduation Address. Edinburgh, April 20, 1881. Printed for private circulation.

219. Optical notes. 1882. (1) On a singular phenomenon produced by some old window panes. (2) On the nature of the vibrations in common light. *R.S.E.*, XI. *S.P.*, LXII.

220. Knots. 1882. Article (first part) in *Encyclopaedia Britannica*. Vol. XIV.

221. Light. 1882. Article in *Encyclopaedia Britannica*. Vol. XIV.

222. Note on the Preceding Paper (a paper on the lowering of the maximum density point of water by pressure, by D. H. Marshall, C. M. Smith, and R. T. Omond). *R.S.E.* XI. Reprinted in *Challenger Reports*, Vol. II, Part IV. 1888. *S.P.*, LXI.

223. Review of Stallo's "Concepts and Theories of Modern Physics." 1882. *Nature*, XXVI.

224. On a method of investigating experimentally the absorption of radiant Heat by gases. 1882. *Brit. Ass. Reports*. *Nature*, XXVI. *S.P.*, LXIII. (The experiments were made and described by J. G. MacGregor in *R.S.E.*, XII, the conclusion being that absorption of radiant heat by air containing 1·3 per cent. of water vapour is between that of air containing 0·06 per cent. and that of air containing 0·2 per cent. of olefiant gas.)

225. Proof of Hamilton's Principle of Varying Action. 1877. *Royal Irish Academy Proceedings* III, 1883.

226. State of the atmosphere which produces the forms of mirage observed by Vince and by Scoresby. *Nature*, XXVIII, 1883. A popular account of the mirage paper of 1881 Reprinted in this volume.

227. On the Laws of Motion. Part I. 1882. *R.S.E.*, XII. *Phil. Mag.*, XVI. *S.P.*, LXIV. (Translation in a German Mathematical journal under the title, *Ueber die Bewegungsgesetze*.)

228. Note on the compressibility of water. 1882. *R.S.E.*, XII. See *Second Challenger Report*. *S.P.*, LXI.

229. Notice of J. B. Listing. 1883. *Nature*, XXVIII. *S.P.*, LXV.

230. Note on the compressibility of water, sea-water, and alcohol, at high pressures. 1883. *R.S.E.*, XII. *Second Challenger Report*. *S.P.*, LXI.

231. Review of Jordan's "The New Principles of Natural Philosophy." 1883. *Nature*, XXVIII.

232. Biographical Note on J.-A.-F. Plateau. 1883. *Nature*, XXVIII.

233. Further note on the maximum density point of water. 1883. *R.S.E.*, XII. *Second Challenger Report*. *S.P.*, LXI.

234. Professor Stokes' Works. 1883. *Nature*, XXIX.

235. Mechanics. 1883. Article in *Encyclopaedia Britannica*. Vol. XV.

236. Listing's Topologie. *Phil. Mag.*, 1884. *S.P.*, LXVI.

237. On radiation. 1884. *R.S.E.*, XII. *S.P.*, LXVII.

238. On an equation in quaternion differences. 1884. *R.S.E.*, XII. *S.P.*, LXVIII.

239. On Vortex Motion. 1884. *R.S.E.*, XII. *S.P.*, LXIX.

240. On knots. Part II. 1884. *Trans. R.S.E.*, XXXII. *S.P.*, XL.

241. Note on reference frames. 1884. *R.S.E.*, XII. *S.P.*, LXX.

242. Further note on the compressibility of water. 1884. *R.S.E.*, XII. *Second Challenger Report*. *S.P.*, LXI.

243. On various suggestions as to the source of atmospheric electricity. 1884. *Journal of the Scottish Meteorological Society*, VIII. *Nature*, XXIX, March 27. *S.P.*, LXXI.

244. Review of Stokes' "Light." 1884. *Nature*, XXIX.

245. On an improved method of measuring compressibility. 1884. Described above, p. 85. *R.S.E.*, XIII. See *Second Challenger Report*. *S.P.*, LX.

246. Note on a Theorem of Clerk Maxwell. 1884. *R.S.E.*, XIII. Embodied in papers on Kinetic Theory of Gases. *S.P.*, LXXVII to LXXXI.

247. Theorem relating to the sum of selected Binomial Theorem Coefficients. 1884. *Messenger of Mathematics*, XIII.

248. On knots. Part III. 1885. *Trans. R.S.E.*, XXXII. *S.P.*, XLI.

249. Note on a singular passage in the *Principia*. 1885. *R.S.E.*, XIII. *S.P.*, LXXII. Reprinted also in Tait's *Properties of Matter* as an Appendix.

250. Note on the necessity for a condensation nucleus. 1885. Points out that the presence of Aitken's condensation nuclei greatly reduces the range of instability indicated by James Thomson's speculations as to the abrupt change from liquid to vapour or from vapour to liquid.

251. Review of Williamson's "Dynamics." 1885. *Nature*, XXXI.

252. On evaporation and condensation. 1885. Continues last note by considering possibility of preserving geometrical continuity and physical stability in the region considered by J. Thomson. This was done by making both the values of the pressure infinite along a vertical asymptot. Suggests experiment with mercury under tension in a vacuum. *R.S.E.*, XIII.

253. Hooke's anticipation of the Kinetic Theory and of synchronism. 1885. *R.S.E.*, XIII. *Nature*, XXXI, p. 546. *S.P.*, LXXVI. Quoted also in Tait's *Properties of Matter*, Chapter XIV.

254. Review of Clifford's "Exact Sciences." 1885. *Nature*, XXXII.

255. Note on a plane strain. *Proc. Edin. Math. Soc.*, 1885. *S.P.*, LXXIII.

256. Summation of certain series. *Proc. Edin. Math. Soc.*, 1885. *S.P.*, LXXIV.

257. Guesses at the Truth (signed G. H.). 1885. *Nature*, XXXII, p. 152.

258. On certain integrals. *Proc. Edin. Math. Soc.*, 1885. *S.P.*, LXXV.

259. Review of Stokes' "Light." 1885. *Nature*, XXXII.

260. On an application of the atmometer. 1885. Proposal to use the instrument in its peculiarly sensitive state to measure humidity of the air. Experiments were tried at Ben Nevis, but not with much success. *R.S.E.*, XIII. Reprinted as Appendix to *Second Challenger Report*. *S.P.*, LXI.

261. On the Foundations of the Kinetic Theory of Gases. 1885. The first of an important series. *R.S.E.*, XIII; subsequently incorporated in the *Transactions* paper, No. 264 below. *Phil. Mag.*, XXI.

262. On the Partition of Energy among Groups of colliding spheres. 1886. *R.S.E.*, XIII.

263. Note on the collisions of Elastic Spheres. 1886. *R.S.E.*, XIII.

264. On the foundations of the Kinetic Theory of Gases. 1886. *Trans. R.S.E.*, XXXIII. *S.P.*, LXXVII.

265. On the foundations of the Kinetic Theory of Gases; with a continuation containing numerical and other additions. 1886. *R.S.E.*, XIV. Both are short statements of what is subsequently proved in full in Part II, No. 274 below.

266. Quaternions. 1886. Article in *Encyc. Brit.* Vol. XX. *S.P.*, CXXIX.

267. Radiation and Convection. 1886. Article in *Encyc. Brit.* Vol. XX. *S.P.*, CXXX.

268. On even distribution of points in space. 1886. *R.S.E.*, XIV.

269. Review of Anderson's "On Heat and Work." 1887. *Nature*, XXXV.

270. On the general effects of molecular attraction of small range on the behaviour of a group of smooth impinging spheres. 1887. *R.S.E.*, XIV.

271. Weight and mass. 1887. *Nature*, XXXV.

272. On the effects of explosives. 1887. *R.S.E.*, XIV. *S.P.*, LXXXII.

273. Review of Stokes' "Light." 1887. *Nature*, XXXVI.

274. On the foundations of the Kinetic Theory of Gases. Part II. 1887. *Trans. R.S.E.*, XXXIII. *S.P.*, LXXVIII. *Phil. Mag.*, XXIII.

275. The assumptions required for Avogadro's Law. 1887. *Phil. Mag.*, XXIII.

276. On the value of $\Delta^n o^m/n^m$, when m and n are very large. *Proc. Edin. Math. Soc.*, v. 1887. *S.P.*, LXXXIII.

277. Notice of G. R. Kirchhoff. 1887. *Nature*, XXXVI. *S.P.*, CXXVII.

278. Note on Milner's Lamp. *Proc. Edin. Math. Soc.*, v. 1887. *S.P.*, LXXXIV.

279. The unwritten chapter on golf. *The Scotsman*, Aug. 31, 1887. *Nature*, XXXVI.

280. An exercise on logarithmic tables. *Proc. Edin. Math. Soc.*, v. 1887. *S.P.*, LXXXV.

281. Notice of Balfour Stewart. 1887. *Nature*, XXXVII.

282. On glories. 1887. *R.S.E.*, XIV. *S.P.*, LXXXVI.

283. On the compressibility of water and of different solutions of common salt. 1887. *R.S.E.*, XV.

284. On the thermoelectric properties of iron. 1888. Attention is called to Battelli's direct measurement of the Thomson Effect in various metals and particularly in iron. His results are in accordance with Tait's early conclusion that the thermoelectric line of iron is not straight. *R.S.E.*, XV.

285. Review of Fleeming Jenkin's "Papers, Literary and Scientific," etc. 1888. *Nature*, XXXVII.

286. On Mr Omond's observations of fog bows. 1888. *R.S.E.*, XV.

287. Reply to Professor Boltzmann. 1888. *R.S.E.*, XV. *Phil. Mag.*, XXV.

288. Preliminary note on the duration of impact. 1888. *R.S.E.*, XV. *S.P.*, LXXXVII.

289. On the mean free path and the average number of collisions per particle per second in a group of equal spheres. 1888. *R.S.E.*, XV. Included in *Transactions* paper, No. 293 below.

290. Quaternion Notes. 1888. *R.S.E.*, XV. *S.P.*, XC.

291. Engineers v. Professors and College Men. 1888. *Nature*, XXXIX.

292. On the Foundations of the Kinetic Theory of Gases. Part III. 1888. *Trans. R.S.E.*, XXXV. *S.P.*, LXXIX.

293. Address to the Graduates. 1888. (Printed for private circulation.)

294. Religion and Science. 1888. *The Scots Observer*.

295. On Laplace's Theory of the internal pressure in liquids. 1888. *R.S.E.*, XV. See Appendix to *Second Challenger Report*. *S.P.*, LXI.

296. Report on some of the physical properties of fresh-water and of sea-water. 1888. From the "Physics and Chemistry" of the voyage of H. M. S. Challenger. Vol. II. Part IV. *S.P.*, LXI.

297. Thermodynamics. 1888. Article in *Encyclopaedia Britannica*. Vol. XXI. *S.P.*, CXXXI.

298. Wave. 1888. Article in *Encyclopaedia Britannica*. Vol. XXIV.

299. On the virial equation for molecular forces, being Part IV of a paper on the foundations of the Kinetic Theory of Gases. 1889. *R.S.E.*, XVI. Included in later *Transactions* paper, No. 318 below.

300. Obituary notice of Balfour Stewart. *Proc. Roy. Soc. London*, 1889. *S.P.*, XCI.

301. The relation among four vectors. 1889. *R.S.E.*, XVI. *S.P.*, XCII.

302. Note on Mr McAulay's paper on differentiation of a quaternion. 1888. *R.S.E.*, XVI.

303. Additional remarks on the virial of molecular forces. 1887. *R.S.E.*, XVI.

304. On the relation among the line surface and volume integrals (quaternionic). 1889. *R.S.E.*, XVI. *S.P.*, XCIII.

305. Quaternion note on a geometrical problem. 1889. *R.S.E.*, XVI. *S.P.*, XCIV.

306. Note on Captain Weir's paper on a new azimuth diagram. 1889. *R.S.E.*, XVI. *S.P.*, XCV.

307. On the relations between systems of curves which, together, cut their plane into squares. 1889. *Proc. Edin. Math. Soc.*, 1889. *S.P.*, XCVI.

308. On the importance of quaternions in physics. *Phil. Mag.*, 1889. *S.P.*, XCVII.

309. Glissettes of an ellipse and of a hyperbola. 1889. *R.S.E.*, XVII. *S.P.*, XCVIII.

310. Graphic records of impacts. 1890. *R.S.E.*, XVII.

311. On impact. 1890. *Trans. R.S.E.*, XXXVI. *S.P.*, LXXXVIII.

312. Note on a curious operational theorem. *Proc. Edin. Math. Soc.*, 1890. *S.P.*, XCIX.

313. Note on ripples in a viscous liquid. 1890. *R.S.E.*, XVII. The complete mathematical theory is given in connection with C. M. Smith's experiments. *Proc. R.S.E.*, XVII. *S.P.*, C.

314. Review of Thurston's translation of Carnot's Essay, "Reflexions on the Motive Power of Heat." 1890. *Nature*, XLII, p. 365.

315. Some points in the physics of golf. 1890. *Nature*, XLII.

316. The pace of a golf-ball. *Golf*, Dec. 1890.

317. On the foundations of the Kinetic Theory of Gases. Part IV. 1891. *Trans. R.S.E.*, XXXVI. *S.P.*, LXXX.

318. Quaternions and the Ausdehnungslehre. 1891. *Nature*, XLIV. *S.P.*, CXXIX (Appendix).

319. Note on the isothermals of ethyl oxide. 1891. *R.S.E.*, XVII. *S.P.*, CI.

320. Some points in the physics of golf. II. 1891. *Nature*, XLIV.

321. Van der Waals' treatment of Laplace's pressure in the virial equation—two letters to Lord Rayleigh. 1891. *Nature*, XLIV.

322. Note appended to Dr Sang's paper on Nicol's polarizing eyepiece. 1891. Mainly a simplification of the mathematics as given by Sang in his early paper of 1837, which was not published by the Royal Society of Edinburgh till 1891. The ingenious suggestion made by Sang to construct a polarizer by placing a thin slice of Iceland Spar between two glass plates was made independently much later by Bertrand in 1884. *R.S.E.*, XVIII. *S.P.*, CII.

323. Hammering and driving. *Golf*, Feb. 1892.

324. On the virial equation for gases and vapours. 1892. *Nature*, XLV.

325. Review of Poincaré's "Thermodynamics," with two letters in reply to Poincaré. 1892. *Nature*, XLV.

326. On the foundations of the Kinetic Theory of Gases. Part V. 1892. *R.S.E.*, XIX. *S.P.*, LXXXI.

327. On impact. II. 1892. *Trans. R.S.E.*, XXXVII. *S.P.*, LXXXIX.

328. Note on Dr Muir's solution of Sylvester's elimination problem. 1892. *R.S.E.*, XIX.

329. Note on the thermal effect of pressure on water. 1892. *R.S.E.*, XIX. *S.P.*, CIV.

330. Note on the division of space into infinitesimal cubes. 1892. *R.S.E.*, XIX. *S.P.*, CV.

331. Note on attraction. *Proc. Edin. Math. Soc.*, 1893. *S.P.*, CVI.

332. Carry. *Golf*, Aug. 1893.

333. Carry and run. *Golf*, Sept. 1893.

334. Some points in the physics of golf. III. 1893. *Nature*, XLVIII.

335. On the compressibility of liquids in connection with their molecular pressure. 1893. *R.S.E.*, XX. *S.P.*, CVII.

336. Vector Analysis. 1893. *Nature*, XLIX.

337. Preliminary note on the compressibility of aqueous solutions in connection with molecular pressure. 1893. *R.S.E.*, XX. *S.P.*, CVIII.

338. Review of Watson's "Theory of Gases." 1893. *Nature*, XLIX.

339. On the path of a rotating spherical projectile. 1893. *Trans. R.S.E.*, XXXVII. *S.P.*, CXII.

340. Quaternions as an instrument of research. 1893. *Nature*, XLIX.

341. Obituary notice of Professor Robertson Smith. 1894. *Nature*, XLIX.

342. On the compressibility of fluids. 1894. *R.S.E.*, XX. *S.P.*, CIX.

343. Note on the antecedents of Clerk Maxwell's electrodynamical wave-equations. 1894. *R.S.E.*, XX. *S.P.*, CXIV.

344. On the application of Van der Waals' equation to the compression of ordinary liquids. 1894. *R.S.E.*, XX. *S.P.*, CX.

345. The initial pace of a golf ball. *Golf*, July 1894.

346. On the electromagnetic wave surface. 1894. *R.S.E.*, XXI. *S.P.*, CXV.

347. On the intrinsic nature of the quaternion method. 1894. *R.S.E.*, XX. *S.P.*, CXVI.

348. Systems of plane curves whose orthogonals form a similar system. 1895. *R.S.E.*, XX. *S.P.*, CXVII.

349. Note on the circles of curvature of a plane curve. *Proc. Edin. Math. Soc.*, 1895. *S.P.*, CXVIII.

350. On the path of a rotating spherical projectile. Part II. 1896. *R.S.E.*, XXI. Abstract of *Transactions* paper, No. 351 below.

351. On the path of a rotating spherical projectile. Part II. 1896. *Trans. R.S.E.*, XXXIX. *S.P.*, CXIII.

352. Note on centrobaric shells. 1896. *R.S.E.*, XXI. *S.P.*, CXIX.

353. Long driving. 1896. *Badminton Magazine*.

354. Note on Clerk Maxwell's law of distribution of velocity in a group of equal colliding spheres. 1896. *R.S.E.*, XXI. *S.P.*, CXXV.

355. On the linear and vector function. 1896. *R.S.E.*, XXI. *S.P.*, CXX.

356. On the linear and vector function. 1897. *R.S.E.*, XXI. *S.P.*, CXXI.

357. Note on the solution of equations in linear and vector functions. 1897. *R.S.E.*, XXI. *S.P.*, CXXII.

358. On the directions which are most altered by a homogeneous strain. 1897. *R.S.E.*, XXII. *S.P.*, CXXIII.

359. On the generalization of Josephus' Problem. 1898. *R.S.E.*, XXII. *S.P.*, CXXVI.

360. Note on the compressibility of solutions of sugar. 1898. *R.S.E.*, XXII. *S.P.*, CXI.

361. Queries on the reduction of Andrews' measurements on Carbonic Acid. Short letter appended to K. Tsuruta's letter in *Nature*, LIX, Feb. 2, 1899.

362. The experimental bases of Professor Andrews' paper on the continuity of the gaseous and liquid states of matter (*Phil. Trans.*, 1859). Title only in *R.S.E. Proc.* for Feb. 20, 1899; but the original MS prepared for this paper is included in " Andrews' measurements of the compression of carbon dioxide and of mixtures of carbon dioxide and nitrogen," edited by C. G. Knott, who, by request of the Council of the Roy. Soc. Edin., completed the work begun by Tait. *R.S.E.*, Vol. XXX, 1909–10.

363. On the linear and vector function. 1899. *R.S.E.*, XXII. *S.P.*, CXXIV.

364. On the claim recently made for Gauss to the invention (not the discovery) of quaternions. 1900. *R.S.E.*, XXIII.

365. Quaternion Notes. 1902. *R.S.E.*, XXIV. Published posthumously in *facsimile* from a sheet of foolscap on which Tait had jotted down some quaternion formulae two days before his death; with commentary by C. G. Knott.

LIST OF BOOKS WHOLLY OR PARTLY WRITTEN BY P. G. TAIT.

1. *A Treatise on Dynamics of a Particle.* By P. G. Tait and W. J. Steele. Macmillan and Co., London. 1856, 1865, 1871, 1878, 1882, 1889, 1900.

2. *Sketch of Elementary Dynamics.* By W. Thomson and P. G. Tait. Maclachlan and Stewart, Edinburgh. 1863. A pamphlet of 44 pages, afterwards incorporated in the large type part of the *Treatise*, No. 3 in this list, and also in Nos. 5 and 7.

3. *A Treatise on Natural Philosophy.* Vol. I. By Sir W. Thomson and P. G. Tait. The Clarendon Press, Oxford. 1867. Second Edition in Two Parts, The University Press, Cambridge. 1878, 1883.

4. *An Elementary Treatise on Quaternions.* By P. G. Tait. The Clarendon Press, Oxford. 1867. Later and enlarged editions, The University Press, Cambridge. 1873, 1890.

5. *Elementary Dynamics.* By Sir W. Thomson and P. G. Tait. The Clarendon Press, Oxford. 1867. Second issue [not published], 1868. Afterwards enlarged to *Elements of Natural Philosophy.*

6. *Historical Sketch of the Dynamical Theory of Heat.* [Not published.] Thomas Constable, Edinburgh. 1867. The first and second chapters of the *Sketch of Thermodynamics* (1868).

7. *Elements of Natural Philosophy.* By Sir W. Thomson and P. G. Tait. The Clarendon Press. 1873. Second Edition, Cambridge. 1879.

8. *Sketch of Thermodynamics.* By P. G. Tait. Edmonston and Douglas, Edinburgh. 1868. Second Edition, David Douglas, Edinburgh. 1877.

9. *Life and Letters of J. D. Forbes.* By J. C. Shairp, P. G. Tait and A. Adams-Reilly. Macmillan and Co., London. 1873.

10. *Introduction to Quaternions.* By Philip Kelland and P. G. Tait. Macmillan and Co., London. 1873, 1881, 1904 (the last edited by C. G. Knott).

11. *The Unseen Universe.* By B. Stewart and P. G. Tait. Macmillan and Co., London. 1875, 1875, 1875, 1876, 1876, 1878, 1879, 1881, 1882, 1885, 1886, 1888, 1890.

12. *Recent Advances in Physical Science.* By P. G. Tait. Macmillan and Co., London. 1876, 1876.

13. *Paradoxical Philosophy.* By B. Stewart and P. G. Tait. Macmillan and Co., London. 1878.

14. *Miscellaneous Scientific Papers by W. J. Macquorn Rankine,* with a memoir of the Author by P. G. Tait. Edited by W. J. Millar. Charles Griffin and Co., London. 1881.

15. *Heat.* By P. G. Tait. Macmillan and Co., London. 1884.

16. *Light.* By P. G. Tait. Adam and Charles Black, London. 1884, 1889, 1900.

17. *Properties of Matter.* By P. G. Tait. Adam and Charles Black, London. 1885, 1890, 1894, 1899, 1907 (the last edited by W. Peddie).

18. The *Scientific Papers* of the late Thomas Andrews, with a Memoir by P. G. Tait and A. Crum Brown. Macmillan and Co., London. 1889.

19. *Dynamics.* By P. G. Tait. Adam and Charles Black, London. 1895.

20. *Newton's Laws of Motion.* By P. G. Tait. Adam and Charles Black, London. 1899.

21. *Scientific Papers.* By P. G. Tait. Vol. I. The Cambridge University Press. 1898.

22. *Scientific Papers.* By P. G. Tait. Vol. II. The Cambridge University Press. 1900.

Of these books, Nos. 4, 8, 11, and 12 were translated into French; Nos. 3, 4, 12, 15, and 17 into German; and No. 12 into Italian.

INDEX

Names of individuals are indexed in heavy type; names of books in italic.

T.

Printed in the United States
By Bookmasters

Printed in the United States
By Bookmasters

Printed in the United States
By Bookmasters